PRINCIPLES OF MOLECULAR CENETICS

By

DR. S. SUNDARA RAJAN
Principal
Reva Institute for Science and Technology Studies
R.T. Nagar
Bangalore – 560 024

ANMOL PUBLICATIONS PVT. LTD.
NEW DELHI - 110 002 (INDIA)

ANMOL PUBLICATIONS PVT. LTD.
4374/4B, Ansari Road, Daryaganj
New Delhi - 110 002
Ph.: 3261597, 3278000
Visit us at: www.anmolbooks.com

Principles of Molecular Genetics

First Edition, 2003

ISBN 81-261-1388-X

PRINTED IN INDIA

Published by J.L. Kumar for Anmol Publications Pvt. Ltd., New Delhi - 110 002 and Printed at ARORA Offset Press, Delhi.

CONTENTS

1. INTRODUCTION **1-10**

Earlier concept of heredity — Perfomation and epigenesis — Epigenesis theory — Pangenesis theory — Germplasm theory of Weismann — Environmental variations — Genotype - Phenotype — Phenocopy — Microbial genetics — Beginnings of biochemical genetics — Some milestones in the History of Genetics

3. THE CELL **11 – 67**

History – Prokaryotic and Eukaryotic cells-comparison – Prokaryotic cell – Eukaryotic cell-cell wall – Plasma membrane-Chemical composition – Membrane functions – Structure of the membrane-membrane models – Bilayer models – Fluid mosaic model – Micellar models – Cytoplasm – Nucleus – Nucleolous – Nucleic acids-DNA – Molecular model of DNA structure – RNA– Mitochondria – Chloroplasts – Chlorophylls – Endoplasmic reticulum – Ribosomes – Golgi complex – Vacuoles – Lysomes – Centrosomes – Peroxisomes – Glyoxysomes.

3. MOLECULAR BASIS OF HEREDITY **68-84**

Bacteriophages and chemical basis of heredity — Nucleic acids — DNA _ Chemical composition — Molecualr model — Alternative forms of DNA helices — RNA — molecular structure — Types of RNA

4. GENEOME IN PROKARYOTES **85 – 95**

The chromosomes of *E .coli* - Detailed structure of the chromosome — Chromosomes of other bacteria — Genome of cyanobacteria — The viral genome .

5. REPLICATION OF DNA **96 – 113**

Types of replication — Evidences for semiconservative replication — Meselson and Stahl's experiment — carin's autoradiographic experiment— Mechanism of replication — Enzymes involved in DNA replication — Models for DNA replication in prokaryotes – Jacob and Brunner's model — Cairn's model — Yoshikawa model —Rolling circle model .

6. **GENETIC CODE** 114– 129

Properties of the code — Codons , anticodons, and Initiation codons — Nonsense codons — Deciphening the code — Colinearity of genetic code — New Gennetic code in mitochondra and ciliate protozoa — Evolution of genetic code .

7. **PROTEIN SYNTHESIS** 130– 145

Protein structure and classification — The central dogma — Wobble hypothesis — Initiation of polypeptide chain synthesis _ Chain elongation — Chain termination .

8. **GENETIC REGULATION** 146– 164

Induction and Respression — The operon concept — Types of regulatory mechanism — Negative control — Positive control — Catabolic respression — Feedback inhibition — The lac operon The gal operon — Tryptophan operon — Arabinose operon — Ga l gene in yeasts — regulation of transcriprion in phage DNA — genetic regulation in eukaryotes

9. **GENETIC RECOMBINATION** 165– 184

Types of genetic material transfer — Conjugation — Mechanism of conjugation — Interrupted conjugation — Interrupted conjugation and chromosome mapping — Sexduction — genetic transformation —Mechanism of transformation -Transduction — Types — Generalized transduction — Specialized transduction — Growth phase of transducing phages — Lytic cycle — Lysogenic cycle.

10. **GENE MUTATION** 185– 205

What are mutations — History — Spontaneous mutations - Induced mutations — Radiation induced mutations — Chemical mutagens — CLB method for detecting mutations — The Muller 5 metod — Balanced lethals — Biochemical mutations — Phenylketonunia — Tyrosine metabolism - Sickcle cell anaemia — Biochemical mutations in microorganisms — Replicating technique — Reserve mutations —Mutations at the molecular level — Frame shift mutations -Substitution mutations — Transversions — Missense mutations — Nonsense mutations — Silent mutations — Genetic suppression of mutation - Practical application of mutations

11. **CELL DIVISION** 206 – 233
Amitosis – Mitosis – Interphase – Prophase – Metaphase – Anaphase – Telophase – Cytokinesis – Theories of Furrowing – Mitotic apparatus – Frequency and duration of mitotis – Significance of mitosis – Role of mitosis in cancer – Meiosis – Definition – Interphase – Leptotene – Zygotene – Pachytene – Diplotene – Diakinesis – Metaphase I – Anaphase I – Telphase I – Meiosis II – Synaptonemal complex – Significance of meiosis – Molecular mechanism of crossing over – Comparison between mitosis and meiosis.

12. **LINKAGE AND CROSSING OVER** 234 – 267
Copuling and repulsion hypothesis – Morgan's concept of linkage – Linkage in maize – Linkage in *Drosophila* – Linkage in mammals – Linkage in bacteria – Crossing over – Meiotic cross over – Mechanism of crossing over – Cytological demonstration of crossing over – Types of cross over – Mitotic crossing over – Significance of crossing over – Chromosome mapping – Linkage map in Maize.

13. **CHROMOSOMAL ABERRATIONS** 268 – 302
Structural aberrations – Deletions – Duplications – Inversions – Translocations – Evolutionary implication of translocations – Numerical aberrations – Aneuploidy – Monosomy – Nullisomy – Trisomy – Haploids – Polyploids – Autopolyploids – Allopolyploids – Role of polyploidy in evolution – Polyploidy and speciation – Triticale a man made cereal.

14. **CYTOPLASMIC INHERITANCE** 303 – 318
Plastid inheritance in *Mirabilis* – Plastid inheritance in Maize – Male sterility in plants – Cytoplasmic inheritance in *Chlamydomonas* – Petite characters in Yeasts – Kappa particles in *Paranmecium* – Coiling of shells in snails – Merogenic hybrids in urodeles – Cytoplasmic inheritance in *Ephestia* – Other instances of cytoplasmic inheritance.

Preface

One of the unique characters of life is its ability to propagate and produce replicas with minimal changes. Genetics is a branch of Biology that studies all the parameters as to how and in what manner an organism transfers its characters to the offspring . Genetic mechanisms are studied form various stand points such as classical genetics , biochemistry, molecular biology etc.

PRINCIPLES OF MOLECULAR GENETICS aims to study the genetic phenomena with a molecular approach. Aspects of classical genetics are also given as without them it is impossible to understand the mechanism of heredity . All along ,however emphasis has been laid on molecular mechanisms. As such the book selects examples from both eu and prokaryotes.In its 14 chapters the book deals with the properties of the genetic material keeping it in the framework of its physics and chemistry, replication, function, mutation, recombination (both eu and prokaryotic), regulation etc.

The book is meant to be a text book for students of life science, pharmacy, agriculture etc at the graduate and post graduate level .

I am thankful to Mr J.L Kumar, M.D of Anmol Publications, New Delhi for publishing this book.

Dr. S. Sundara Rajan
Principal
Reva Institute For Science and Technology Studies
R.T. Nagar
Bangalore - 560 024

1

INTRODUCTION

"Like begets like" and "Chip of the old block" are two of the adages quoted often whenever we come across a situation to compare parents an children. These popular adages are indeed so commonplace that we may not be able to appreciate the complete truth behind it. Indeed why should a human being give birth to a human being and not to any other organism. A mango seed sown in soil produces only a mango tree and not a guava tree. What is it that is hidden in the mango seed which makes it grow into another mango tree only. When children are born we often hear statements like "the nose is like the father, the eyes are like the mother, but the mouth and hair are different". This shows that the new born in addition to having parental characters has its own haracters too. again what is the mystery behind the fact, that the child has sufficient similarity with the parents and yet sufficiently different from the also?

Genetics is a branch of biology that aims to find answers to the above questions. In fact, genetics is still trying to unravel the total mystery behind the statement, like begets like, even though most of the facts both at the morphological and at the molecular level have been understood.

The factg that children resemble their parents is due to heredity and they also differe from their parents to some extent is because of Variation. Heredity may be defined as the potential of an individual to transmit its characters to the offspring while variations are the differences that the individual acquires due to the interaction with the environment. The main aim of Genetics is the study of these seemingly opposite phenomena namely heredity and variation. While heredity tries to maintain uniformity, variation brings in plasticity.

The term Genetics was first used by Bateson in 1906, even though the science of genetics made its firm beginnings in the form of Mendel's epoch making work with pea plants. The rediscovery of Mendel's work independently by

Devries. Tschermark and Correns indeed paved the way for founding the science of Genetics even though the name was coined only in 1906.

Organisms are capable of transforming their characters to the next progeny as a result of reproduction. Indeed ever since life first arose on earth (due to a chance combination of elements), it has maintained itself till to day in myriad forms only because of its capacity of reproduction. Untill 1868 when Mendel working with pea plants came out with coinvincing explanation regarding the mechanisms of heredity, it was largely speculative as to how organisms pass on their characters to the progeny.

Earlier concepts of heredity

Earlier concepts of heredity made their beginning first with the discovery of sexualityin organisms-both plants and animals. The invention of the microscope gave a great impetus to probe into the world of microcosm and get a visual knowledge about the link between one generation and the toher bringing about hereditary transmission of characters.

Harvey (1578-1657) speculated that all animals arise from eggs, with the semen only playing a vitalizing role. R.deGraff (1641-1673) opined that since the offspring has characters of both the parents, both of them should contribute to the heredity. Leeuwenhoek (1677) who studied the sperms in several animals suggested their association with the eggs.

Reproductive organs in plants were reported for the first time by Grew in 1682. It was Camerarius however who for the first time described sexual reproduction in plants. In 1717 Fairchild produced a hybrid having characters of both the parents. The hybrid was called *Fairchild's mule.*

Performation and epigenesis

The discovery of sexuality, revealed the physical link between one generation and another, thus providing a physical basis for the transmission of heredity from parents to progeny.

The homunculus theory of Wammerdam suggested that the development of an organism is a simple enlargement of a minute but completely preformed individual. According to Swammerdam, the preformed individual or homuncullus could be present either in the sperm or in the ovum. During 18th and early part of 19th centuries there were two groups among "Preformationists". Those who attached importance to sperm were called *animalculists.* Subsequent studies however clearly showed, that not only ovists and animalculists were wrong, but the entire preformation theory was nothing but a figament of imagination.

Epigenesis theory

Wolff (1738-1794) who opposed the preformation concept opined that, neither the egg nor the sperm had a preformed organism, but both of them have undifferentiated mass of living substance which developed into the organism after fertilization. According to this theory called epigenesis, new organs and tissues are formed *de novo* (from the mass of living substances in thegametes) due to vital forces.

Pangenesis theory

Hippocrates (400 BC) believed that reproductive material is produced from all parts of the body so that all the characters are represented in theprogeny. Aristole (350 BC) also was a believer of direct inheritence of characters.

Charles Darwin (1809-1882) was also a votary for this type of inheritance. He modified the views of Hippocrates slightly and called it the theory of pangenesis. According to this, all parts of the body productive invisible gemmules or pangenes that are carried into sex organs *via* the blood stream. There the pangenes are assembled into gametes. During fertilization, pangenes of mother and father combine to produce the characters of the offspring.

Lamarckian theory (J.B. Lamarck 1744-1829) of inheritance, often called inheritance of acquired characters opines that any new character acquired during the life time of an individual is inherited. It will be seen from the above the Charles Darwin's pangene theory is nothing but a version of Lamarckian theory.

Germplasm theory of Weismann

August Weismann (1834-1914) disproved the pangene theory with his experiments on rats. He cut the tails of rats and observed the progeny for 22 generations. In every generation the rat progeny continued to have tails in spite of the fact that the parents were tailless. If pangene theory were to be corect, the tailless parents could not have sent the tail pangene to the gametes and consequently the progeny should have been tailless.

In order to account for inheritance, Weismann proposed his germplasm theory, which is essentially correct even today. According to this theory, the body of an organism can be divided into two - **somatoplasm** and **germplasm.** Somatoplasm constitutes thevegetative body of the organism, while the germplasm constitutes the reproductive part. Somatoplasm develops newly in every individual and there is no continuity; while the germplasm is continuous and links the generations. With the information present in germplasm, every individual develops its own somatoplasm. Hence any change or variation occuring in somatoplasm cannot be transmitted to the next generation, while those of germplams can be inherited. The fermplasm theory provided a convincing physical basis for distinction of variations into heriditary and environmental (see later in the chapter).

Variations

Variations appearing in organisms are of two categories. *Hereditary variations* and *environmental variations.* Before the classical explanation of Wismann (germplasm and somatoplasm) it was baffling the biologists as to how variations of some kind are inherited while others are not. The explanation of Weissman that the body consists of germplasm and somatoplasm and only the former constitutes the link between one generation and the other provided for the first time a basis to classify the variations into two categories (named above).

Hereditary variations

These arise in the germplasm and consequently can be passed on to the next generation. These are also called heritable or permanent variations as they go from one generation to the other. It is primarily the hereditary variations that are responsible for the variations between one species and theother.

Environmental variations

These arise in thesomatoplasm due to the influence of environmental factors and consequently cannot be passed on to the next generation. They are called transient variations becausethey will last only for one generation.

Between the two types of variations, from the evolutionary point of view, heritable variations are of some significance as they tend to carry the variation for several generations.

Some interesting experiments have been made by Clausen *et al* in the Californian (USA) plant *Potentilla glandulosa,* a member of Rosaceae, concerning the difference between heritable and environmental variations. It is generally known that the individuals of a plant species growing at diferent altitudes ina mounainous region tend to show morphological difference. The individuals that grow at sea level have long stem and leaves and take nearly six months to produce flowers from the seedling stage. A much taller form is encountered at about 4000-45000 ft above sea level and it takes only 50-60 days to flower in its natural habitat. At about 10,000 ft above sea level (alpine region) there is a dwarf form of *Potentilla glandulosa* which takes about 50-55 days to produce flowers. To what extent these differences encountered are due to environment and to what extend due to heredity? In order to find out an answer, the plant from each locality was cut into there pieces and replanted in three experimental gardens at coasta level, at 4600 feet and at 10,000 feet. The alpine race (form) grown at sea level resembles the original race, but grows to a taller height at 4500 ft. The ming altitude race (4500 ft) becomes dwarf at the coastal level and extremely dwarf at the alpine region. The coastal race grown at 4500 ft altitude is dwarfer and at 10,000 ft the plant dies. The above experiment clearly illustrates that the members of the samestock (clone) react differently to different environments as shown in the case of height. Hence variations of this type are called environmental variations.

Genotype - Phenotype

An individual organism has tow sets of attributes-an endowment which it gets from the parents and an individuality which reflects a reaction of the hereditary potentiality with the surroundings. In order to clearly demarcate the two in genetic terms, Hohansen (1909) a Danish scientist proposed the *genotype - phenotype concept.* Genotype of an individual has been defined as thesum total of all the hereditary potentialities that the organism inherts from the parents; while the phenotype is the morphological and physiological attributes of the organism - like the size shape, colour, metabolic reaction rate etc. Phenotype reflects theinteraction between the genotype and the environment. Since the individuals of a species may grow in different environments their interactions are different. In other words the same genotype may interact variouslyin different environment and produce various phenotypes. In the case of *Potentilla glandulosa,* the same genotype behaved differently in different environments.

To sum up, we can say that as far as an individual is concerned its genotype is permanent (barring extreme circumstances) while its phenotype is transient. Individuals with different genotypes some time show similar phenotypes in a similar environment.

The diverse phenotypes possible for a given genotype under various environmental influences (without causing lethality to the individual) is called the *norm* or *reaction range.* The norm or reaction range extends from one extreme to the other (with reference to all factors or parameters of the environment) to theother (with reference to all factors orparameters of theenvironment) in which a given genotype can interact and survive. The full reaction range of a genotype perhaps can never be understood in its entireity, because in order to assess this, one has to know all parameters of the environment. This is easier said than done.

Hereditary diseases

These are inborn diseases that the individual has in its genetic mechanism. It reflects an inbuilt fault in the genetic machinery. Many of the human diseases are known to be hereditary - Taste blindness to PTC, colour blindness, Diabetes, albinism, Haemophilia, Xeroderma, phenyl Ketonuria, Alkaptanuria, Sickle cell anaemia are all due tto a fault in the genetic machinery and an individual suffers from these whether one likes or not. But there is a distinction in these diseases. Some of them like Haemophilia, colour blindness etc., will affect the individual irrespective of age, physiological condition etc of the individual. On the other hand, diseases like diabetes (not all types of diabetes), even though hereditary may not affect the individual until quite late in life. This means that hereditary diseases also interact with the environment and express themselves only in a suitable environment.

In two other human diseases

Albinism and xeroderma, the former is recognised by complete loss of pigments in skin, hair, iris of the eyes etc; the latter is also a skin ailment in which exposure to sun rays results in freckles on the skin. But both these can be withstood. Albinos can protect themselves by not exposing too much tgo light, the same is the case with xeroderma people. Till now there has been no total cure for a hereditary disease - the remedies are only pelliative. But recent techniques of genetic engineering have brought a ray of hope that the faulty genes can be identified and corrected very early. However this is still in the experimental stage.

Phenocopy

In order to cure a hereditary disease or rectify a genetic abnormality; it will be necessary for us to know the norm or range of reaction of the genotype in all the environmental conditions. Even though it is difficult to fully analyse the entire reaction range of a genotype it has been possible to do so to some extent atleast in some organisms.

In the fruit fly *Drosophila melanogaster,* the normal body colour is brown with black margins. A hereditary mutant for this fly was discovered by Morghan (1`910) where the body colour was yellow. This was a genotypic character in that the body colour was constant in both the flies in all environments. In 1939, Rapport discovered that, if larvae of these flies are fed on silver salts, they develop into yellow flies irrespective of the fact whether they were genetically brown or yellow. The yellow bodies fly which is genotypically brown is a variant of the original yellow bodies fly. This is now called a phenocopy. It is easy to distinguish between the original yellow body and its phenocopy. The offspring of a phenocopy, would have brown body, if they are not fed again silver salts. The temporary development fo body colour (yellow) in brown flies is an example of norm of its genotype.

MICROBIAL GENETICS

Gentetics studies during the early post mendelian period were concentrated mainly on higher organisms. The reasons for this are not far to seek. Manipulation of organisms for breeding and study of the chromosomes can be easily carried out in higher organisms. Further, phenotypic variations are discreet and offsprings after mating could be easily analyzed for genetic varaitions. Visual observation of segregation of characters also can be done easily. In microorganisms however the picture is totally different. Organisms themselves are too small to handle. Further, when the entire organism is made up of a single cell what sort of a phonotypic variation one can visually observe in microbes? Genetic studies in microbes indeed call for an entirely different approach than the conventional methods used for higher organisms.

In microbes where no morphologically variable traits are available within the individuals of a species for the purpose of hybridization and genetic analysis, the physiological traits are the only dependable characters. What is a physiological trait? When microbes such as bacteria are cultured on a medium continuously, they undergo mutation and this can be detected by their nutritional requirements. For instance if a bacterial colony does not require the addition of biotene to the medium even though it is required for its growth, the assumption is, it is able to synthesize biotene on its own. If such a colony sometimes does not grow on a medium which does not have specially added biotene the conclusion is, the colony has lost the ability to synthesize biotene due to mutation. Such a colony can we grown on a medium which has biotent in it. By this analysis we now have two genetic variations of a bacterial colony (I) colony that can synthesize biotene and (ii) colony that can not synthesize biotene. These two can now be used for genetic studies as they have two 'phenotypic traits'. The traits however are not morphological but nutritional. In other words the two colonies mentioned above are biochemical variants. In microbial genetics the phenotypic traits are nothing but biochemical traits and genetics of microbes is essentially biochemical genetics.

Beginnings of biochemical genetics

Earliest cases of biochemical mutants in metabolic pathways were studied by A.E. Garrod in 1909 which has been published in his classical book *'Inborn errors of metabolism'*. He studied three abnormalities associated with the metabolic breakdown of phenylalanine. The three abnormalities are *Phenyl ketonuria, Alcaptonuria* and *Albinism*.

The next significant step in biochemical genetics pertaining to microbes was that of G.W. Beadle and E.L. Tatum who were awarded Nobel prize for their outstanding work on the biochemical genetics of fungus *Neurospora.* Beadle and Tatum discovered several biochemical mutants in *Neurospora* and proposed their celebrated concept - **one gene, one enzyme hypothesis.** According to this hypothesis, if a biosynthetic pathway involves several steps, each step is controlled by a specific enzyme, which in turn in governed by a particular gene. Mutation in any one of the genes would alter the biochemical pathway. In addition, elucidation of biosynthetic pathways also will be made easy by the study of mutations. This is particularly true for microbes. For the study of biochemical mutations and mutations that bring about phage resistance or drug resistance, microbial systems are admirably suited. Large populations of microbes can be handled easily and mutations may be detected. Thus biochemical genetics of microbes came to be recognized as microbial genetics.

Genetic notations used for microbes. In order to discuss genetics one must be able to distinguish phenotypically one cell from the other. This is hardly possible in microbes as they do not manifest any morphological traits.

In microbes a phenotype constitutes a biochemical trait. Thus a cell which can synthesize methionine is called *Met*$^{+}$ and if it cannot synthesize methionine it is called *Met*$^{-}$. Obviously a Met^{-} cell must have a defective (mutated) gene which prevents the synthesis of methionine.

A typical microbe such as a bacterium has only one set of genes; it is haploid. Enkaryotic organisms on the other have two sets of genes i.e., they are diploid. A diploid cell might have normal copy of the gene and one mutated copy. In this case the heterozygote will have both copies and the gene symbols are separated by a diagonal line Met + Met^{-}. Alternatively a cell may have both defective copies or both normal copies. To summarize, the following conventions are used (Friefelder, 1987).

1. Abbreviations of phenotypes contain three Roman letters (the first capitalized) with a superscript + or - to denote the presence or absence of a designated character and a non superscript + or - for drug resistance.

2. The genotypic designation is always lower case, except for superscripts, all components of the symbol are itacilized.

The above conventions, however has not been adopted for bacteriophages, for which one and two letter symbols are widely used.

Some milestones in the History of Genetics

Name	Year	Contribution
Empedocles-		Semen theory
Aristotle	200BC	Semin is highly purified blood
Harvey	1651	Living beings originate from egg.
Hartsoikes and	1694	Supported preformation theory
Datempotius	1692	"Observed" miniature human being in a sperm's head
Charles Darwin	1809-1882	Pangenesis theory
August Weismann	1889	Germplasm theory
Gregor Mendel	1822-1884	Laws of Heredity; regarded as father of Genetics
Hertwig	1875	Discovered fertilization
Flemming	1879	Described longitudinal splitting of chromosomes
Benden E.W.	1883	Gametes of round worm have half the number of chromosomes as the body cells
Waldeyer	1888	Coined the term chrosome
Tschermark Devries and Correns	1900	Rediscovering of Mendelian laws
Landsteiner	1900	Blood groups
Devries	1901	Mutation in Oenothera
McClung	1902	Sex chromosomes
Bateson. W	1906	Coined the term genetics
Bateson and Punnet	1906	Linkage in peas
Johanssen	1909	Coined the word gene
T.H.Morghan	1910	Sex linkage in Drosophila
Bridges and Tatum	1941	Biochemical genetics in Neurospora
Stern	1931	Cytological proff of crossing over
Avery, Mcclead and McCarty	1944	DNA - genetic material
Muller	1946	Radiation genetics
Chargaff	1950	Base pairs are equal in proportion in DNA
Watson and Crick	1953	Molecular model of DNA
Otchoa and Kornberg	1959	DNA synthesis
Jacob and Monad	1961	Operon concept
Nirenberg, Holley		

and Khorana	1968	Genetic code
Calude and Palade	1974	Ultra structure of cells.

2

THE CELL

One of the marvels of nature is the creation of tiny yet potent biological compartments called cells. There is virtually no living being in the biological world save for Viruses which is not cellular. Either one, two or many millions of cells go to make the body of the living beings. A cell may be defined as the 'structural and functional unit of a living being. It is the minimal biological unit capable of maintaining and propagating itself. The study of cell comes under the perview of cytology, physiology and anatomy; while anatomy deals mainly with the structural aspects, the other two lay emphasis both on structure and function.

History

The word cell has been derived from the Latin word *cellula* meaning a small compartment. The term in the present sense was first used by Robert Hooke (1665) in his book *Micrographic.*

There has been some mention about cell in ancient Indian literature also. Parasara (1st century BC) in his book *Vrkshayurveda* refers to invisible compartments in the leaf and names them *Rasakosha.* He further opines that each cell has a sap *(Rasa)*, not visible to the naked eye *(Anavah)* and has two boundaries *(Kalavestitena)*, some of the cells have a colouring matter *(Ranjakayukta)*.

The credit however of discovering the cell goes to Robert Hooke (1965), since it s difficult to establish the discovery to the ancient Indians in the absence of complete text of Parasara and the general apathy among Indians to their scientific heritage. Robert Hooke who constructed the first compound microscope observed the sections of cork and opined that they contain honey comb like compartments. German Biologists M.J. Schleiden and T.S. Schwann(1838) established the '*cell theory*' that all organisms are made up of cells.

Dutrochet in 1824 separated the cells from *Mimosa* plant. He boiled the leaves in nitric acid to isolate the cells. Turpin and Meyers (1830) observed that every tissue is an assemblage of cells.

One of the significant discoveries of the cell came from Robert Brown (1830). He discovered the presence of a spherical body in the centre of every cell, which he named *'nucleus'*

In the year 1835-37, Purkinje and Mohi independently discovered that protoplasm is an important constituent of every cell and it plays an important role in every cell activity including division. Golgi (1838) discovered the golgi apparatus. Balbiani (1881) discovered chromosomes in the salivary glands of *Chironomus.* At about the same time, Flemming (1882) studied cell division in detail and gave the name *Mitosis.*

Endoplasmic reticulum was discovered by Porter in 1945, while Benda gave the name mitochondria to organelles originally discovered by Hemming. Lysosomes were discovered in 1955 by de Duve.

Shape of Cell

The shape of the cell may be variable as in the case of *Amoeba* or fixed. When the shape is fixed or permanent, all conceivable shapes are found-they may be spherical, rectangular, flattened, polygonal, oval, triangular, cone like, columnar etc. As in *Acetabularia,* the body of the plant made of a single cell is differentiated into a rhizoid like base, a long and slender stalk and an umbrella like cap. Certain cells, like the Ascospores of *Claviceps* are acicular (needle like).

Size

There is a great range of variation among cells in size also. Some of them are not visible to the naked eye, some are barely visible, while some are macroscopic This smallest cell size can be encountered in *Coccus* bacteria (0.2 to 0.5m), while the largest size of cell is seen in Ostrich egg (nearly 15 cms). The table given below summarises the variations in cell size.

Table 1 (Adopted from U. Sinha and S. Sinha, 1976)

Cell type	Size
Fibre cell of *Linum*	550mm (length)
Egg of ostrich	170x35mm
Egg of Hen	60x45mm
Human egg	0.1mm=100μm
Amoeba	100μ
Sea urchin egg	70μ
RBC	7μ
Typhoid bacillus	2.4x0.5μ

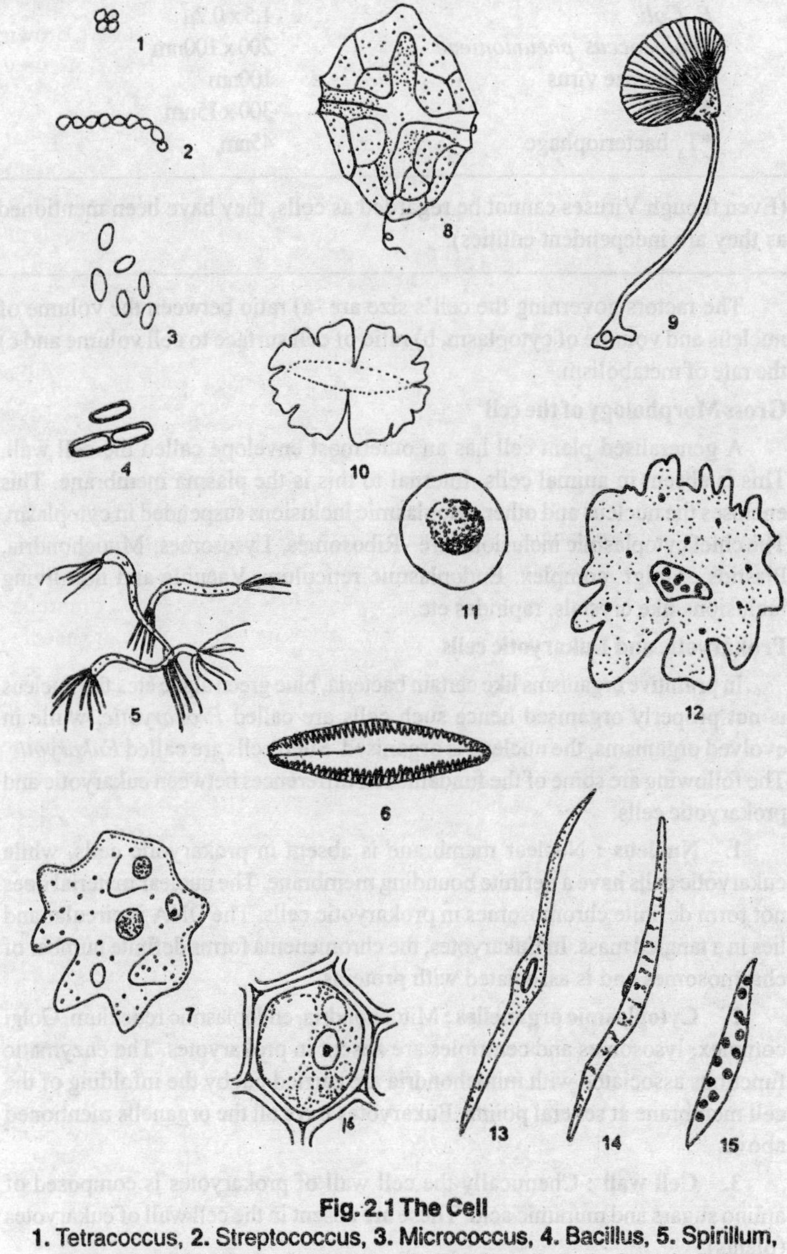

Fig. 2.1 The Cell

1. Tetracoccus, 2. Streptococcus, 3. Micrococcus, 4. Bacillus, 5. Spirillum, 6. Navicula (Diatom), 7. Amoeba, 8. Dinoflagellate, 9. Acetabularia, 10.Desmid,11. Human Erythrocyte, 12. Melonocyte, 13. Smooth muscle cell,14. Striated musclecell, 15. Trachied from a vascular plant, 16. A parenchyma cell

E. Coli	1.5 x 0.7μ
Diplococcus pneumonieae	200 x 100nm
*Influenze virus	100nm
*TMV	300 x 15nm
*T_3 bacteriophage	45nm,

(Even though Viruses cannot be regarded as cells, they have been mentioned as they are independent entities).

The factors governing the cell's size are -a) ratio between the volume of nucleus and volume of cytoplasm, b) ratio of cell surface to cell volume and c) the rate of metabolism.

Gross Morphology of the cell

A generalised plant cell has an outermost envelope called the cell wall. This is absent in animal cells. Internal to this is the plasma membrane. This encloses the nucleus and other cytoplasmic inclusions suspended in cytoplasm. The chief cytoplasmic inclusions are -Ribosomes, Lysosomes, Mitochondria, Plastids, Golgi complex, Endoplasmic reticulum, Vacuole and non living inclusions like crystals, raphides etc.

Prokaryotic and Eukaryotic cells

In primitive organisms like certain bacteria, blue green algae etc., the nucleus is not properly organised hence such cells are called *Prokaryotic*, while in evolved organisms, the nucleus is organised. Such cells are called *Eukaryotic*. The following are some of the fundamental differences between eukaryotic and prokaryotic cells.

1. **Nucleus :** Nuclear membrane is absent in prokaryotic cells, while eukaryotic cells have a definite bounding membrane. The nuclear material does not form definite chromosomes in prokaryotic cells. The DNA is circular and lies in a tangled mass. In Eukaryotes, the chromenema forms definite number of chromosomes and is associated with proteins.

2. **Cytoplasmic organelles :** Mitochondria, endoplasmic reticulum, Golgi complex, lysosomes and centrioles are absent in prokaryotes. The enzymatic functions associated with mitochondria are carried out by the infolding of the cell membrane at several points. Eukaryotes have all the organells mentioned above.

3. **Cell wall :** Chemically the cell wall of prokaryotes is composed of amino sugars and muramic acid. These are absent in the cell wall of eukaryotes (plants).

4. **Flagella :** The flagella of eukaryotes have eleven fibres with (2 + 9 arrangement) 2 central and 9 peripheral fibres, while prokaryotic flagella do not exhibit this arrangement.

5. Plastids : These are absent in prokaryotes. In photosynthetic prokaryotes the chlorphyll is associated with the lamellae, but the lamellae are not enclosed by any membrane. In contrast to this, in all eukaryotes, the chlorophyll pigment is found in definite organelles called chloroplasts.

In addition to the above differences, the streaming movement of cytoplasm may also be added, which is present in eukaryotes, but absent in prokaryotes.

Table 2. Comparison between Prokaryotic and Eukaryotic Cells
(Adopted from De Robertis Et al 1980)

	Character	Prokaryotic	Eukaryotic Cells
1.	Nuclear membrane	Absent	Present
2.	DNA	Naked and circular	Combined with proteins
3.	Chromosomes	Single	Multiple
4.	Nucleolous	Absent	Present
5.	Cell division	amitosis (fisson)	Mitosis or meiosis
6.	Ribosomes	70s (50s + 30s)	80s (60s + 40s)
7.	Endomembrane system	Absent	Present
8.	Mitochondria	Definite organelle absent enzymes of respiration and photosynthesis stored in the folds of plasma membrane	
9.	Choloroplast	Absent	Present
10.	Cell wall	Non cellulosic (present in plants only)	Cellulosic
11.	Exocytosis and endocytosis	Absent	Present
12.	Flagella	No definite arrangement of fibrills	9 + 2 fibrillar arrangement

PROKARYOTIC CELL

The prokaryotic cells are generally smaller and vary in size in different members. In mycoplasma it is about 0.12)μm while in *Oscillatoira,* a filamentous blue green the size is 40 x 5\μ. A great majority of them, however are about μ in size. Among the bacteria the smallest are to be found among cocci (0.1 μ) while the largest are the spirilla (60 x 6~m). The various components of the prokaryotic cell are described below.

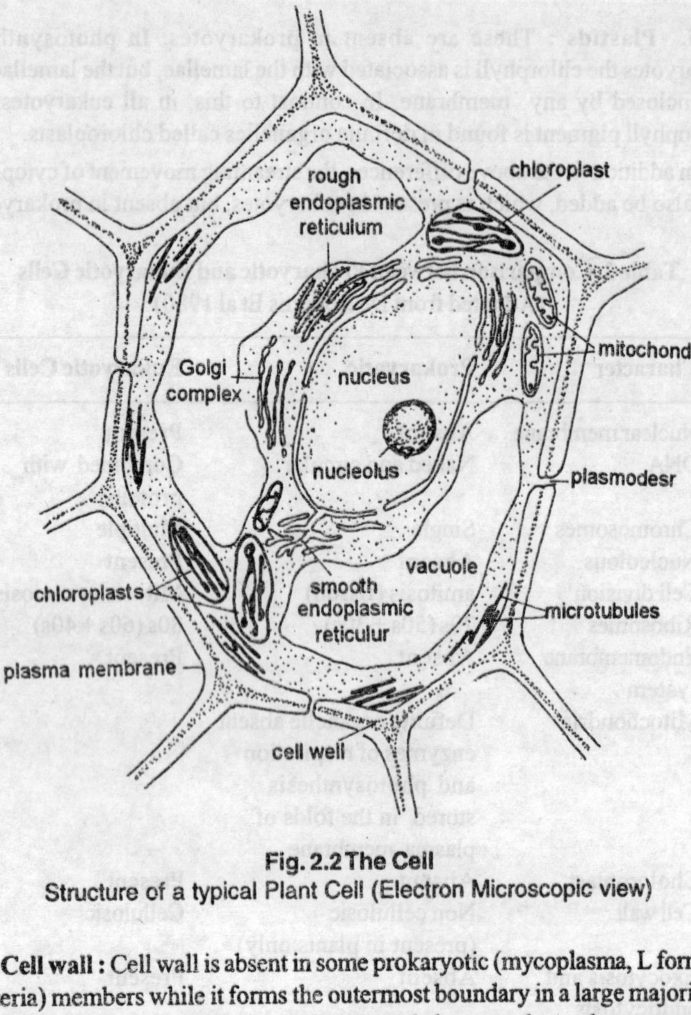

Fig. 2.2 The Cell
Structure of a typical Plant Cell (Electron Microscopic view)

Cell wall : Cell wall is absent in some prokaryotic (mycoplasma, L forms of bacteria) members while it forms the outermost boundary in a large majority of them. While the cell wall is non living and does not have the property of selective permeability it can act as a seive as it is porous.

In some prokaryotes the space between the cell membrane and cell wall is filled with some enzymes and metabolites consituting the periplasm.

Chemically the cell wall is made up of strong fibres of hetropolymers called *mucopeptides ofpeptidogycans*. They are also refered to as *glycopeptide, muropeptide, murein etc.*

The peptidoglycans consist of alternating units made up of N acetyl glucosamine and N acetyl muramic acid with 1-4 linkages. This type of a linkage is common to all Prokaryotes. Attached to the muramic acid is a short peptide chain consisting of alanine, glutamic acid and either lysine or diaminopimelic acid (DAP) and muramic acid. DAP and some of D-amino acids are found only in prokaryotic cell walls.

Cross linkage of these polymers assures structural rigidity with the type and extent of linkage being species specific. For example in some bacteria a pentaglycine links the tetrapeptide side chains extending from the muramic acid units while in others the terminal D- alanine of one tetrapeptide may be covalently linked to the DAP of an adjacent tetrapeptide. The peptidoglycans of gram negative bacteria have more cross linkage than those of gram positive bacteria.

In addition to peptiodglycans the wall has many other chemicals. The wall of gram negative bacteria shows as many as five layers under the electron microscope. These are-the periplasm space, Peptidoglycan layer, an intermediate zone, the outer membrane and lipopolysaccharides.

Cell membrane : This is the bounding layer of cytoplasm and is about 6-8nm thick. Unlike in the case of eukaryotic cells it carries out multiple functions. Some of these are -

(i) Presence of enzymes of biosynthetic pathways leading to the synthesis of wall components

(ii) Occurrence of *permease* enzymes responsible for the transport of materials across the membrane

(iii) Membranes of aerobic bacteria have the electron transport chain and oxidative phosohorylation

(iv) The purple bacteria have their photosynthetic apparatus in the membrane

(v) The membrane has attachment sites for chromosomal DNA (also plasmid) during DNA partitioning

(vi) Regulatory mechanism for chemotaxis is located in the membrane.

Structurally the membrane is triple layered consisting of two electron dense layers surrounding an electrontranslucent layer. In otherwords it is similar to a unit membrane. The architecture and chemical coposition is similar to the fluid mosaic model proposed by Singer and Nicolson (1972) (For details see later in the chapter).

Cytoplasm : Cytoplasm does not appear to have an organized structure when observed under electron microscope. It may be regarded as a concentrated solution containing a host of metabolites, enzymes, amino acids, inorganic ions, sugars, vitamins etc. Cytoplasm plays an important role in intermediary metabolism. There is no cytoskelcton.

Flagella (see bacteria)

Capsule (see bacteria)

Pili (see bacteria)

Surface layers : These are present in all gram negative and gram positive bacteria. Three main types of S layers have been reported in prokaryotic cells (bacteria) The S layers are composed of single homogenous peptides with occasional carbohydrates as minor components. These layers are believed to act as macromolecular sieves preventing the entry of toxic molecules and also to protect the peptidoglycans of the wall from the action of lytic enzymes.

Nucleoid : This represents the genetic material of prokaryotic cells. It is similar in all prokaryotes. (for details see - bacteria) The molecular structure of DNA is same as in eukaryotic cells.

Inclusion bodies : These are storage granules and other particles distributed in the cytoplasm. There are two kinds of inclusion bodies-membrane enclosed and membraneless (Shiviey 1974).

Membrane enclosed bodies : These are enclosed by a non unit type of membrane 2-4nm thick and made up of proteins. The inclusion bodies encountered in the prokaryotic cells are chlorosomes, carboxysomes, magnetosomes, gas vacuoles, polyhydroxybutyrate (PHB) granules, sulphur granules and glycogen granules.

PHB granules occur in cyanobacteria, sulphur granules are found in Thiorhodaccae and glycogen granules in *Clostridium* spp.

Chlorosomes have been demonstrated in photosynthetic bacteria belonging to chlorlobiaceae. (Cruden and Stanier, 1970). These are different from chromatophores in being non-membranous.

Carboxysomes are seen in cyanobacteria (Murphy et al 1974).

Magnetosomes are particles which help the bacterium in orienting itself in relation to the magnetic behaviour of the surroundings. Reported to occur in *Aquaspirillum magnetotacticum* (Balkwill et al 1980), the magnetosomes allow the bacteria to punise the most efficient aerotactic behaviour.

Non membranous Inclusion bodies : Phycobilisomes, cyanophycin granules (found in cyanobacteria), polyglucoside granules and several crystalline substances belong to this group.

In addition to the above there are R - bodies (retractile stuctures) demonstrated in bacteria by Lalucat and Mayer (1978). These are supposed to be similar to kappa particles of *Paramecium*.

Ribosomes (see bacteria)

Mesosomes (see bacteria)

Chromatophores : These are present in the photosynthetic bacteria belonging to the families Rhodospirillaceae, Chromatiaceae and Cyanophyceae.

Chromatophores vary in form and have vesicles, tubes, bundles, stacks or thylakoids etc.

EUKARYOTIC CELL

Cell Wall

The cell wall is the characteristic feature of the plant cells. It is made up of non living matter and is believed to be secreted by the plasma membrane. The structure and composition of the cell wall not only varies in different plants, but in different types of cells of the same plant. Often, there might be variety in the cell wall composition at different stages of growth of the cell. The cell wall is very rigid and maintains the shape of the cell acting as a boundary. A unique feature of the cell wall is its permeability which is due to the presence of small pores.

Gross structure

Three layers are distinguishable in the cell wall. these are the *primary cell wall, middle lamella* and the *secondary cell wall.* The primary cell walls of adjacent cells are separated by middle lamella. Secondary wall is deposited after a certain period of development.

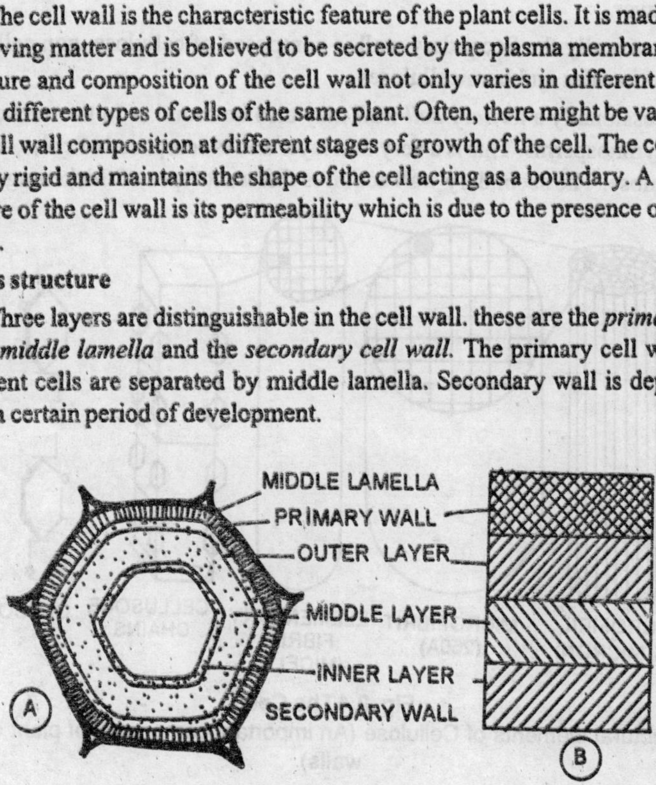

Fig. 2.3 The Cell
Arrangement of wall layers. A. Cross section, B. Arrangement of fibrils

The primary wall is secreted by the plasma membrane. In some instances like the leaves, fleshy stems, roots etc., the cells have only the primary wall and the middle lamella. The primary wall is formed during the early stages of growth and is about 1 to 3nm in thickness. Chemically the primary wall is composed of cellulose, hemicellulose and pectic compounds. The young primary wall is plastic and is prone to changes. Often it may be lignified.

The Middle lamella is the intercellular substance binding together the primary and the secondary walls. The middle lamella is formed during cell division and is amorphous and colloidal in nature. Chemically the middle lamella is composed of pectin, cellulose, calcium and magnesium pectate.

The secondary wall is deposited on the primary wall after cell maturation. It is very rigid and does not alter its shpae. The thickness varies from 5 - 10nm. The secondary wall usually has three layers viz., *outer layer, middle layer* and *inner layer*.

Chemically the Secondary wall is composed of cellulose, non cellulosic polysaccharides and hemicelluloses.

Ocassionally a *tertiary cell wall* may also be found as in the tracheids of some gymnosperms. This is a very thin layer and is composed of *xylan,* instead of cellulose. The secondary wall acquires different types of thickenings.

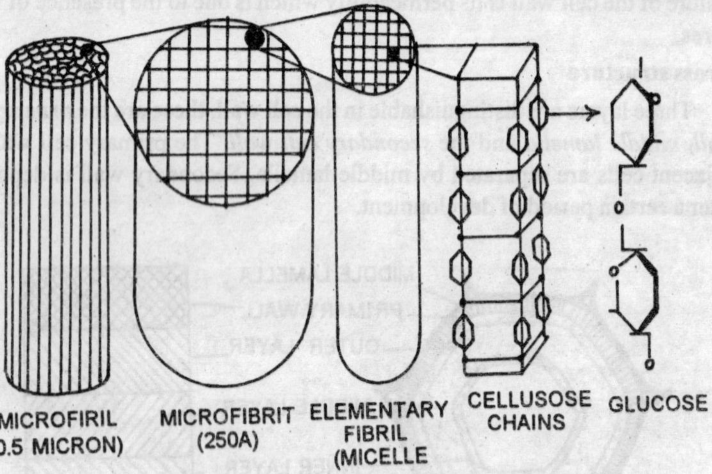

Fig. 2.4 The Cell
Structural elements of Cellulose (An important component of plant cell walls)

Ultra structure of the cell wall

The cell wall is mainly composed of cellulose. Cellulose is a polymer of glucose molecules. Approximately each molecule of cellulose consists of about 3000 glucose residues joined together by 1-4 oxygen bridges. These cellulose chains are oriented along the axis of the cell. About 100 of these cellulose chains bundle together to form *micellels* or *elementary fibrils*. Each micelle is about 100Å in diameter. Many hundreds of these micelles of these stack together to form *microfibrils*. Each microfibril is 200Å to 300Å in diameter. The micorfibril is visible only under eletron microscope. About 2000 of these microfibrils join together to constitute the *macrofibril* which is about 0.5μ in width. The macrofibrils are visible under the light microscope.

In secondary walls, the microfibrils are more orderly and compact than in the primary walls. The space between fibrils is filled by an interfibrillar substance usually the matrix. The matrix is filled by non cellulosic polysaccharides - such as pectin, hemicellulose, xylans etc.

The primary cell wall of dicotyledons contains many complex polysaccharides such as *xyloglucans, arabinogalactans* and *rhamnogalacturonans.* These are linked to the cellulose fibril and to one another by covalent linkages.

In certain walls, a glycoprotein called extensin is present. It is similar to collagen in being rich in 4 - hydroxyproline.

Extensin has a long protein chain, to the hydroxy groups of which are linked, residues of *arabinose* and *galactose.*

Plasmodesmata : Protoplasts of adjacent cells are connected to one another by means of thin cytoplasmic threads called plasmodesmata. These pass through the small pores (pit fields) present in the cell wall. Tubules which are in continuity in the endoplasmic reticulum are seen within plamodesmata thus maintaining an effective intercellular communication.

Origin of cell wall : At the completion of *Karyokinesis,* small vesicles of endoplasmic reliculum *(phragmoplast)* accumulate along the equatorial line to form the cell plate. The cell plate becomes thick by the addition of material from the golgi complex. Plasmodesmata are left open in the cell plate to maintain communication. This cell plate forms the lamella. Later, on either side of the middle lamella, cytoplasmic fibrils start accumulating to form the wall.

Plasma membrane : The cytoplasm in all the cells (both plant and animal) is surrounded by a living membrane called the *plasmalemma* or *plasma membrane.* This membrane is vital to the cell as it controls the exit and entry of all substances to and from the cell. The plasma membrane, as demonstrated by McAlear and Edwards (1958) in fungi, invaginates and maintains connection with the nucleus. It is also believed that many of the cell and their bounding membrane is same as the plasma membrane These organelles are nucleus, mitochondria, chloroplasts etc.

The presence of the cell membrane is proved by the following evidences.

1. On injury to a cell, when protoplasm flows out, it remains as a discreet unit.
2. Observations under electron microscope have revealed the presence of a boundary membrane over the surface of cytoplasm.

Size of the membrane : It is about 75Å thick and is composed of three concentric layers. The middle layer is about 35Å in thickness, while the outer layers on either side or 20Å thick.

Chemical composition : Membranes from different sources show variations in chemical composition. Essentially however all membranes are lipoproteinanceous. The components are held together in a thin sheet by non convalent bonds. The ratio of protein to lipids varies depending on the type (Golgi membrane, plasma membrane etc) of membrane, type of organism (eukaryote, prokaryote, etc.) and the type of cell (epidermis, phoelm etc.). In adition to these two, certain membranes also have carbohydrates.

Membranes have three dfferent categories of proteins. These are - structural *proteins, functional proteins* (enzymes) and *carrier proteins.* Structural proteins maintain the structural frame work of the membranes. The average molecular weight of structural proteins found in membranes in $3x10^4$. Enzymes catalyzing various physiological activities are found in the membrane of mitochondria, endoplasmic reticulum etc. *Carrier proteins* are known to form a complex with the substances and help them to pass through the membrane.

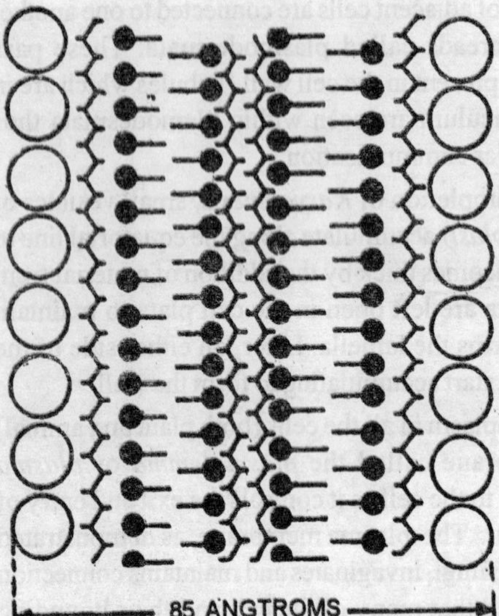

Fig. 2.5 The Cell
Diagrammatic representation of a cell membrane

The protein composition of different membranes vary (as has been pointed out already). For instance plasma membranes have 50% while mitochondrial membranes have 75% proteins.

The proteins present in plasma membrane are categorized into two viz. *integral proteins* and *peripheral proteins*. While integral proteins form the back bone of the membrane, peripheral proteins have a weak association maintained by weak electrostatic interaction. The integral proteins are hydrophobic in nature as they have to penetrate the hydrophobic lipids to form the basic structure of the membrane. These proteins do not easily disassociate from the membrane and it requires harsh chemical procedures to accomplish it. The integral proteins have a hydrophilic portion also which protrudes out of lhe lipoid bilayer. Peripheral proteins are associated either with the hydrophilic head of lipids or hydrophilic part of the integral proteins.

The Lipids of membranes are of several types, but all of them share a common character in being *amphipatic* i.e. having both *hydrophilic* and *hydrophobic* portions within a molecule. There are three categories of membrane lipids-they are phosopholipids, glycolipids and sterols. While most membrane lipids are phospholipids, cholesterol does not have phosphates.

Phospholipids are of two categories viz. *phosphoglycerides* and *sphingophospholipids*. Phosplioglycerides have a glycerol back bone, while sphingophospholpids have the back bone derived from sphingosine an aminoalchobol having a long hydrocarbon chain.

Glycolipids contain one or many monosaccharide units attached to the lipids based on *sphingenine* or *ceramide*. There are three kinds of glycolipids viz., *Cerebrosides, sulphatides* and *gangliosides*.

Cerebrosides contain a single monosaccharide unit (galactose or glucose). Sulphatides are sulphuric acid esters of cerebrosides. Gangliosides are glycosphingolipids having *sialic acid*.

Sterols are steroid alcohols and are of three types - cholesterols (present in animal tissues), phytosterols (present in plants) and ergosterols (present in eukaryotic microorganisms).

Among the glycerophospholipids *lecithin* and *cephalin* are most abundant in higher plants and animals.

Membrane functions

The following are the major functions of membranes.

1. **Compartmentalization:** Membranes constitute the living boundary for cytoplasm and cell contents. As unbroken sheets they divide the living mater into self sustaining units to effectively coordinate and regulate the activities.

2. **Regulation of movement:** In a biologically active cell many susbstances enter and many others go out of the cell. However, if the entry and exit is unregulated, it is deleterious to the cell. Hence plasma membrane regulates and decides which substances are to be allowed and which should not. This unique property of the cell membrane called selective permeability or differential permeability is responsible for most of the physiological activities of the cell.

3. **Communication:** All the membranes have certain special substances (on their surface) called receptor molecules which mediate and control the message from one to another. The receptor molecules vary in different membranes. These molecules combine with substances (ligands) of different types and generate signals (in the form of cyclic AMP) directed to the interior of the cell. The best studied ligands are the hormones and neurotransmitters.

4. **Interaction between cells:** It is the responsibility of the plasma membrane as the living boundary between cells to effectively coordinate actions between cells since an organism consists of differnt types of cells and many a time the actions have to be coordinated to achieve harmony.

5. Energy transformations : Membranes are intimately associated with the inter conversion of energy. For instance, in green plants the light energy has to be converted into chemical energy. Similarly the chemical energy stored in carbohydrates has to be converted into energy currency i.e. ATP molecules. The machinery for energy capture and conversion is hidden in the membranes of mictochondria and chloroplasts.

6. Locus for biochemical functions : Membranes provide the structural framework for the location of various enzymes. The scaffolding of the membrane interconnecting the various organelles helps in the coordination of activities. Such an effective coordination of activities is best illustrated in the mitochondrial membrane where the various components of the electron transport chain work together as they are located in an orderly fashion.

7. Other activities : Membranes also function in various other capacities such as providing mechanical strength, involving in cell movement secretion etc., and also acting as electrical insulators.

Structure of the membrane : The molecular structure of the membrane has been interpreted in various ways by different cell biologists in order account for their known function. Various models have been proposed for the molecular structure of the membrane. These fall into two basic categories, viz., bilayer models and micellar or subunit models. The bilayer models are believed to have protein and lipid to be arranged in the form of layers, while the subunit models assume the presence of a number of independent subunits. We shall discuss a few of these models. The chart below summarises the list of various models.

MEMBRANE MODELS

- Bilayer models
  - Lipid membrane Model
  - Sandwich Models
    - (a) Danielli Davson model
    - (b) Unit membrane model
    - (c) Greater membrane odel
    - (d) Kavanau's model
- Micellar models
  - Models in which protein penetrate into the lipid layer
    - (a) Benson's model
    - (b) Lenard and Singer' model
    - (c) Mosaic membrane model
    - (d) The fluid mosaic model

BILAYER MODELS

Lipid membrane model : Overtoil (1895) discovered that fat soluble subsatances easily pass through the membrane indicating their affinity to the lipids. Hober (1910) and Fricke (1925) found out that the low electrical conductivity of the Iving cell is due to the presence of lipids. The experiments of Mudd and Mudd (1931) in placing an oil droplet which adheres to the cell surface provided the presence of lipids in the cell membrane. It was Gorter and Grendel (1925) who suggested a structural model based on lipids. They studied the erythrocyte (RBC) membrane and concluded that it is composed of two layers of liqid molecules whose polar groups are situated on the outside of the membrane.

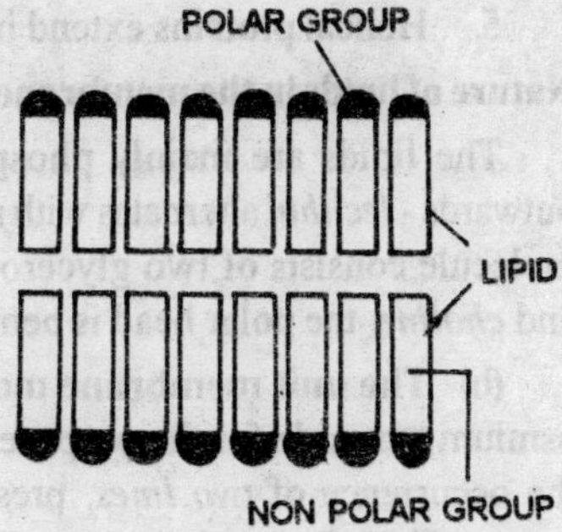

Fig. 2.6 The Cell
The lipid membrane model of Forter

Sandwich models : In these models the lipid layers are supposed to be sandwiched between the two protein layers on either surface. The polar groups of the lipids adhere to the proteins on either side. The following are some of the models based on sandwich principle.

(a) Danielli Davson model : The study of surface tension of cells by Daniell and Harvey (1935) and Harvey and Cole (1931) indicated that there must be more to the structure of the membrane than merely being made up of lipids. In 1935, Hugh Davson and James Denielli proposed that the plasma membrane is a lipid layer sandwiched between layers of globular proteins. In a way the Danielli model was the first structural interpretation to account for the physiological properties of the membrane.

Over the years various modifications have been proposed for the model. Danielli (1938) opined that there could be two kinds of globular proteins on the surface and diagonally arranged proteins in contact with the lipid layer. In order to account for the permeability, certain small pores were visualised in the lipid layers with an inner lining of protein molecules providing an inner hydrophilic surface.

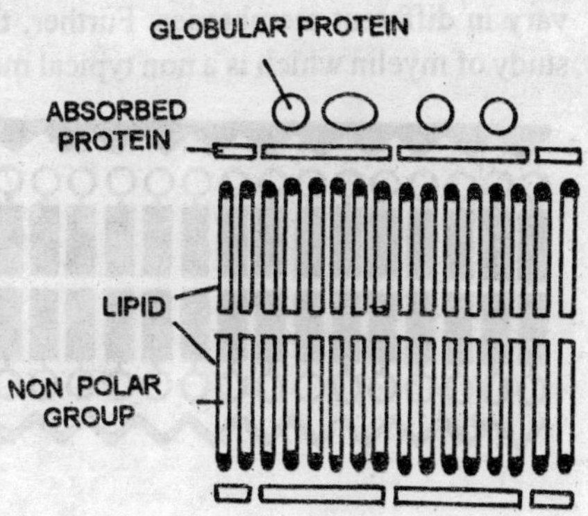

Fig. 2.7 The Cell
The Danielli-Davson model

The following are some other variations in the protein lipid arrangement.

1. Proteins are in the form of folded chain.
2. Proteins are in coiled from on both sides of the lipid layer.
3. Proteins are asymmetrically arranged i.e., one side has globular proteins and the other side a folded β chain.
4. Globular proteins are found on either side.
5. Helical proteins extend into the pores between the lipid molecules.

Nature of lipids in the membrane

The lipids are mainly phospholipids with their polar groups pointing outwards - *lecithin* alternates with a steroid molecule (*cholesterol*). Each lecithin molecule consists of two glycerol chains with the part containing *phosphate* and *choline*, the polar head is bent like a walking stick.

(b) The unit membrane model : The specimens of membrane fixed in osmium tetroxide for the purpose of electron microscopic observation reveal the occurrence of *two lines*, presumably indicating their double membrane nature. This observation originally made in 1950's by Humberto Fernandez Moran and David Robertson led to the unit membrane model. Robertson used a new fixative potassium permanganate. He discovered that each of the dense line possessed a sub structure having a trilaminar appearence. Each membrane is composed of two thin eletron dense lines of 20Å, with the intervening space of about 35Å. The inner and outer layers are non identical. The outer is a mucoid protein, while the inner surface is a non mucoid protein. The unit membrane model is similar to Danielli Davson model except that protein layers are asymmetrical. Membrane of cell organelles, plasma membrane and endoplasmic reticulum are thought to be similar in havingo the unit membrane structure. But current studies have shown that arrangement of lipids and proteins vary in different membranes. Further, the Robertson model is based on the study of myelin which is a non typical membrane.

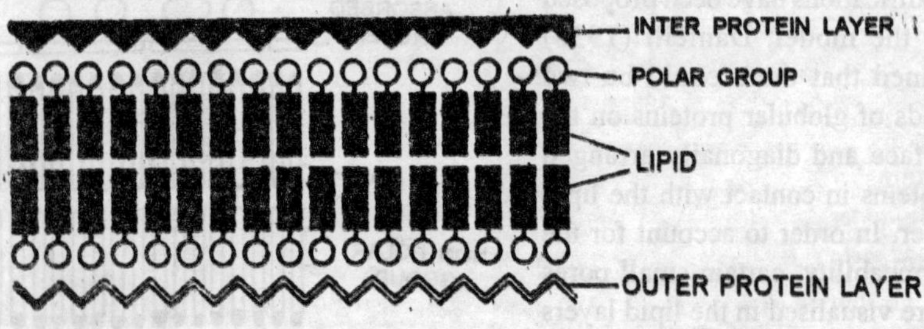

Fig. 2.8 The Cell
The unit membrane model of Robertson

(c) The Greater membrane model : This is similar to the trilaminar sandwich model except that the outer and inner surface proteins are different. The inner surface has an unconjugated structural protein adhering to polar heads of lipids. The outer surface also has structural proteins, on which is superimposed a *glycoprotein*. To this glycoprotein are attached side chains of oligosaccharides.

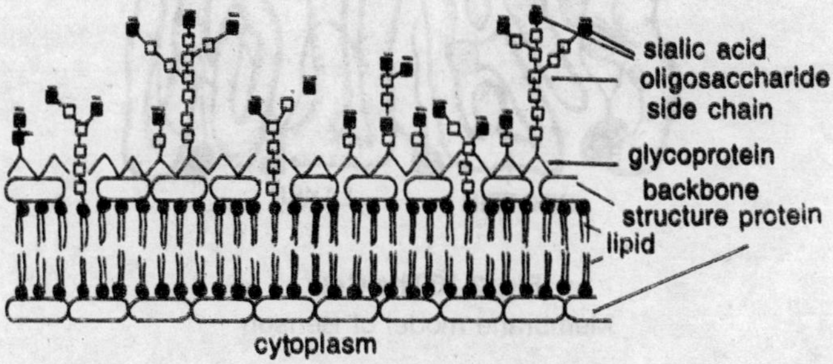

Fig. 2.9 The Cell
The greater membrane model

(d) Kavanau's lipid pillar model : According to Kavanau J.L. (1965), the lipid layer exists in two forms. Kavanau's model is essentially similar to Danielli Davson model. The two types of lipid structures are *pillars* and *flattened discs*. Pores are present between pillars to faciliate the passage of ions. Membranes with pillar lipids are thicker than those with discs. Internally the pillars have non polar tails, while the surface is made up of polar phospholipids. Proteins are present on either side of lipids.

Models based on protein penetration of lipids

(a) Lenard and Singer's model : This is based on the protein arrangement in the membrane. About 25 to 30% of proteins are helical while the rest are irregular coils.

(b) Benson's model : This model was proposed by Benson (1966) based on the study of membranes of chloroplasts. In this, it was postulated that lipid tails are bound to the corresponding hydrophobic points within the coils of proteins. The polar heads of the lipids lie on the surface.

(c) Mosaic membrane concept : According to this model proposed by Baum (1967), the membrane consists of units, each about 80Å in diameter. The wedge shaped protein is enclosed on either side and at edges by the lipids. The polar heads protrude out combining with the hydrophobic areas of proteins.

(d) The Fluid mosaic model : The molecular model of the plasma membrane proposed by Singer and Nicolson (1972) emphasises on the dynamism of membrane organisation and visualises it to be a semifluid like structure with a mosaic arrangement of lipids and proteins. The fluid mosaic model, though essentially based on bilayer models, rejects the idea of continuous bilayer.

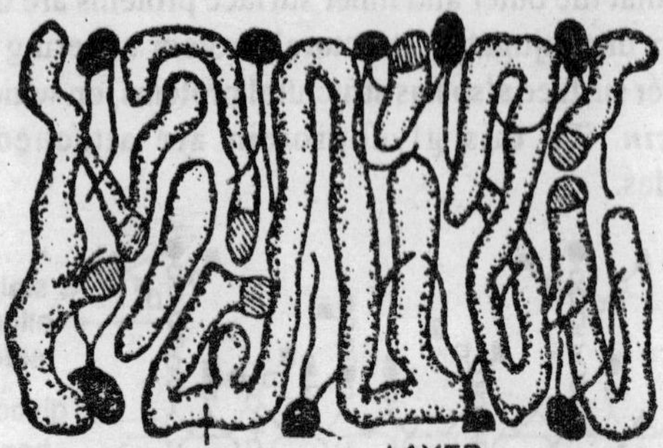

Fig. 2.10 The Cell
Membrane model of Benson

Instead of forming continuous external layer adhering to the hydrophilic heads of the phospholipids, as in the Danielli model; the proteins constitute discontinuous particles often penetrating deeply into the lipid layer. In a way the protein units could be compared to floating incebergs in a sea of phospholipids. The countinuity of the bilayer is not maintained because of complete or incomplete interruption by proteins. While the Danielli model assumes a hydrophilic bonding between lipids and proteins, the Singer Nicolson model regards it to be hydrophobic.

The membrane proteins in the fluid mosaic model are of two types distinguishable by their intensity of lipid association. These are the *integral* or *intrinsic proteins* and the *extrinsic* or *pheripheral proteins*. ITie integral proteins penetrate deep into the lipid layer and are the hydrophobic as they have to coexist in the hydrophobic environment of fatty acids. The integral proteins are *amphipatic*. While their tips are hydrophilic; the central portion is hydrophobic. These proteins are difficult to extract from the membranes.

The peripheral proteins which are weakly associated whith the membrane can be easily extracted. They are hydrophilic and never penetrate the interiors of the lipid bilayer. They form weak bonds either with the polar heads of lipids or with the hydrophilic surfaces of integral proteins. One of the best evidences for a broken surface layer of proteins comes with the using of phospholipases. These phosphate degarding enzymes very easily percolate into the membrane, which would have been impossible, if there were to be an unbroken surface envelope of proteins.

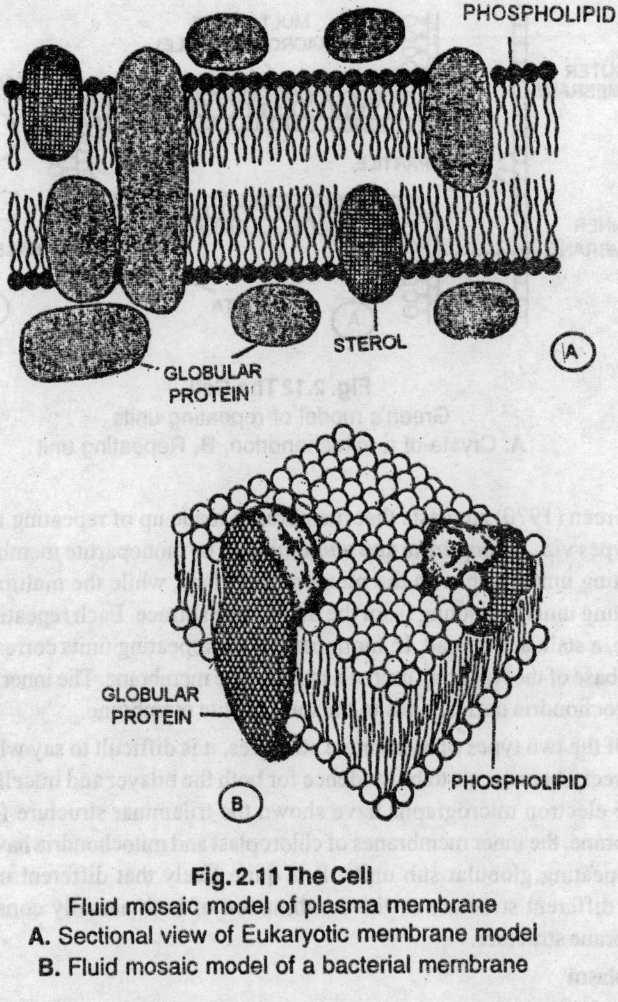

Fig. 2.11 The Cell
Fluid mosaic model of plasma membrane
A. Sectional view of Eukaryotic membrane model
B. Fluid mosaic model of a bacterial membrane

The fluid mosaic model is by far the best structural interpretation of the membrane which satisfactorily explains all its (membrane's) known functions.

MICELLARMODELS

According to Fernandez Moran (1963) the membrane is corpuscular in nature consisting many repeating units. These units called *globular units* or elementary particles range in diamiter form 40 - 70Å and form closely packed repeating units. The membrane structure may be globular or unit membrane in different organelles or under different metabolic conditions. According to some cell biologists, the corpusular nature is an artifact.

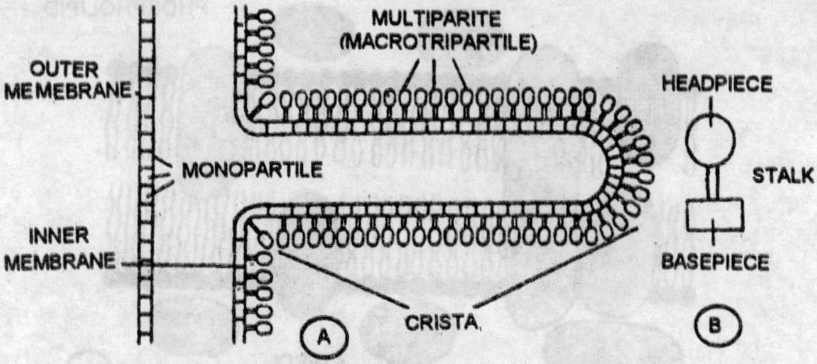

Fig. 2.12 The Cell
Green's model of repeating units
A. Crysta of a mitochondrion, **B.** Repeating unit

Green (1970) suggests that membranes made up of repeating units are of two types viz., *monopartite* and *multipartite* The monopartite membranes have repeating units flush with the membrane surface, while the multipartite have repeating units projecting from the membrane sufrace. Each repeating unit has a base, a stalk and a head. In monopartite, the repeating units correspond only to the base of the repeating unit of a multipartite membrane. The inner membrane of mitochondria corresponds to the multipartite membrane.

Of the two types of membrane stuctures, it is difficult to say which model is correct There seems to be evidence for both the bilayer and micellar models. While electron micrographs have shown the trilaminar structure for the unit membrane, the inner membranes of chloroplast and mitochondria have revealed the repeating globular sub units. It is quite likely that different membranes show different structure or the combination of both actually constitutes the membrane structure.

Cytoplasm

The colloidal substance in the cell, internal to the plasma membrane (without the inclusions) is called cytoplasm. The cytoplasm is usually granular and has many substances either dissolved or suspended in it The suspended particles may be in the solid or semi solid condition. There are several theories regarding the structure of the cytoplasm. Some of these are-

Reticular theory : Cytoplasm is made up of network of fibrils or particles.

Alveolar theory : According to Butschill (1892), cytoplasm consists of a number of droplets or alveoli. It gives an appearence of oil droplets floating on water.

Granular theory : According to Altmann (1893), Cytoplasm is composed of a number of minute granules suspended in a matrix.

Fibrillar theory : This theory was proposed by Fisher (1894) and Flemming (1897). Electron microscopic studies have proved the hypothesis of Fisher. According to this view, cytoplasm is made up of a number of fibrils.

Properties of cytoplasm : Cytoplasm has a number physical and biological proerties.

Physical properties

The colloidal nature of cytoplasm is responsible for the various physical properties A colloidal system may be defined as a system of liqid medium containing suspends particles varying in diameter from 1/1,000,000 to 1/10,000 nm. Some of the physical properties are follows.

Phase reversal : Cytoplasm may be differentiated into a *sol state* and *gel state*. Sol state is the liquid state and the gel state is the semi solid state. These two phases are interchangeable. This capacity is known as phase reversal.

Elasticity : Depending on the circumstances, the cytoplasm can extend or contract subject to a certain limit. This is known as elasticity.

Cohesiveness : The suspended particles in cytoplasm have mutual attraction, thus exhibiting cohesiveness.

Contractility : It is the capacity of peripheral cytoplasm to absorb or remove water from cell to the exterior. This is manifested very well by guard cells.

Viscosity : The suspended particles of cytoplasm are responsible for its viscous nature.

Brownian movement : The suspended particles of cytoplasm are in a state of to and fro movement called Brownian movement

In addition to this, cytoplasm also exhibits movements (seen in the plasmodium of slime molds) and 'cyclosis or streaming movements (seen in the leaf cells of *Elodea)*.

Biological properties

Growth, respiration, excretion, nutrition, reproduction, metabolism, irritability etc., are some of the important biological activities of cytoplasm.

Chemical composition of cytoplasm : 90% of cytoplasm is water. The remaining 10% is constituted by organic and inorganic compounds.

About 40 or more inorganic constituents are found in cytoplasm. These are found in the form of ions (either cations or anions). Some of the common inorganic constituents are - Oxygen (62%), Hydrogen (10%), Carbon (18%) and Nitrogen (3%). The less common inorganic contituents are Calcium (0.25%), Phosphorus (1.2%), Sodium (0.1%), Magnesium (0.7%), Iodine (015%) and Chlorine (0.15%).

Among the various organic substances proteins, carbohydrates, lipids, hormones, vitamins and nucleic acids are most important.

Cytoplasmic Inclusions : Cytoplasm has a number of living and non living inclusions. The living inclusions are - Nucleus, mitochondria,. plastids, endoplasmic reticulum, golgi complex, lysosome, centrioles, microbodies, etc.

Nucleus

This is the most important organelle of the cell which regulates all its activities. It was discovered by Robert Brown in 1831.

Position : Usually it is present at the centre of the cell, but in some instances as in mature parenchyma cells, the nucleus is peripheral. In *Acetabularia,* the nucleus occupies various positions, though mainly present at the basal region.

Shape, size and number : The shape is variable depending on the cell type. Though the nucleus is generally spherical, ellipsoidal or flattened ones also may be seen in certain cells. The surface is generally smooth but as in leucocytes it may have small infoldings.

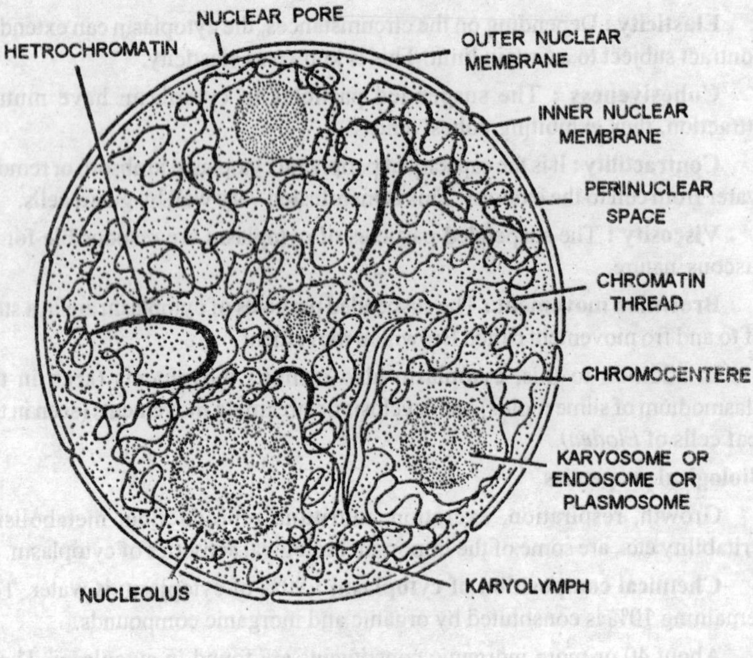

Fig. 2.13 The Cell
Structure of the nucleus

The size of the nucleus is also variable depending on the cell; the size of the nucleus being directly proprolional to that of cytoplasm. The following formula or Heywig may be used to determine the size of the nucleus in a particular cell.

$$NP = Vn/Vc - Vn$$

When *NP* = nucleoprotein index, *Vn* is the volume of the nucleus and *Vc* is the volume of the cell.

Most of the cells are uninucleate, while some *(Paramecium)* may be binucleate or multinucleate *(Cladophora)*. In certain fungi like *Mucor,* where there are no septa dividing the plant body into cells, numerous nuclei occur indicating a coenocytic condition.

Ultrastructure of the nucleus : Ultrastructurally the nucleus has the following parts - 1. Nuclear membrane, 2. Karyolymph or Nuclear sap, 3. Chromonemata, 4. Nucleolus and 5. Endosomes.

Nuclear membrane : In all eukaryotic organisms, a limiting membrane envelops the nucleus. This membrane helps in effective communication between nucleus and cytoplasm. The elements of endoplasmic reticulum contribute to the nuclear envelop during cell division.

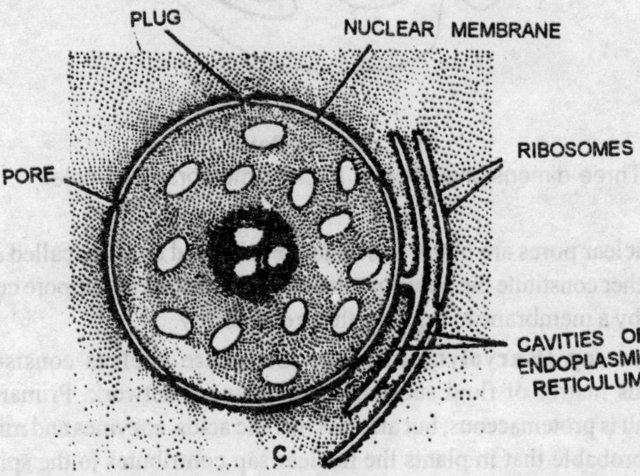

Fig. 2.14 The Cell
Diagrammatic representation of nuclear membrane and pore

Electron microscopic studies have revealed the nuclear membrane to be a double membrane with a number of pores. According to Watson (1959) these pores called *nuclear pores* are octogonal in shape. The outer nuclear membrane is called endokaryotheca. Each membrane is about 75 - 90Å thick and they enclose a space between them about 100 - 150Å (De Robertis et al 1970). This space is called the *perinuclear space* or *cisterna.* The outer membrane is rough due to the presence of RNA particles. Internally the nuclear membrane is followed by a supporting membrane called *fibrous lamina* (300Å thick).

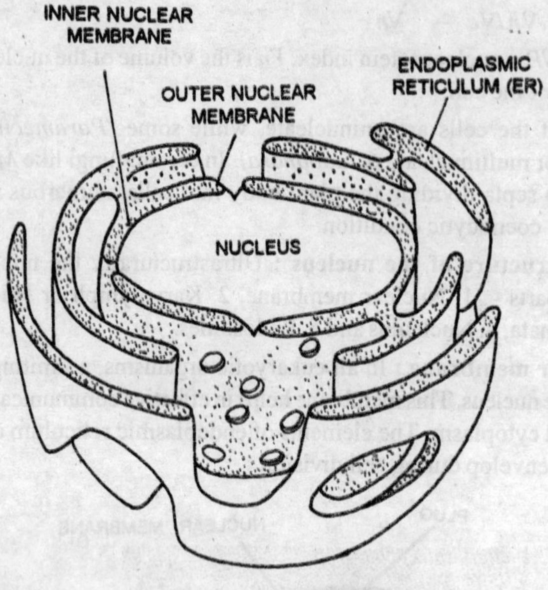

Fig. 2.15 The Cell
Three dimensional view of nuclear membrane complex

The nuclear pores are 600Å wide and are enclosed by rings called *annuli*. These together constitute the pore complex. In some instances, the pore complex is covered by a membrane to regulate the premeability.

Nuclear sap (Karyolymph) : The interphase nucleus consists of a homogenous matrix of fluid substances enclosing a network. Primarily the Karyolymnh is proteinaceous, but also has nucleic acids, enzymes and minerals. It is quite probable that in plants the nuclear sap contributes to the spindles.

Chromonemata : Enclosed in the Karyolymph and visible in the interphase nucleus are found a number of fibrillar structures constituting a network called *chromonemata* or *chromatin fibrils*. These stain positively with basic fuchsin stain. Some coarse granules are deposited on the chromatin network. These are a called *chromocentres* and constitute the points of condensation of chromosomes. During cell division the chromatin network breaks up into specific number of chromosomes. Two regions can be identified in the chromatin material. These are *heterochromatic* region and *euchromatic* region.

The heterochromatic region stains darkly and shows numerous bead like structures called *chromomeres*. The heterochromatic region has less DNA. This region is believed to be genetically and metabolically inert. The light staining region of the chromatin is called the euchromatic region. This region contains more of DNA and is supposed lo be genetically active.

Nucleolus : This was first discovered by Fontana (1874). A spheroidal body, situated either in the central or pheripheral position, the nucleolus is supposed to regulate the synthetic activity of the nucleus. Usually two or more chromosomes are associated with the nucleolus (this can be seen during late prophase) and these are called *nucleolar organisers* as they play a role in the reappearence of the nucleolus after cell division. The number of nucleoli per nucleus varies form one to two or three.

Chemically the nucleolus is rich in RNA. According to Palade (1963), the nucleous of liver cells contain 3 - 5% RNA, while pea embroys have 10 - 20% of nuclear RNA. Orthophosphates helping in RNA synthesis have been reported in nucleolus. Besides this many enzymes (acid phosphatase, nucleoside phosphorylase, RNA methylase etc) have been reported in the nucleolus.

Types of nucleoli : According to Wilson, there are two kinds of nucleous viz., Plasmosomes and Karyosomes (Ocata). The plasmosomes are positive to acidic stains and have a transparent outer region with a dense interior. The karyosomes stain positively with basic dyes. Three kinds of karyosomes have been identified-*Netknot* type (like nodules attached to threads), *Chromosome nucleoli* (occuring only in gametocytes) and *Karyospheres* (spheriodal bodies with basic chromatin).

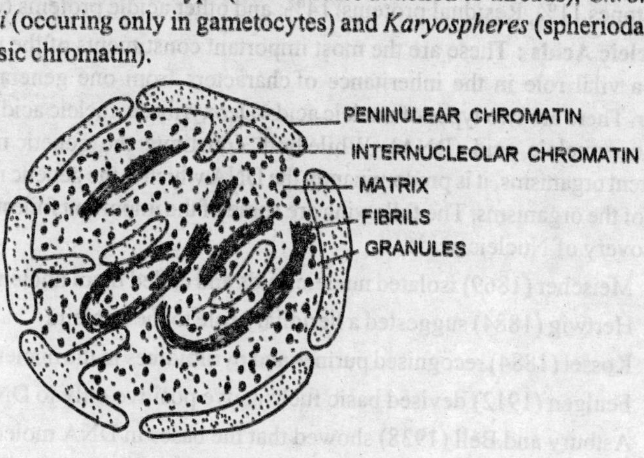

Fig. 2.16 The Cell
Schematic representation of Ultra structure of nucleus

In some instances, plasmosomes and karyosomes combine to form amphinucleoli; these are common in molluscan eggs.

Fine structure of the nucleolus : Detailed analysis of the nucleolus has revealed the presence of three regions. These are granular region, fibrillar region and amorphous region. The granular region occuring at the peripheral region of the nucleolus is composed of granules of 150-250Å thick. This region has ribonucleic acid and proteins. The fibrillar region also known as nucleolonema is composed of fibrils 50-80Å in length. The amorphous region has low electron density and is the one to disappear first at the time of cell division.

There is no limiting membrane to the nucleolus and the organisation is maintained by the calcium ions.

Functions of nucleolus

1. It is the active site of RNA synthesis
2. It is the source of ribosoinal RNA
3. It produces precursors of ribosomes
4. According to Maggio (1990), protein synthesis takes place in the nucleolus. But this may not be totally true.

Endosomes : These are granular structures present in the karyolymph and are smaller in size than the nucleolus.

Chemical composition of Nucleus : The main constituents of nucleus are nucleoprolcins, besides enzymes and inorganic salts. The following gives the total distribution of nucleoproteins in the mammalian liver cells-DNA 9%, RNA 1%, Histones 11%, Residual proteins 14%, and other acidic proteins 65%.

Nucleic Acids : These are the most important constituents of the nucleus playing a vilal role in the inheritance of characters from one generation to-the other. There are two types of nucleic acids-Deoxyribose nucleic acid (DNA) and Ribose nucleic acid (RNA). While both constitute the genetic material in different organisms, it is predominantly the DNA which is the genetic material in most of the organisms. The following are some of the important milestones in the discovery of Nucleic acids.

1. Meischer (1869) isolated nucleic acids and called them nuclein.
2. Hertwig (1884) suggested a heriditary role to the nuclein.
3. Kossel (1884) recognised purines and pyrimidines in the nucleic acids.
4. Feulgen (1912) devised basic fuchsin dye positive only to DNA.
5. Astbury and Bell (1938) showed that the bases in DNA molecule are arranged one above the other.
6. Avery, Macleod and Macarty (1944) presented evidence for DNA to be the genetic material.
7. Chargaff (1950) showed that DNA contains equal amount of puriness and pyrimidines.
8. Wilkins et al (1950) identified the molecular distances between the bases.
9. Furberg (1952) opined that DNA is a coil of single chain.
10. Pauling and Corey (1952) suggested that DNA is a helix of three chains.
11. Waston and Crick (1953) demonstrated the double helix model for DNA molecule.

12. Meselson and Stahl (1958) demonstrated the semi conservative replication of DNA molecule.
13. Jacob and Monad (1961) postulated the presence of RNA in protein

Deoxy Ribose nucleic Acid (DNA) DNA is the main genetic material in almost all organisms except for certain Viruses.

In prokaryotes like *Escherichia* coil the genetic material is a single giant molecule of DNA. While DNA is mainly concentrated in the nucleus, it has also been reported in cytoplasmic organelles like chloroplast, mitochondria etc.

Chemical composition of DNA : DNA extract from many living organisms has revealed it to be a complex macromolecule. It is a long chain polymer composed of monomers (units) called *nucleotides.* Each nucleotide in turn is composed of three sub units. These are a) nitrogen base, b) sugar and c) phosphate.

Nitrogen bases : The nitrogen bases are of two kids viz. *pyrimidines* and *purines.* Both pyrimidines and purines are heterocyclic compounds, with a basic six membered ring in which one or the other carbon position is replaced by nitrogen.

The pyrimidines have a basic six membered ring of benzene in which, one or the other carbon position is replaced by nitrogen. There are two kinds of pyrimidines-*cytosine* and *thymine.* Structurally cytosine has a single six membered ring with a hydroxyl group at the 2' position and an amino group at the 6' position (*2 hydroxy, 6 amino pyrimidine*). Thymine also has a single six membered ring with hydroxyl groups at the 2' and 6' position .and a methyl group attached to the 5' position (*2.6' hydroxy, 5 methyl pyrimidine*).

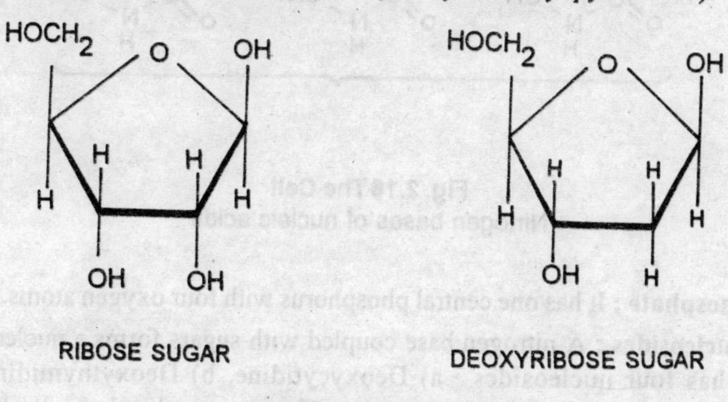

Fig. 2.17 The Cell
Pentose sugars of nucleic acids (left, robose; right deoxyribose)

The purines have a double ringed structure. Attached to the six membered benzene ring is a 5 membered imidazole ring. The imidazole ring is joined to the hexagonal ring at 4' and 5' positions. The benzene ring has nitrogen in the 1' and 3' position (replacing C).

There-are two types of purines viz., *Adenine* and *Guanine*. Adenine has an amino group at the 6' position (*6 aminopurine*), while guanine has an aminogroup at the 2' position and a hydroxyl group at the 6' position (*2 amino, 6 hydroxy purine*).

Sugar : The sugar molecule in DNA is basically a ribose sugar. The ribose sugar is a pentose sugar. It is a 5 membered ring. The ring consists of four carbon atoms while the 5th C atom is outside the ring. Position 1 '.3' and 5', in the ring have OH groups while the 2nd carbon position has only H. There is no O; hence the sugar is said to be Deoxyribose (no oxygen in the second carbon position).

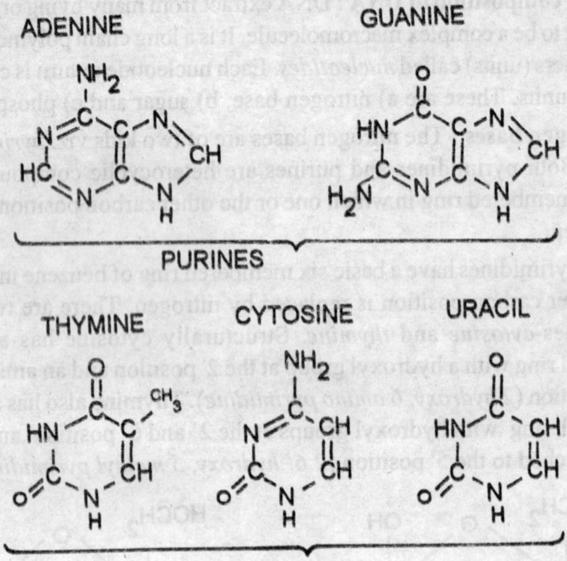

Fig. 2.18 The Cell
Nirtogen bases of nucleic acids

Phosphate : It has one central phosphorus with four oxygen atoms.

Nucleosides : A nitrogen base coupled with sugars forms a nucleoside. DNA has four nucleosides : a) Deoxycytidine, b) Deoxythymidine, c) Deoxyadenosine and d) deoxyguanosine. The sugar molecule is attached to the nitrogen base to the 3' position of the imidazole ring in the case of purines.

Nucleotides : A nucleotide is derived from the nucleoside by the addition of phosphoric acid molecule to the latter. The phosphate group is usually linked to the two sugar molecules above and below. The linkage is with the 3rd C atom of one deoxyribose sugar and with the 5th C atom of another. DNA has four nucleotides- Deoxycytidilic acid, Deoxy thymidilic acid, Deoxy adeniylic acid and Deaxyguanylic acid.

Polynucleotide : DNA is said to be a polynucleotide chain with a number of nucleotides joining together with the help of DNA. polymerase.

Molecular model of DNA structure : A molecular model of DNA structure first proposed by J.D. Waston and F.H.C. Crick (1953) has been universally accepted. They were awarded Nobel prize for their discovery. The work of Crick and Watson has been supported by X-ray diffraction picture obtained by Wilkins and his associates.

Fig. 2.19 The Cell
F.H.C. Crick (Left) and J.D. Watson (Right)

According to this model, a DNA molecule is composed of two chains that are spirally coiled and form a double helix, much like a spiral stairway. The distance between the two chains is uniform and it is maintained by base pairing. The back bone of the chain is made up of alternating sugar and phosphate molecules. At specific , intervals, the chains are joined by 'steps.' Each step is made up of two nitrogen bases. Each nitrogen base is attached to the sugar molecule of the opposite chain. Between themselves, the nitrogen bases of a 'step' are joined by hydrogen bonding. The pairing of the nitrogen bases is always specific. It is made up of a purine and a pyrimidine i.e., a double ring and a single ring. Two purines would occupy too much space and can not be accommodated between two chains: similarly two pyrimidines would form a short step and leave a gap between two chains. Hence the pairing is always

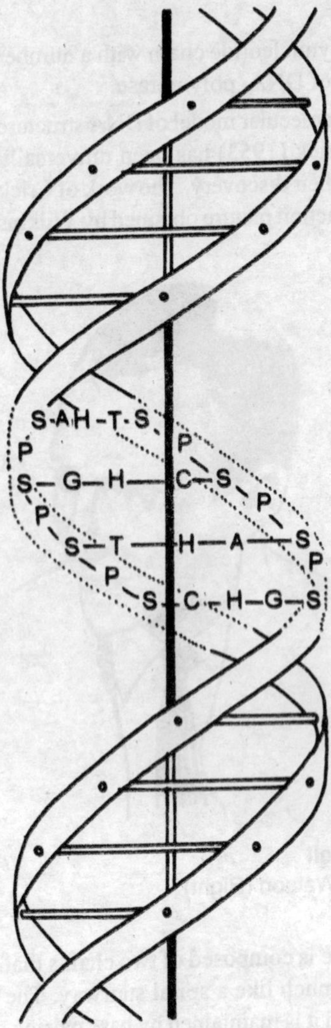

Fig. 2.20 The Cell
The double helix model of DNA

between a purine and pyrimidine. Because of this pairing, in a DNA molecule total purine content is always equivalent to total pyrimidine content

The purine pyrimidine base pairing is further specific. It is always Adenine with Thymine and Guanine with Cytosine. Adenine and Thymine are joined by two hydrogen bonds through atoms attached to positions 6' and 1'. Cytosine and Guanine are joined by three hydrogen bonds at position 6', I' and 2'. The bonding of hydrogen is weak; but this facilities easy separation of two strands during replication. According to Crick and Waston, the helix has a diameter of 20Å and there is one complete turn of the helix at every 34Å and has a stack of 10 nucleotides at every turn.

One unique character of DNA is its unity and yet infinitesimal variety. The DNA structure mentioned above is uniform in all organisms, whether it is man or virus. At the same time it is also true that DNA is the genetic material that decides the characters of individual organisms. Then how does DNA decide the individuality. The answer lies in the base pair sequence. The combination of A/T, G/C is uniform, but their sequence is not; for instance in one individual it may be AT, GC, AT, AT, AT, in another-GC, GC, AT, GC, AT etc. There is no limit to the variety in which base pair sequences are arranged. This is the secret of individual differences. In no two individuals, thebase pair sequence is same. **Unusual bases In DNA :** The normal bases in DNA are A, T, G, and C. In some viruses like PBSIand PBS2 uracil replaces thymine. In some bacteriophages C is replaced by 5 hydroxy methyl cytosine.

Single strand DNA : Although DNA in most organisms is double stranded in some phage viruses (f x 174), single stranded DNA is found. The following table summarises the differences between double stranded DNA and single stranded DNA.

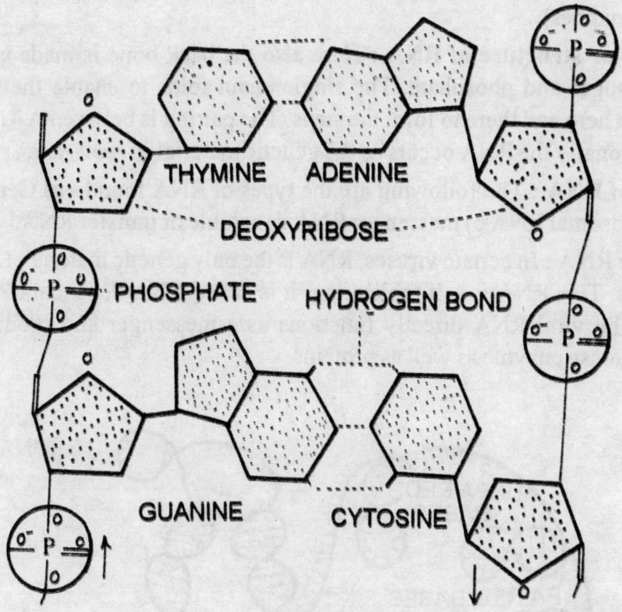

Fig. 2.21 The Cell
Arrangement of nucleotides in a segment of the DNA model

Table 3

	Character	Double Strand DNA	Single Strand DNA
1.	U.V. absorption	constant from 10-80^{0}C and then rises rapidly	steady increase from 2^0 to 90^{0}C.
2.	Action of formaldehyde	Resistant	not resistant
3.	Ratio of nucleotides	Equal, A = T and G = UC	A: T: G: C is 1.33:0:98:0.75
4.	Shape	Linear	Circular
5.	Replication	two strands separate and replication begins	becomes double stranded (replicate from) temporarily)

Ribonucleic acid (RNA) : The RNA is also a polymer of nucleotides. These are called ribonucleotides. Each ribonucleotide has three parts viz., a pentose sugar, a nitrogen base and a phosphate group.

The pentose sugar is a ribose sugar much like that of DNA, but with oxygen (present) attached to the 2' carbon position of the ring.

The nitrogen bases are - Adenine and guanine (purines) and cytosine and uracil, (pyrimidines)

Molecular structure of RNA : Here also the back bone is made up of alternating sugar and phosphate. The single chain folds to enable the base pairs to meet here and there to form the steps. The pairing is between A-U and C-G. Occasionally the RNA occurs as the genetic material in *reoviruses*.

Types of RNA : The following are the types of RNA found - a) Genetic RNA b) Ribosomal RNA c) messenger RNA d) soluble or transfer RNA.

Genetic RNA : In certain viruses, RNA is the only genetic material, DNA being absent. This RNA is self replicating; it is called RNA dependant RNA synthesis. The viral RNA directly functions as a messenger and produces RNA polymerase enzyme as well as proteins.

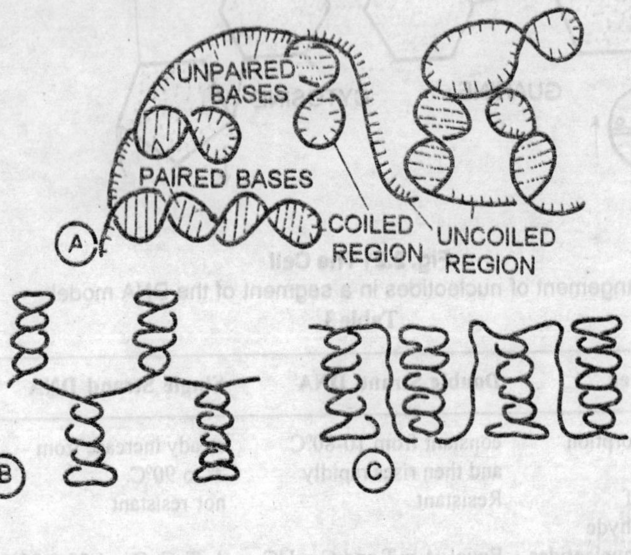

Fig. 2.22 The Cell
Secondary structure of ribosomal RNA (Diagrammatic)
A. Extended strand, B. Negative interaction showing repulsion,
C. Positive interaction showing association

Ribosomal RNA (rRNA) : This is found in the ribosomes. It constitutes about 80% of total RNA of the cell. The base pairs of rRNA are complementary to that part of the nuclear DNA from where they are produced. rRNA consists of a single strand folding upon itself at some regions. The folded regions have paired bases, while the other regions have unpaired bases. Due to this in rRNA there is no purine pyrimidine equality.

There are three types of rRNA based on sedimentation and molecular weight. These are a) rRNA with molecular weight over a million, eg. 21s - 29s rRNA, b) rRNA with molecular weight less than a million, eg. 12s- 18s RNA, c) rRNA with very low molecular weight (40,000) eg 5s rRNA.

Messenger RNA (mRNA) mRNA carries the message from DNA to ribosomes during protein syntheses. It is about 3 - 5% of total RNA content of the cell.

mRNA has a molecular weight of about 5,00,000 with a sedimentation coefficient of 8s. This however is highly variable, mRNA is always single stranded and there is no base pairing, even though random coiling may be observed. The mRNA molecule is produced from the coding strand and its base sequence is complementary to the coding strand. The following are some of the general features of mRNA.

1. **Cap :** This is a blocked methylated structure found at the 5' end of RNA.
2. **Non coding region :** This follows the cap and consists of 10 - 100 nucleotides. There are no codes for proteins in this region.
3. **Initiating codon :** This region begins the coding.
4. **Coding region :** Codes for proteins and has about 1500 nucleotides.
5. **Terminating codon :** The coding region is followed by a terminating codon, which stops translation.

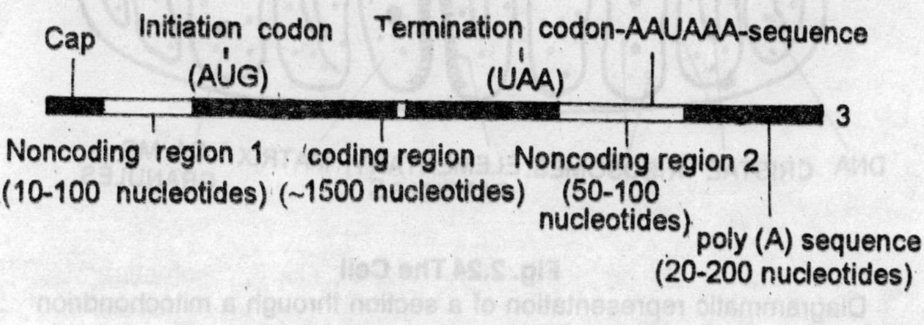

Fig. 2.23 The Cell
General features of eukaryotic mRNA

Transfer RNA (tRNA or sRNA) : The transfer RNA also called soluble RNA, is next in quantity to rRNA. It has a low molecular weight of about 25,000 - 30,000 and the sedimentation coefficient is 80s. It is made up of about 70 - 90 nucleotides.

tRNA is synthesised in the nucleus on a DNA template. Only a minute fraction of DNA (0.025%) codes for tRNA. The tRNA molecule consists of a single strand looped around itself with the 3' end always terminating in a C-C-A sequence. Some bases are paired.

The function of tRNA is to carry aminoacids to mRNA during protein synthesis, Each amino acid couples with a specific tRNA. Since the protein molecule consist of of about 20 types of amino acids, there must at least be 20 types or tRNA. But there could be many more, since there are two tRNAS for every amino acid in several, cases.

Mitochondria: (Gr. *mito* = thread, *chondiron* = granule)

Popularly called 'power house' of cells, these granular or filamentous organelles are present in all eukaryotes. They are characterised by their special structure, and unique physiological function concerning intracellular energy transformation.

Historically, mitochondria were first observed by Altmann (1894) who called them *bioblasts*. The present name was coined by Benda (1897) who developed the crystal violet staining. Bensely and Hoerr (1934) first isolated the liver cell mitochondria. The discovery that mitochondria are seats of cellular respiration was made by Hogeboom *et al* in 1948.

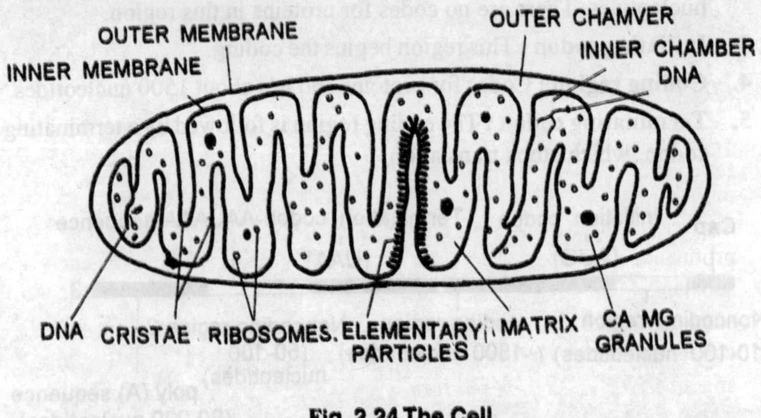

Fig. 2.24 The Cell
Diagrammatic representation of a section through a mitochondrion (ETPs are not shown on all christae)

It is difficult to observe mitochondria in living cells due to their low refractive index, but they can be studied under phase contrast microscope after staining with a dilute solution of janus green. Mitochondria may be isolated from cells by centrifugation between 20,000 g -40.000 g.

Shape : Generally mitochondria are granular or filamentous, even through the shape is variable. During certain conditions a long mitochondrion may swell at one end lo become club shaped or look like a tennis racket .(De Robertis et al 1970). Mitochondria may also become vesicular.

Size : Size also is variable. The width is about 0.5μm and the length may reach a maximum of 7μm. The si/e depends on the osmotic pressure.

Distribution : In general, mitochondria are distributed evenly in the cytoplasm, but in some instances they may accumulate around the nucleus or at the periphery of the cytoplasm. During cell division, mitochondria accumulate near the spindles. The distribution of mitochondria within the cell is in relation to the energy demand. For example in skeletal muscles they are grouped like rings or braces around the I-band of the myofibril.

Orientation : Mitochondria have more or less a specific orientation. In cylindrical cells, their long axis is parallel to the long axis of lhe cell. In leucocytes, mitochondria are arranged radially in respect to the centrioles.

Number : Mitochondria are absent in prokaryotic cells. The quantity of mitochondria per cell varies, but in liver cells it is estimated that mitochondria contains 30-35% of the total protein content of the cell. A normal liver cell contains about 1000-1500 mitochondria. Some oocytes coniain an many as 300000 mitochondria.

Structure of mitochondria : The mitochondrion is enveloped by two membranes viz, *inner membrane* and *outer membrane.* The outer membrane is 6nm thick; internal to this is the inner membrane. Separating these two is a space (outer chamber) filled wilh a watery fluid. This space is 40-70Å thick. The inner membrane is folded into finger like projections called *cristae.* This membrane encloses the inner chamber or matrix, usually filled by dense protinaceous material. The matrix may be homogenous or may have a granular substance. The cristae of the inner membrane facing matrix is called the M side, while the other side is called the C side. The cristae have an inner space called the intercristae space which is in continuity with the intermembrane space.

Within the mitochondrial matrix are small ribosomes and a circular DNA. The ribosomes are of 70s type and similar to those of bacteria. These can synthesise proteins (Tyier 1973). The DNA called mDNA may be circular as in most of the animal cells or linear as in plant cells. Rabinowitch (1968) has demonstrated that the mitochondrial DNA has more G-C content than the nuclear

DNA.

The mitochondrial DNA can replicate itself as it has DNA ploymerase. It is due to this, that mitochondria can replicate themselves.

The inner membrane : The inner membrane has numerous nail like or tennis racquet like projections protruding into the matrix. Each particle is about 70-100A⁰ in diameter. The particles are placed apart a distance of 100A⁰ and each is attached to the inner membrane by a short stalk,of 35-50A⁰ long. The number of these particles per mitochondrion ranges from 10^4-10^5. These are called elementary particles, F_1 *particles* or *subunits.* It was formerly believed that these particles contain all the enzymes necessary for electron transport and oxidative phosphorylation. Hence they were named *electron transport panicles* (ETP). Each particle was thought to have four complexes, with complex I and II in base, III in stalk and IV in the head piece. According to current thinking, the head piece has ATPase, the stalk consists of a special protein (oligomycin sensitivity confering protein) and the piece contains the proton channel.

The inner membrane has the respiratory chain consisting of five complexes. Of these, four are involved in the electron transport, while the fifth one is involved in the synthesis of ATP.

Mitochondria play a vital role in respiration. Glycolysis takes place in the *cytosol* (cytoplasm) while the enzymes of the Kerbs cycle are found in the matrix, those of oxidative phosphorylation are found in the inner membrane.

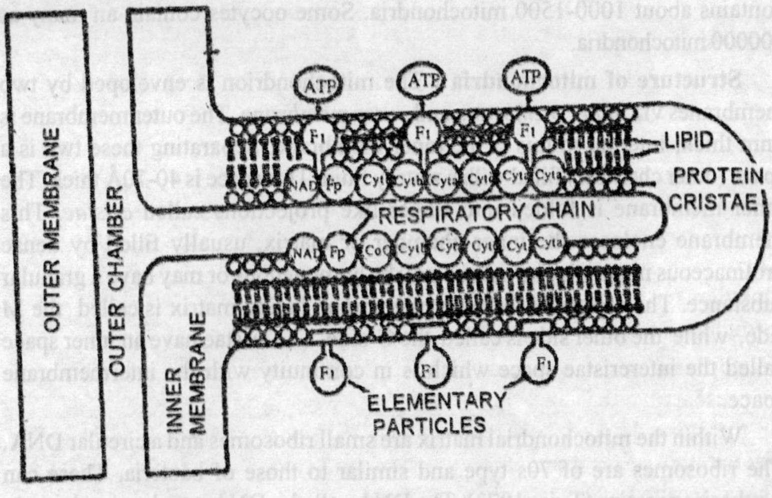

Fig. 2.25 The Cell
Arrangement of complexes (I-IV) in the ETP (old view)

Biogensis of mitochondria : Each mitochondrion lasts for about a week or little more. There are three views regarding the origin of mitochondria-a) origin from cell membranes, b) Division of pre existing mitochondria and c) *de novo* origin.

According to Robertson (1964), mitochondria in muscle cells arise by the infloding of the plasma membrane. This infolding joins another piece and forms a vesicular structure. In the nerve fibres of cray fish, endoplasmic reticulum may form mitochondria.

According io Luck and Rich (1964), mitochondria are capable of self feproduction. DNA, protein synthesis, DNA polymerase and all the other essential, features for self repliciation are present in mitochondria.

Harvey (1951) supports a *de novo origin* for mitochondria. Centrifugal separation of sea-urchin eggs into two halves, a nucleated half (without mitochondria) and a non nucleated half (with mitochondira) showed that the nucleated half produced the mitochondria.

Are mitochondria symbolic organisms? The many similarities between a bacterial cell and mitochondria made many people (Lloyd 1974; Borst 1977, O'Brien 1977 believe that the mitochondrion was once an autonomous entity which later entered into a symbiolic association with an anaerobically respiring host cell. The host cell derived its energy only from glycolysis. Subsequently with the association, the host cell started respiring aerobically. De Robertis (1970) opines that in evolutionary terms it is possible that a symbiotic relationship could have evolved into the present semiautonomus condition of the mitochondria.

The same theory has been Proposed for chloroplasts also, as they also have all (the independent attributes of mitochondria.

PLASTIDS : These are living cytoplasmic inclusions found in most of the plants except for some bacteria and some fungi.

The plaslids are of three categories viz., chromoplasts, leucoplasts and chloroplasts.

Chromoplasts : These are pigmented plastids. The pigments are non chlorophyllous like, carolenes, xanthophylls, fucoxathin, phycoerythrin etc. In photosynthetic bacteria, the chromoplasts have bacteropchlorophyll.

Leucoplasts : They are colourless plastids. They lack pigments and are usually present in cells which do not receive direct light. Leucoplasts may be seen in the storage leaves of onion. Leucoplasts that store starch are called *amyloplasts*, those that store oil are called elaioplasts and the ones storing proteins are called *aleuroneplasts*.

Chloroplasts : These are by far, the commonest and the most plastids. As the primary sites for trapping and converting solar energy they are very vital for the existence of not only the green plants, but for the whole living world.

Shape : Chloroplasts have varied shapes. The algal chloroplasts exhibit an array of shapes. They may be cup shaped *(Chlamydomonas)*, girdle shaped *(Utothrix)*, reticulate *(Hydrodictyon)*, stellate *(Zygnema)*, ribbon shaped *(Spirogyra)* or discoid. They are generally spherical or ovoid in higher plants.

Size : There is a great variety in siz.e. Normally they are about 1μm thick and 4-6μm in length. Chloroplasts ofpolyploid cells are generally larger than in the diploid cells. Cells in the shaded area have longer chloroplasts but with less intensity in colouration.

Distribition : They arc uniformly distributed all over cytoplasm, but in some instances they cluster towards the nucleus. The concentration of chloroplasts will also depend on light intensity.

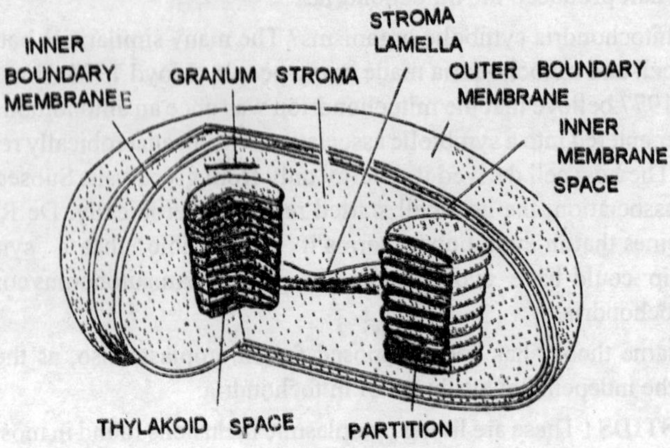

Fig. 2.26 The Cell
Model of Unltrastructure of chloroplasts

Number : The number of choloroplasts per cell varies. In some algae like *Chlamydomonas*, there is only one chloroplast per cell or two as in *Zygnema*. In higher plants, cells have a number of chloroplasts. The leaf of *Ricinus communis* has nearly lour lakh chloroplasts per square millimeter area.

Che·nical composition : The chloroplasts contain carbohydrates, lipids, proteins chlorophyll, carotenoid and xanthophylls. The protein lipid ratio is 40:30. Besides the above, nucleic acids have also been demonstrated in the chloroplasts.

Ultra structure of the chloroplasts : The chloroplast has a covering of two membranes with an inner membrane space. These membranes are smooth and there are no perforations or particles. The membranes are differentially permeable (Mudrak).

A section of the chloroplast reveals an intricate system of membranes enclosed in a granular matrix. These membranes are called lamellae and the surrounding matrix-the *stroma*. In a sectional view, the lamellae can be seen packed and these stacks are called *thylakoids*. In the chloroplasts of higher plants, the thylakoids themselves form highly compact bundles called *grana*. Some thylakoids of granum extend into the stroma and maintain contact with other grana. These are *called-Stroma thylakoid* or *stroma lamellae* or *intergrana*.

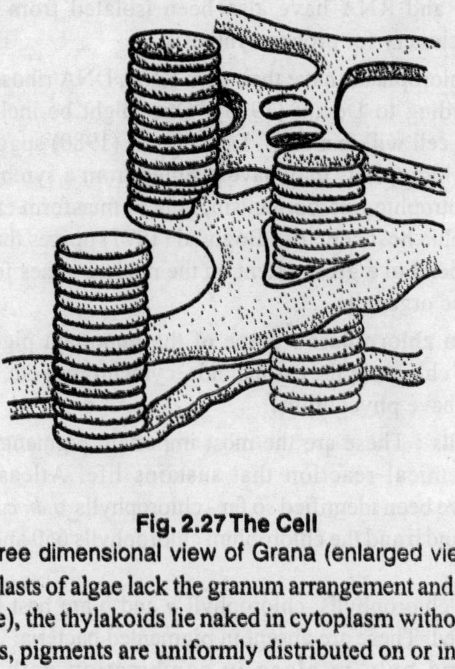

Fig. 2.27 The Cell
Three dimensional view of Grana (enlarged view)

The chloroplasts of algae lack the granum arrangement and in cyanobacteria (blue green algae), the thylakoids lie naked in cytoplasm without any envelope. In such instances, pigments are uniformly distributed on or in the lamellae.

The stroma has in addition to the lamellae - granules (globuli), lipid droplets, starch grains and vesicles.

Structure of the lamellae : All the photosynthetic pigments are concentrated in the chloroplast lamellae. The lamellae as well as the outer membranes are composed of lipoprotein sub units, each sub unit according to Weir and Benson has a protein core surrounded by a lipid sheath. The chloroplast membranes have a single layer of sub units, whereas in thylakoids it is double due to the juxtaposition of the membranes. The appressed areas are called partitions. These are hydrophobic in nature The space between the two membranes of thylakoid (fret membrane) is called *fret channel*. The space between two thylakoids is called *loculus* while the ends of disc shaped thylakoids are called *margins*. These areas (loculus, fret channel) are hydrophilic. The chlorophyll molecules are concentrated in the fret channel ie., in the space between the two membranes of thylakoids. The chlorophyll molecules may be entirely in the space or the *heads* may be in the fret channel while the *tails* are buried in the sub units, occasionally the entire chlorophyll molecule may be found in the fret membrane. Other pigments like cytochromes, carotenoids, etc., are also found in the fret membranes.

Four sub units of the partition constitute a photosynthetic unit or a *quantosome*.

Ribosomes and RNA have also been isolated from the chloroplasts indicating a machinery for protein synthesis.

Since the chloroplasts, have their own RNA, DNA ribosomes and protein synthesis, according to Deviln (1975) "One might be inclined to view the chloroplast as a cell within a cell". De Robertis (1980) suggests "It has been suggested that chloroplasts may have resulted from a symbiotic relationship between an autotrophic microorganism, able to tmasform energy from light, and a heterotrophic host cell" But Bogorad (1977) opines that the chloroplast components depend to a great extent on the nuclear genes for their assembly into an integrated organelle.

Pigments in chloroplast : Some of the important pigments present in chloroplasts are chlorophylls, carotenoids, cytochromes etc. In algal cells the thylakoids also have phycobilins.

Chlorophylls : These are the most important pigments involved in the basic photochemical reaction that sustains life. Atleast nine types of chlorophylls have been identified so far - chlorophylls *a. b. c. d* and *e*. Bacterio Chlorophylls a and b and the chlorobium chlorophylls 650 and 660μm.

Among the chlorophylls, chlorophyll *a* and *b* are best known and most widely distributed. These are absent in pigmented bacteria. Chlorophylls *c, d* and e are found only in algae in combination with chlorophyll a. Bacteriochlorophylls and chlorobium chlorophylls are found only in photosynthetic bacteria.

The chlorophyll molecule is a complex structure basically made up of a 'head' and 'tail' resembling a tennis racquet. The 'head' is a *porphyrin* structure made of four *Pyrrole rings* attached to each other at the centre by an isocyclic ring containing Mg atom at the centre. Extending from one of the pyrrole rings is the 'tail' - the alcoholic chain *(phytol)*. The emperical formula for the chlorophyll molecule is CH, O,N Mg. The phytol chain is esterified on the C atom of one of the pyrrole rings and has only one double bond.

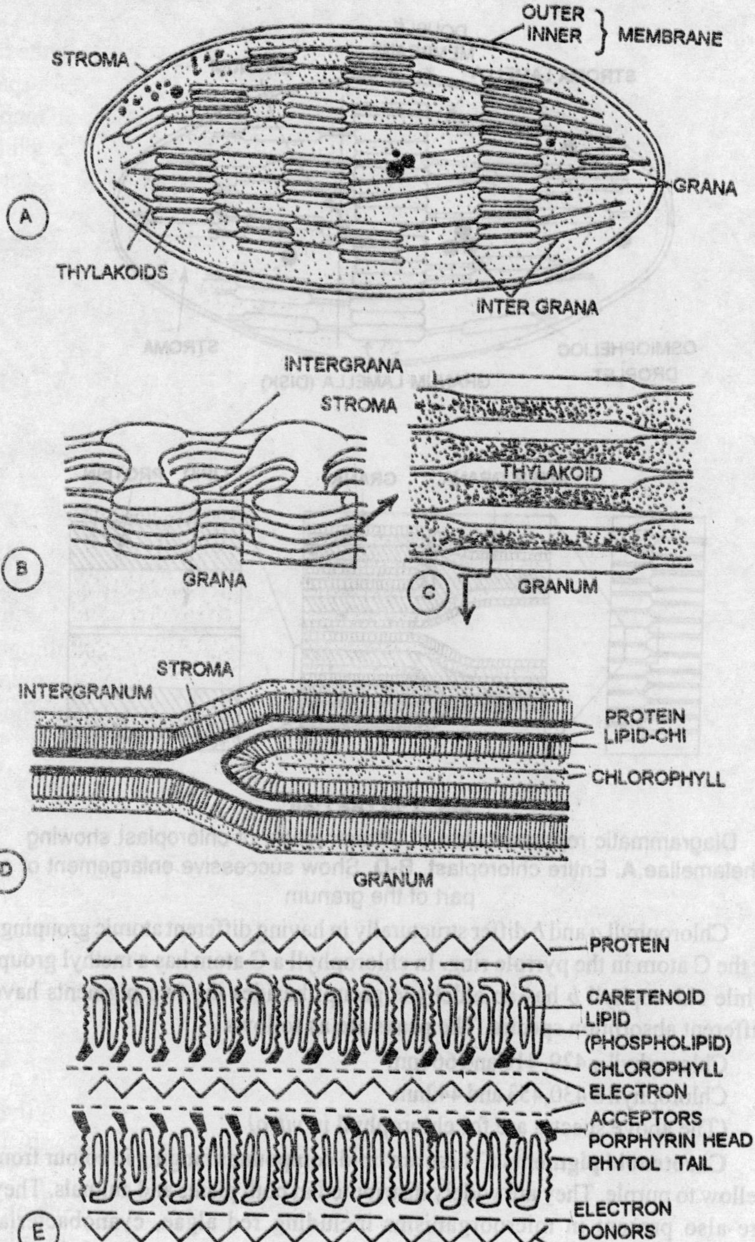

Fig. 2.28 The Cell
Cell structure of Chloroplast

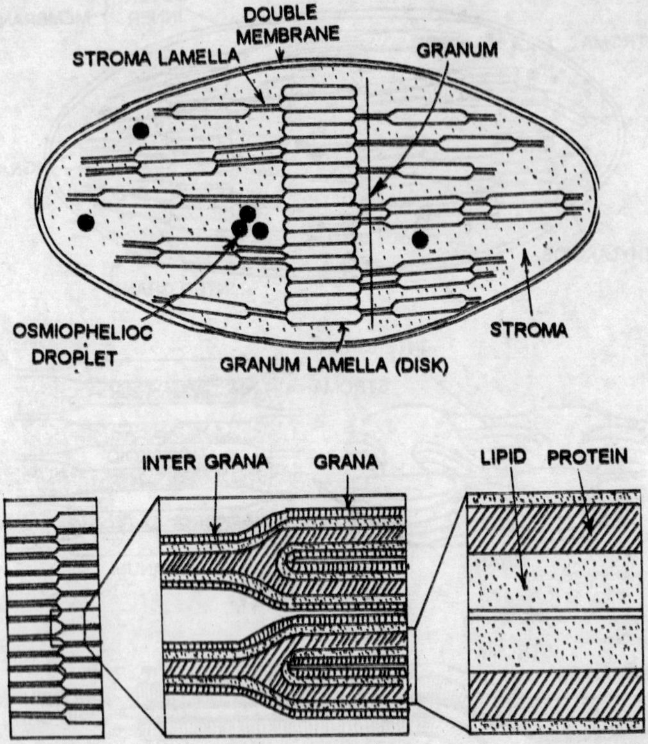

Fig. 2.29 The Cell

Diagrammatic representation of Ultrastructure of chloroplast showing thelamellae. **A.** Entire chloroplast, **B-D.** Show successive enlargement of a part of the granum

Chlorophyll *a* and *b* differ structurally in having different atomic groupings at the C atom in the pyrrole ring. In chlorophyll a C atom has a methyl group, while chlorophyll *b* has an aldehyde group; besides the two pigments have different absorption spectra. The peaks are as follows

Chlorophyll *a* 429, 410 and 660nm

Chlorophyll *b* 430,453 and 442nm

(The above spectra are for chlorophyll in *vitro)*

Carotenoid pigments : These are lipid compounds ranging in colour from yellow to purple. They are widely distributed in both plants and animals. They are also present in microorganisms including red algae, cyanobacteria, photosynthetic bacteria, fungi, etc. Wackenroder (1931) isolated the first carotenoid - carotene from tissue.

Carotenoids are derivatives of *lycopene* a red pigment found in many plants. They (carotenoids) are known to have eight isoprene ($CH_2 = C[CH_3]$ - $CH = CH_2$) like residues.

The main carotenoid found in plants is the orange yellow coloured β carotene with some quantity of α carotene.

The carotenoids are also located in the chloroplast. According to Goodwin (1960), carotenoids and chlorophylls may be combined with the same protein to form a complex known as *photosynthein*.

Carotenoids protect the chorophyll form photooxidation and transfer the light energy they absorb to chlorophyll *a*.

Phycobilins: These are found only in algae. There are two types of pycobilins - the *Phycoreythrin* (red) and *Phycocyanin* (blue) These pigments are strongly associated with a protein and consequently it is difficult to isolate the pigments in the pure state.

Like carotenoids phycobilins are also involved in the transfer of light energy (they absorb) to the chlorophyll.

The absorption spectra of phycobilins presents an intersting study considering the fact that they act as accessory pigments in photosynthesis. R - phycoerythrin has peaks at 495, 540 and 545nm, while R - phycocyanin has peaks at 550 and 615nm.

ENDOPLASMIC RETICULUM : The cytoplasm in the cells of most of the plants and animals is traversed by a network of anastamosing fibrillar structures called endoplasmic reticulum (ER). Originally known as ergastoplasm, the current name ER was given by Porter and Kaallman (1952). The ER is connected to the outer nuclear membrane with the ER space opening into the perinuclear space between two membranes. The ER is known to be connected with the plasma membrane.

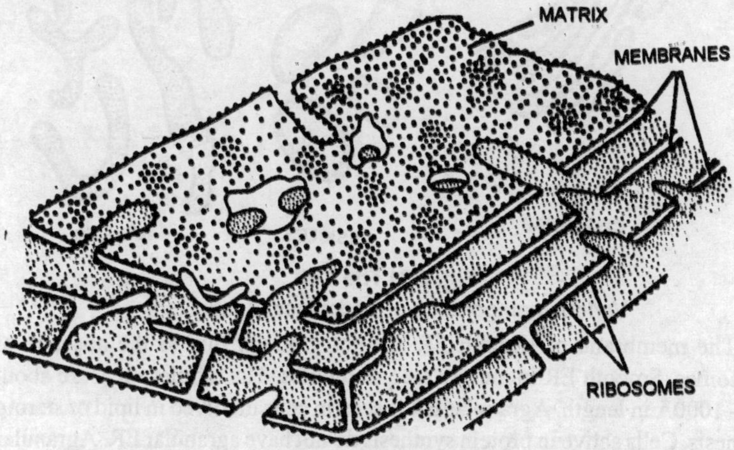

Fig. 2.30 The Cell
Three dimentional view of endoplasmic reticulum

Morphologically the ER consists of three types of structures viz., *Cistemae, Vesicles* and *Tublllues.*

The **Cisternae** *are* elongated cyllindrical unbranched structures arranged parallel to one another respresenting a lamellated structure. They are 40 - 50μm in diameter. Cisternae are generally found in cells active in synthesis.

The **Vesicles** or sacs or oval bodies, attached to the membrane and have a diameter of 25-500μm. These are abundant in pancreatic cells.

The **Tubulues** are branched irregularly and are of diverse shapes having a size of 50 - 100μm. The tubules are abundant in cells which are active in the synthesis of steroid compounds. They are also found in the epithelical cells of retina.

Ultra structure : All the three components of ER are bounded by a unit membrane 50 - 60Å thick. The membrane is trilamenar like the other unit membranes. Palade (1956) has observed some secretory granules in the cavity of endoplasmic reticulum. The membrane has two outer layers of protein enclosing two layers ofphospholipids. However the ER membrane is symmetrical unlike in other organelles (mitochondria).

Types of ER : Two types of ER have been observed in the cells (same or different). These are - smooth ER (agranular) and rough ER (granullar).

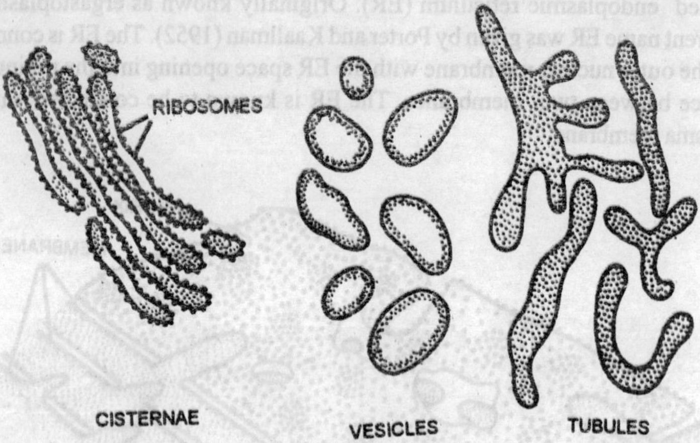

Fig. 2.31 The Cell
Various components of endoplasmic reticulum

The membrane of smooth ER are not granular due to the absence of ribosomes. Smooth ER in mainly seen in tubules. The tubules measure about 500 - 1000Å in length. Agranular ER is seen in cells involved in lipid or steroid synthesis. Cells active in protein synthesis do not have agranular ER. Agranular reticulum of muscle cells is called *sarcoplasmic reticulum* and it forms lace like sleeve around myofibrils. In the pigmented cells of retina, the agranular ER occurs in the form of tightly packed tubules and vesicles called *myeloid bodies.*

The rough ER has a granular surface due to the presence of a large number of ribosomes attached to it. Since ribosomes are actively involved in protein synthesis rough ER is conspicuous in cells that are actively involved in protein synthesis e.g. meristematic cells, plasma cells, goblet cells etc. The rough ER stains positively by basic dyes.

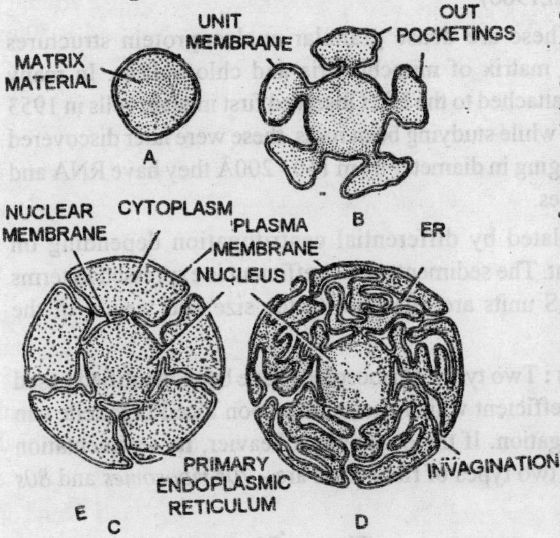

Fig. 2.32 The Cell
Stages in the origin of Endoplasmic reticulum

The cells (Spermatocytes) of certain vertebrates show an unnsual type of ER. Usually ER membrane does not have pores or openings. But in the instance mentioned above pores or annuli have been reported resembling the ones found in the nuclear membrane. There is a diaphragm across the octogonal symmetry (Manul 1968).

The ER performs many vital cellular functions. These are

(a) Enzymatic activities
(b) Transport of synthetic products
(c) Storage of metabolities
(d) Formation of nuclear envelope
(e) Intracellular impulse conduction etc.

Microsomes : These are structures that result due to the homogenization of the membranes. The cell membranes break up into fragments, round off and form microsomes. Microsomes are actually fragments of ER. They can be isolated by high speed centrifugation (100,000g) when they settle after nuclear and ribosomal fractions. Observed under electron microcope they appear as membrane bound vesicles of 500- 1500 Å in diameter. It should be understood however that microsomes are not natural structures found in the intact cell. They are the result of homogenization of cell. The microsome fraction contains fragmented membranes.

Origin of ER : The origin is uncertain, invagination of the plasma membrane might give rise to a canalicular system. But evidences for this hypothesis are yet to come (Dallmes et al,1966).

RIBOSOMES : These are dense granular nuclecoprotein structures occuring in cytoplasm, matrix of mitochondria and chloroplasts. In many instances ribosomes are attached to the ER. Observed first in plant cells in 1953 by Robinson and Brown while studying bean roots, these were later discovered in animal cells also. Ranging in diameter from 150 - 200Å they have RNA and protein in equal quantities.

Ribosomes are isolated by differential centrifugation depending on sedimentation coefficient. The sedimentation coefficient is expressed in terms of *svedberg units.* The S units are related with the size and weight of the ribosome molecules.

Types of ribosomes : Two types of ribosomes have been identified based on the sedimentation coefficient which shows how soon a cell organelle can sediment under centrifugation. If the organelle is heavier, its sedimentation coefficient is more. The two types of ribosomes are - *70s ribosomes* and *80s ribosomes.*

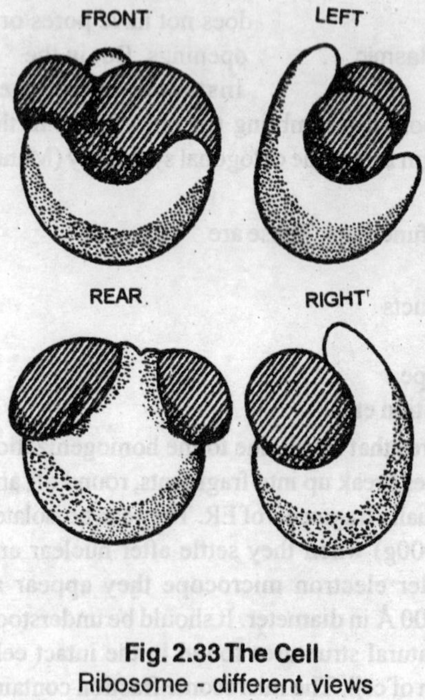

Fig. 2.33 The Cell
Ribosome - different views

The 70s ribosomes are comparatively smaller and have a molecular weight of 2.7×10^6 daltons (Dalton is a unit of molecular weight; one dalton = weight of one H atom = 1) According to Huxley and Zubay (1960) the size of a 70s particle is 170 x 170 x 200Å. These ribosomes occur in prokaryotic cells as well as in chloroplasts of eukaryotic cells. The 70s ribosomes have two sub units - a large 50s sub unit and a small 30s sub unit.

The 80s ribosomes have a moleculer weight of 4×10^6 daltons, and are predominantly found in eukaryotic cells. There are two sub units in an 80s ribosome. These are - 60 and 40s.

Ribosomes may occur singly as isolated units when they are called *monosomes.* When they occur in clusters or groups, they are called *polyribosomes. The* polyribosomes may have a sedimentation coefficient of 100s - 600s.

The number of ribosomes per cell varies, it may be 10,000 (bacterial cell) or up to 10 million (eukaryotic cell).

Ribosomes of chloroplasts and mitochondria have their own protein synthesis. They have a sedimentation coefficient of 55s with two sub units 40s and 30s.

Ultra structure of ribosomes : Ribosomes are oblate or spheroidal structures having two sub units (a large and a small). The large sub unit is dome like and the smaller subunit is placed above like a cap. The 70s ribosome has two units 50s and 30s as already mentioned. The 50s sub unit has a diameter of 140 - 160A^{0}and is pentagonal in shape. In the center of this sub unit occurs a round area of 40 - 60A^0. According to Fernando (1968), there is a pore like transparent area in the centre which prohibits the entry of proteolytic enzvmes.

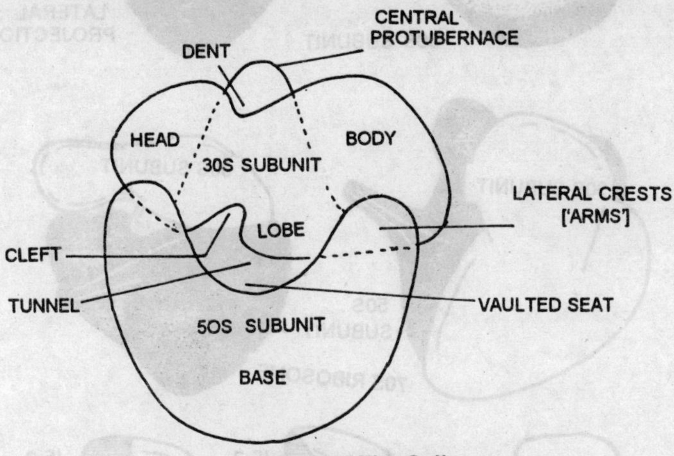

Fig. 2.34 The Cell
Witlmann's model of prokaryotic ribosome

The pores have been found in 50s sub units of 80s ribosomes also. The smaller sub units of both the types of ribosomes are often divided into two portions which remain connected to each other by a strand of 30 - 69A^0 Thickness.

Models for ribosomes structure : There are two models, viz., Wittmann's quasi symmetrical model and Lake's asymmertrical model.

According to Wittmann's model, the 30s sub unit is an elongate slightly

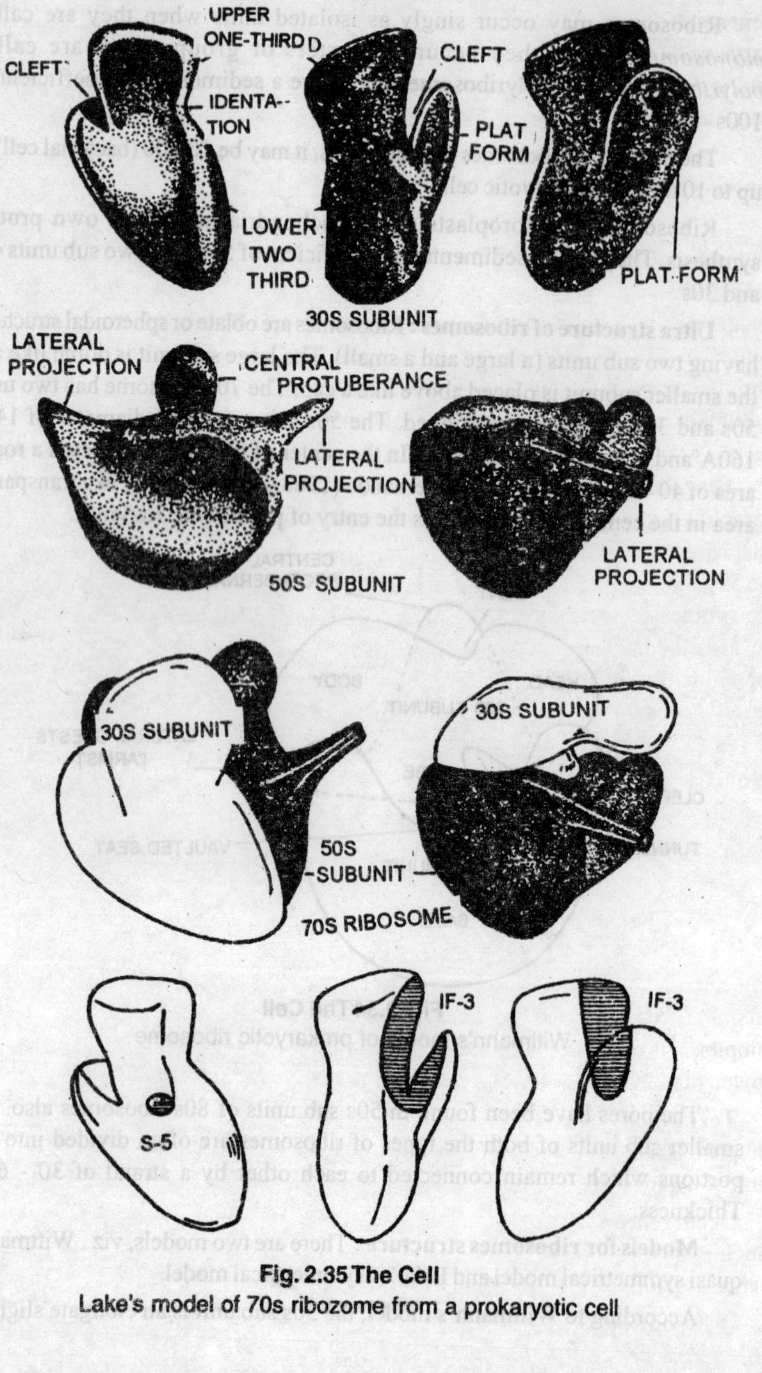

Fig. 2.35 The Cell

Lake's model of 70s ribozome from a prokaryotic cell

bent stucture, having a head and a larger body with a cleft in between; with two parts protruding uniequally. It appears like a telephone receiver. The 50s sub unit is variously shaped - round shaped, maple leaf .shaped etc. The 50s sub unit has three protruberances.

The two sub units of 70s ribosomes have four areas of contact. The mRNA binding. region is located on the right side of the *30s* sub unit.

Lake's model regards the 30s sub unit to be totally asymmetrical. The sub unit is divided into two parts (1/3 and 2/3) unequally. Extending from the lower 2/3 region is a portion called the *platform.* This cleft is important in the sense of location of codon - anticodon interaction. The major difference between the Wittmanns model and Lake's model for the 30s sub unit is that the latter regards the 30s sub unit as totally asymmertrical. According to Lake, even the 50s subunit is asymmetrical.

Biogensis of ribosomes : The synthesis of eukaryotic ribosome is a complicated process involving several areas of the cell. The genes coding for 18s, 28s and 58s RNA are in the nucleolar organizing part of the chromosomes; the 5s genes are localized eleswhere in the chromosome. The ribosomal proteins are synthesised in the cytoplasm. The proteins and 5s gene products move towards the nucleolus, where they are assembled into ribosomes. Evidence for nucleolus being the site of ribosome biogeneis has been obtained by observations on the oocytes of *Xenopus laevis.*

In bacteria, the ribosomes (70s) are synthesised in the cytoplasm as the cistron portion produces a ribosomal RNA which is immediately relased.

Functions of ribosomes : As is very well known, ribosomes (polyribosomes) are the sites of portein systheses. The polyribosomes serve as a platform for the asembly *of* amino acids (on a mRNA template) brought together by specific tRNA from cytoplasm.

GOLGI COMPLEX : Described first by Camillo Golgi in 1890 in the nerve cells of barn owl, the Golgi complex has attracted the attention of cell biologists not only by its complex structure but also by its varied functions. In fact Golgi complex was observed earlier by St George in 1865 itself. Nassonov (1923) and Bowen (1929) established the secretory function in the Golgi apparatus.

Terminoloy : Various terms have been employed to refer to the Golgi complex. Baker (1953) described them as Lipochondria as they have a high lipid content. Golgi complex found in plant cell are often refered to as *Dictyosomes.*

Occurrence and distribution : All the eukaryotic cells have a golgi complex which is absent in prokaryotic cells. Even in eukaryotic cells of some fungi, spenrmatozoids of bryophytes etc. golgi complex is absent.

Distribution : In higher plants, golgi complex is scattered all over the cytoplasm apparently without any localization (Hall 1974). As a contrast in animal cells, the golgi complex is localized into an organelle. In animal cytoplasm, usually there is only one golgi apparatus, but in some cases as in *Storeomyxa* there may be many complexes. In shape, the Golgi complex is variable depending on the cell type; it varies from a compact mass to dispersed filamentous reticulum.

In size, the golgi complex varies from large (nerve cells) to small (muscle cells).

Morphology : The morphology is variable depending on the source. In plants the golgi bodies are about I - 3nm in length and 0.5nm thick. Each golgi body consists of the following parts - *cisternae, tubules, vesicles* and golgian vacuoles.

The *cisternae* are flattened sacs filled with a fluid. Each golgi body consists of 4 - 8 flat structures which are slightly curved and are arranged in stacks. Each cisterna is enclosed by the trilaminar membrane 6nm in thickness and is separated from the other by a 100 - 150Å wide space. In some instances the intercisternal space is filled by fibres called intercistemal elements. The membranes of the cisternae are *fenestrated* (porous). The pores may be localized or present all over the cistemae. Occasionally cisternae may form reticulate structures.

Tubules: Arising from the peripheral region of the cisterane, the tubules form an anastomosing network of 300 - 500Å in diameter.

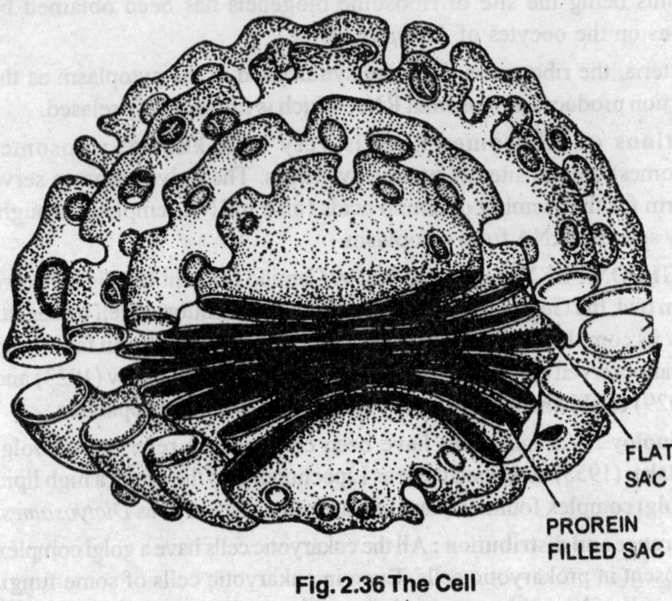

Fig. 2.36 The Cell Golgi Complex

Vesicles : These are goblet like structures attached to the tubules. There are two types of vesicles *smooth vesicles* and *coated vesicles.* The smooth vesicles are 20 - 80nm in diameter and are often called secretory vesicles. These are pinched off from the ends of cistetnal tubes. The coated vesicles are spherical outgrowths about 50nm in diameter. The function of the coated vesicles is not known.

Vacuoles : These are large spaces occuring at the distal end of the cisternae. They represent the cisternae whose membranes are widely separated.

In the case of Dictyosomes (isolated golgi bodies) of plants, cisternae are extensively fesnestrated and have rough and smooth vesicles attached to them. Parallel fibres fill the intercisternal place.

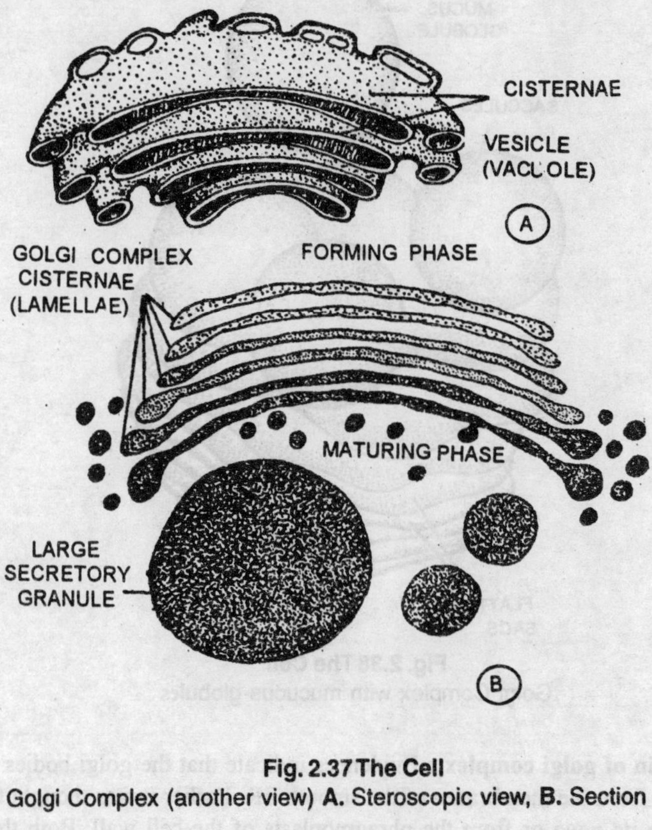

Fig. 2.37 The Cell
Golgi Complex (another view) **A.** Steroscopic view, **B.** Section view

Chemical composition of golgi complex : Proteins and phospholipids are in equal concentration. Golgi material in the nerve cells consists of cephalin and lecithin. Giroud has reported Vitamin C in the golgi complex.

Functions of the golgi complex : The following are some of the functions attributed to the golgi complex -

1. Absorption of compounds
2. Sites of enzyme production
3. Sites of hormonal production
4. Sites of protein storage
5. Forms acrosome during sperm maturation
6. Forms intracellular crystals
7. Formation of plant cell wall - the golgi bodies synthesise pectin, hemiceullulose and cellulose microfibrils. They also help in the formation of cell plate during mitosis.

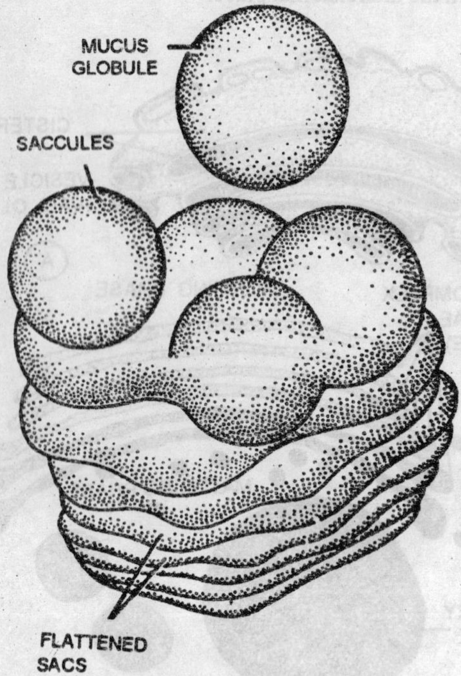

Fig. 2.38 The Cell
Golgi Complex with mucucus globules

Origin of golgi complex : Evidences indicate that the golgi bodies are produced from the membranes of the smooth ER. Earlier it was thought that they arise *de novo* or from the phragmoplasts of the cell wall. Both these theories are found to be incorrect.

Lysosomes : (*Gr.Lyso* = digestive, *soma* = body). First discovered in 1949 as dense round bodies, the lysosomes were called *pericanalicular bodies*. It was de Duve, (1955) who gave the present name.

Occurence : Lysosomes are found in most of the animal cells and few plant cells. The hepatic cells, pancreatic cells, kidney cells etc. posses large number of lysosomes. According to Matile (1969) the lysosomes in plant cell are membrane bound. Generally lysosomes are absent in the prokaryotic cells. Slime molds also have lysosomes.

Lysosomes are pleomorphic as seen in the meristematic cells of the root. The diameter may be 0.4 to 0.8μ; but in some mammalian kidney cells lysosomes as long as 5μ have been found.

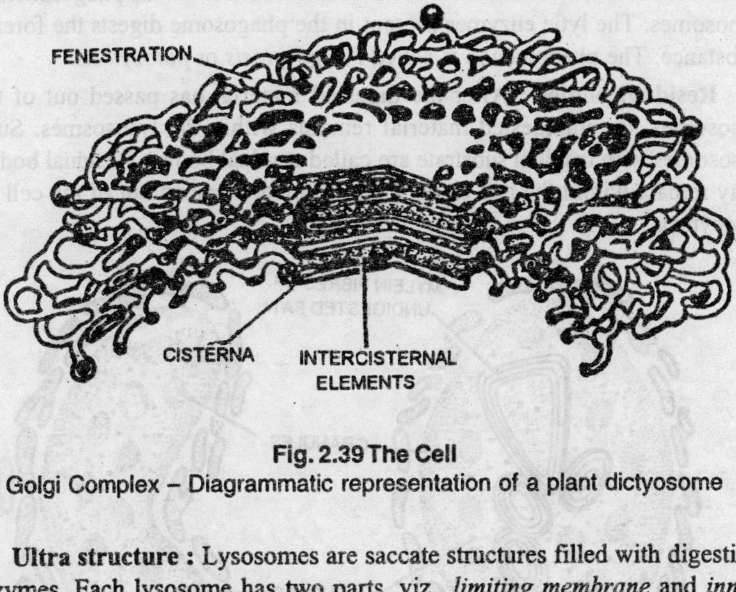

Fig. 2.39 The Cell
Golgi Complex – Diagrammatic representation of a plant dictyosome

Ultra structure : Lysosomes are saccate structures filled with digestive enzymes. Each lysosome has two parts, viz., *limiting membrane* and *inner dense mass.*

The **limiting membrane** is not double membraned like in other organelles. It is single and composed of lipoproteins. The membrane however is bilayered like any unit membrane.

The **inner dense mass** is the core of lysosomes; it is mostly enzymatic. The membrane is impermeable to substrates the enzymes of which are present in the lysosomes. Occassionally certain substances called *labilizers* destabilize the membrane causing the enzymes to come out. Alternatively, there are substance called *stabilizers* which stabilize the membrane. The limited permeability of the membrane prevents excessive digestion of cellular components by the lytic enzymes of lysosomes. The enzymes present in the lysosomes are - Nucleases, Phosphatases, Lipases, Glycosidases, and Sulphatases.

Polymorphism : Lysosomes are ploymorphic assuming various shapes depending on the cell type and situation. Basically there are two types of lysosomes viz. *Primary lysosomes and secondary lysosomes.* The secondary lysosomes are of three categories - *Phagosomes, residual bodies* and *autophagic vaculoes.*

Primary lysosomes : There are sac like structures produced directly from ER from the Glogi bodies.

Phagosomes : These are also known as heterophagosomes or digestive vaculoes. When a foreign substance enters the cytoplasm, the plasma membrane forms a sac around these substances. These are the phagosomes or pinosomes. The lytic enzymes present in the phagosome digests the foreign substance. The phenomenon is called *Phagocytosis* or *pinocytosis.*

Residual bodies : After the digested material has passed out of the lysosomes, the undigested material remains within the lysososmes. Such lysosomes with residual substrate are called residual bodies. Residual bodies may remain within the cell or very often they are pushed out of the cell by exocytosis.

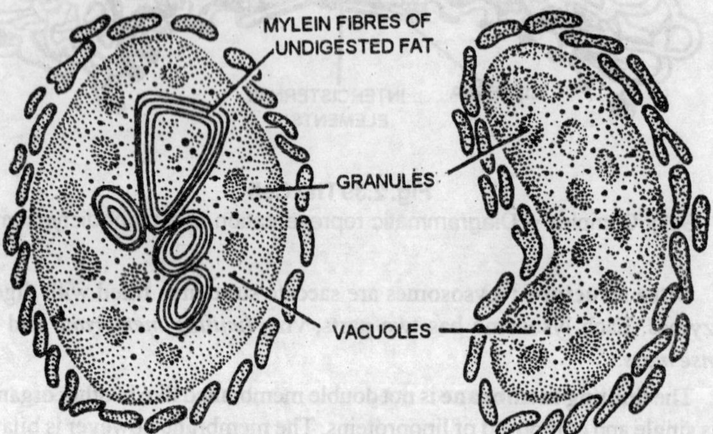

Fig. 2.40 The Cell
Lysosome – Structure

Autophagic vaculoe : Occasionally, a lysosome may digest a part of the cell (ER, mitochondria, etc.). This is a mechanism to digest a part of the cell without destroying the entire cell. Liver cells show autophagic vacuoles or autophagosomes under starvation.

Origin of lysosomes

(a) *Extra cellular origin* : Vacuoles absorbed due to'pinocytosis or phagocytosis may become secondary lysosomes after fusing with primary lysosomes.

(b) *Origin from golgi bodies* : There are enough evidences to show that lysosomes are produced from golgi bodies at the height of their secretory activity.

(c) *Origin from ER* : According to Novikoff (1965), the rough ER gives rise to the lysosomes.

Major functions of lysosomes : The main function of the lysosomes is digestion of intracellular substances. According to Hirsch and Coh (1964), lysosomes help in destroying the dead cells from the tiusse. Recently it has been found out that in some animals (frog), lysosomes play an important role in metamorphosis. Lysosomes also perhaps help in the formation and destruction of bone cells.

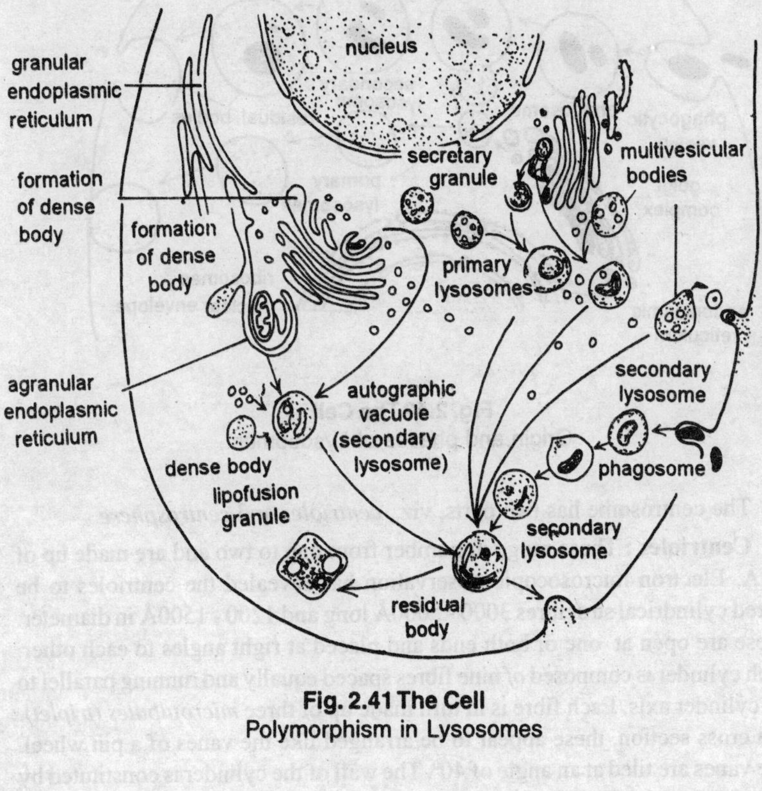

Fig. 2.41 The Cell
Polymorphism in Lysosomes

CENTROSOMES : The centromes are found in the centre of the cell near the nucleus. These are present in all animal cells and lower plants like Dinoflagellates, *Euglena* and *Chlamydomonas,* etc. Van Benden (1887) disocvered centrosomes.

The centrosomes play an important role in organising the mitotic apparatus in flagellated animals, while in flagellated plants they function as basal granules of flagella.

Size : In mature centrosomes the diameter is 1600Å - 2500Å while the length is about 160A⁰. Giant centrioles have been found in spermatocytes or oocytes of certain organisms.

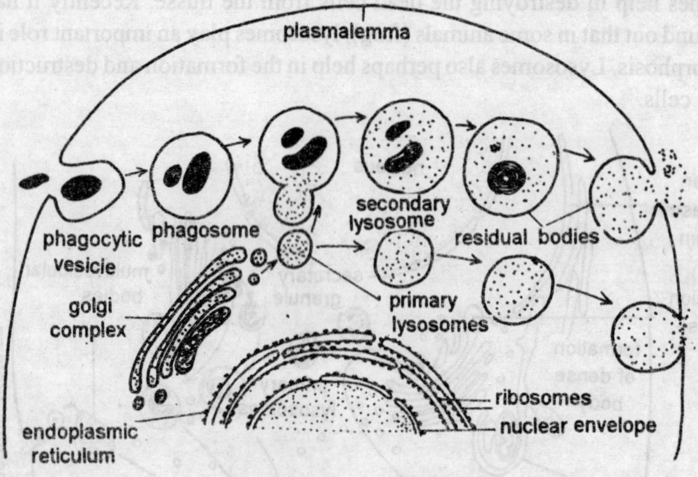

Fig. 2.42 The Cell
Origin and phases of Lysosomes

The centrosome has two parts, viz., *centrioles* and *centrosphere*

Centrioles : These range in number from one to two and are made up of RNA. Electron microsocopic observation has revealed the centrioles to be paired cylindrical structures 3000 - 5000Å long and 1200 - 1500Å in diameter. These are open at one or both ends and placed at right angles to each other. Each cylinder is composed *of* nine fibres spaced equally and running parallel to the cylinder axis. Each fibre is in turn made up of three *microtubules (triplet).* In a cross section, these appear to be arranged like the vanes of a pin wheel. The vanes are tiled at an angle of 40⁰. The wall of the cylinder is constituted by these nine triplet fibres.

The microtubule triplet in a fibre can be designated as A,B and C of which A is innermost. Each microtubule wall is 50A^0 in diameter and the lumen in 120 - 130A^0 (the total diametre varies from 180 - 200A^0). The length of each microtubule is about 560A^0. At the innerside of each triplet fibre a dense line runs and it is called *triplet base* connecting A and C tubules. In addition to this, a dense RNA region called *foot* is present.

Centrosphere : The dense granlar region of cytoplasm surrounding the centrioles is called the centrosphere.

Structure : Structurally the microtubules are lipoproteinaceous (Fulton 1971).

Functions : Centrosomes appear to play an important part in cell division. people believe that centrioles play an important part in the formation of spindle. They however constitute the mitotic pole in higher animals. In primitive animals which are ciliated, centrioles seem to help in the generation of cilia.

Microbodies : The cells of plants, fungi, certain protozoa and liver cells of vertebrates contain certain small granular structures associated with ER, mitochondria and chloroplasts. These spherical bodies ranging in size from 0.2μ to 1.5μ have a central crystalline core and are called *microbodies*.

Recent studies have revealed the presence of two types of microbodies viz., **Peroxisomes and Glyoxysomes.**

Peroxisomes are found in animal cells and also in the leaves of higher plants. These particles participate in the oxidation of substrates resulting in the formation of hydrogen peroxide. In plants, peroxisomes are involved in photorespiration.

Structurally peroxisomes are variable in shape and size but are generally circular. They have an envelope of a single membrane made up of proteins and lipids enclosing a granular matrix. In certain instances, the matrix is crystalline or amorphous. The peroxisomes catalyze the H_2O_2 production. In plants, the peroxisomes contain enzymes of glycolate pathway.

Glyoxosomes are found in the cells of yeast and certain other fungi *(Neurospora)* and also seeds (oil rich) of many higher plants. They are similar to peroxisomes except that their core consists of dense rods of 6.0μm in diameter.

Functionally the glyoxysomes contain enzymes for fatty acid metabolism, involved in the conversion of lipids to carbohydrates during germination. The *glyoxysomes* play an important role in lhe *glyoxylate cycle.* The enzymes *Urate oxidase* and *allantoinase* are also found in glyoxysomes. Urate oxidase converts uric acid to allantonin, while allantoinase hydrates allantonin to allantonic acid during the degrading of purines.

3

MOLECULAR BASIS OF HEREDITY

The cells have diverse chemical components of which most were known to scientists by the first quarter of the nineteeth century. These are classified into three major groups namely–carbohydrates, proteins and lipids. Of these three–proteins are the most complex and diverse and many biologists thought that they may constitute the basis of biological diversity.

Eventhough another group of macromolecular compounds – nuclei acids were discovered in 1869 by Sweedish biochemist Fredrich Miescher, it was only in 1952, their (nuclei acids) role in inheritance was proved.

Meischer collected pus filled hospital bandages and from them he isolated and purified nuclei. On chemical fractionation he found that a cellular chemical content had a very high component of phosphorous compound than what was known before. As the new compound was obtained from nucleus, Miescher named it *nuclein*. Subsequently, nuclein was isolated from many sources and due to its acidic properties, it was named nucleic acid. Upto 1952, however, interest in nuclei acids was limited to only its physiochemical studies. When its role in inheritance was proved nucleic acid became the focus of intensive research. Comparative studies of nuclei acids and proteins revealed that while proteins are of diverse types, there are only two types of nuclei acids. It was hard to believe that nuclei acids which apparently had no diversity could be the basis of heredity and not the proteins.

Presently, however, there is not even an iota of doubt as to the fact that nucleic acids alone constitute the hereditary factor. Three major experiments helped provide an evidence for this.

Bacterial Transformation

Fred Griffith, an English bacteriologist, worked with *Diplococcus pneumococci* and found out that there are two strains in this – *Virulent and Avirulent.* When injected to experimental mice the virulent strains cause the death of the mice, while avirulent strains do not casue any harm. Griffith showed that heat killed virulent strains injected into the mice along with avirulent strains cause the death of the mice and from the bodyof the mice live virulent strains can be obtained. Dawson and Sia (1931), showed that the heat killed virulent strains can transform avirulent strains on a nutrient broth. Alloway (1933), showed that even a lysed cell of virulent strain retains the ability of transformation thereby indicating that some component of the cell rather than the entire cell was involved in transformation.

Nature of the transforming agent

Three scientists – O.T. Avery, Colin Macleod and Maclyn McCarty (1944), went into serious search of the transforming agent from the lysed cells of bacteria. Even after digestion of polysaccharide coat, the cell retained the ability of transformation. This shows that the capsule has no role in transformation. Similarly, protein degradation also had no adverse effect on transformation. With the two important components out of the way, it thus became clear that nucleic acids perhaps might play a role in transformation. It was shown that the DNA fraction of nucleic acids was responsible for transformation. It was also shown that if the cells are treated with DNAse (a DNA degrading enzyme), the transforming ability will be lost. Enzymes which digest RNA do not disturb transformation.

Bacteriophages and chemical basis of heredity

Bacteriophages are viruses that attack a bacterial cell and multiplty inside the (bacterial) cell. The phage ataches itself to the bacterial cell and insects its components into the cell, where replica of phages are produced. Hershey and Chase (1952) conducted an experiment to find out which component of the bacteriophage carries the genetic information necessary for the replication of viruses. They cultured the host bacterium *Escherichia coli* in a medium containing radio isotopes of Sulphur (S^{35}) and phosphorus (P^{32}). When T_2 phages attacked these cells subsequently, the labelled compounds get incorporated in phage DNA which always has radioactive phosphorus but not any sulphur. In the next set of experiment only the protein coat of T_2 phage was labelled with S^{35} and the DNA core of another was labelled with P^{32}. Hershey and Chase separated the radioactive phages by centrifugation and allowed them to infect a population of non radioactive bacteria. A few minutes after initial infection, they separated the phages from bacteria by agitating the medium. After centrifugation, the host and the parasite were separated by centrifugation and their radioactive measured. About 95% of P^{32} was found in bacteria while about 95% S^{35} was seen in phages. This experiment conclusively proved that during infection, only DNA is injected into the medium. This is the reason why the non radioactive bacteria showed labelled P^{32} (as they obtained it from labelled phage DNA). For this work, Hershey was awarded Nobel prize in 1969.

Thus, the experiments of Hersehy and Chase, Avery, Mcleod and McCarty. Griffith etc., clearly proved that the DNA constitutes the chemical basis of heredity.

TMV and Chemical Basis of Heredity

The discovery that DNA is the hereditary material posed some problems to explain the inheritance in instances such as TMV (Tobacco mosaic virus), where there is no DNA (TMV is a RNA virus). But, the absence of DNA has not prevented TMV from transmitting its characters from one generation to the other. TMV is made up of a RNA core and a helical protein coat. The protein RNA component of TMV can be easily separated by shaking the viral particles in a mixture of water and phenol. Frankek Conrad and Wichams (1955), separated the protein and RNA from TMV and reassembled the particles from the individual components thus indicating the purification process (of the components) does not affect their function. Subsequently, Gierr and Shramm (1956), showed that on contact with the host only the RNA can bring about the disease and not the protein coat. In TMV therefore, it is RNA as the hereditary material. It can therefore be concluded that nucleic acids in general, and DNA in particular is the hereditary material in most of the organisms; and in the absence of DNA (as in some viruses) RNA can constitute the genetic material.

Nucleic Acids

These are the most important constituents of the nucleus playing a vital role in the inheritence of characters from one generation to the other. There are two types of nucleic acids – Deoxyribose nucleic acid (DNA), and Ribose nucleic acid (RNA). While both constitute the genetic material in different organisms, it is predominantly the DNA which is the genetic material in most of the organisms. The following are some of the important milestones in the discovery of Nucleic acids.

1. Meischer (1869), isolated nucleic acids and called them nuclein.
2. Hertwig (1884), suggested a heriditary role to the nuclein
3. Kossel (1884), recognised purines and pyrimidines in the nucleic acids.
4. Feulgen (1912), devised basic fushsin dye positive only to DNA.
5. Astbury and Bell (1938) showed that the bases in DNA molecule are arranged one above the other.
6. Avery, Macleod and Macarty (1944), presented evidence for DNA to be the genetic material.
7. Chargaff (1950), showed that DNA contains equal amount of purines and

pyrimidines.

8. Wilkins *et al* (1950), identified the molecular distance between the bases.
9. Furberg (1952); DNA is a coil of single chain.
10. Pauling and Corey (1952); DNA is a helix of three chains.
11. Watson and Crick (1953) demonstrated the double helix model for DNA molecule.
12. Meselson and Stahl (1958), demonstrated the semi conservative replication in DNA.
13. Jacob and Monod (1961), postulated the presence of RNA in protein synthesis.

DEOXY RIBOSE NUCLEIC ACID (DNA)

DNA is the main genetic material in almost all organisms except for certain viruses. In prokaryotes like *Escherichia coli,* the genetic is a single giant molecule of *DNA*. It is mainly concentrated in the nucleus, it has also been reported in cytoplasmic organelles like chloroplast, mitochondria etc.

Chemical Composition of DNA

DNA extract from many living organisms have revealed it to be a complex macromolecule. It is a long chain polymer cmposed of monomers (units), called nucleotides. Each nucleotide inturn is composed of three sub units. These are – (a) nitrogen base, (b) sugar and (c) phosphate.

Nitrogen bases

The nitrogen bases are of two kinds viz., pyrimidines and purines. Both pyrimidines and purines are heterocyclic compounds, with a basic six membered ring in which one or the other carbon position is replaced by nitrogen.

The pyrimidines have a basic six membered ring of benzene in which, position 1' and 3' are replaced by nitrogen. There are two kinds of pyrimidines – *Cytosine* and *thymine.* Structurally, cytosine has a single six membered ring with a hydroxyl group at the 2' position and an amino group at the 6' position (*2 hydroxy, 6 amino pyrimidine).* Thymine also has a single six membered ring with hydroxyl groups at the 2' and 6' position, and a methyl group attached to the 5' position *(2, 6 hydroxy 5 methyl pyrimidine).*

The purines have a double ringed structure. Attached to the six membered benzene ring is a 5 membered imidazole ring. The imidazole ring is joined to the hexagonal ring at 4' and 5' positions. The benzene ring has nitrogen in the 1' and 3' position (replacing Carbon).

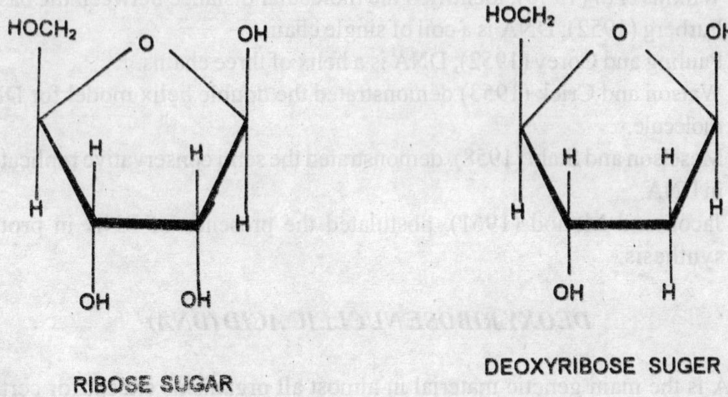

Fig. 3.1 Molecular basis of heredity

Pentose sugars of nucleic acids (left, robose; right deoxyribose)

There are two types of purines viz. *Adenine* and *guanine*. Adenine has an amino group at the 6' position (*6 aminopurine),* while guanine has an aminogroup at the 2' position and a hydroxyl group at the 6' position (2 *amino, 6 hydroxy purine).*

Sugar

The sugar molecule in DNA is basically a ribose sugar. The ribose sugar is a *Pentose Sugar.* It is a 5 membered ring. The ring consists of four carbon atoms while the 5' C atom is outside the ring. Positions 1', 3' and 5' in the ring have OH groups, while the 2' carbon position has only H. There is no O; hence the sugar is said to be Deoxyribose (no oxygen in the second carbon position).

Phosphate

It has one central phosphorus with four oxygen atoms.

Nucleosides

A nitrogen base coupled with sugars forms a nucleoside. DNA has four nucleosides –

(a) Deoxycytidine
(b) Deoxythymidine
(c) Deoxyadenosine and
(d) Deoxyguanosine

The sugar molecule is attached to the nitrogen base to the 3' position in the benzene ring in the case of pyrimidines and to the 9' position of the imidazole ring in the case of purines.

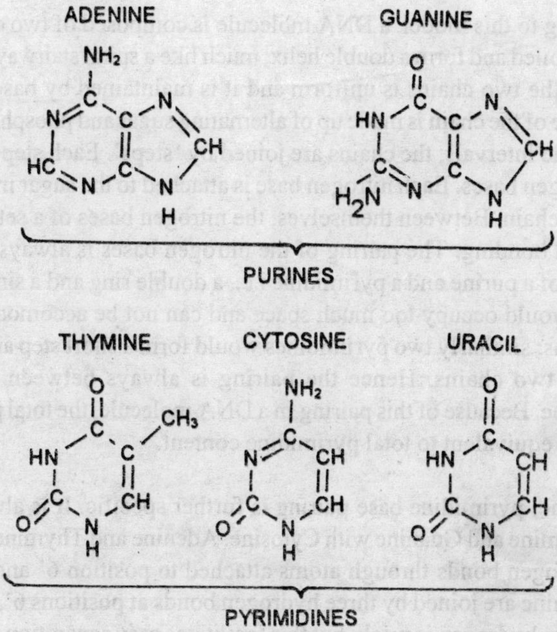

Fig. 3.2 Molecular basis of heredity
Nitrogen bases of nucleic acids

Nucleotides

A nucleotide is derived from the nucleoside by the addition of phosphoric acid molecule to the latter. The phosphate group is usually linked to the two sugar molecules above and below. The linkage is with the 3'C atom of one deoxyribose sugar and with the 5' C atom of another. DNA has four nucleotides – *Deoxycytidilic acid, Deoxy thymidilic acid, deoxy adenylic acid* and *Deoxyguanilic acid.*

Polynucleotide

DNA is said to be a polynucleotide chain with a number of nucleotides joining together with the help of DNA polymerase.

Molecular model of DNA structure

A molecular model for DNA structure first proposed by J.D.Watson and F.H.C

Crick (1953) has been universally accepted. They were awarded Nobel prize for their discovery. The work of Crick and Watson has been supported by X-ray diffraction picture obtained by Wilkins and his associates.

According to this model, a DNA molecule is composed of two chains that are spirally coiled and form a double helix, much like a spiral stairway. The distance between the two chains is uniform and it is maintained by base pairing. The back bone of the chain is made up of alternating sugar and phosphate molecules. At specific intervals, the chains are joined by 'steps'. Each step is made up of two nitrogen bases. Each nitrogen base is attached to the sugar molecule of the opposite chain. Between themselves, the nitrogen bases of a set are joined by hydrogen bonding. The pairing of the nitrogen bases is always specific. It is made up of a purine and a pyrimidine i.e., a double ring and a single ring. Two purines would occupy too much space and can not be accomodated between two chains; similarly two pyrimidines would form a short step and leave a gap between two chains. Hence the pairing is always between a purine and pyrimidine. Because of this pairing,in a DNA molecule, the total purine content is always equivalent to total pyrimidine content.

The purine, pyrimidine base pairing is further specific. It is always Adenine with Thymine and Guanine with Cytosine. Adenine and Thymine are joined by two hydrogen bonds through atoms attached to position 6' and 1'. Cytosine and Guanine are joined by three hydrogen bonds at positions 6', 1' and 2', the bonding to hydrogen is weak; but this facilitates easy separation of two strands during replication. According to Crick and Watson, the helix has a diametre of 20 A^0 and there is one complete turn to the helix at every 34 A^0 and has a stack of 10 nucleotides at every turn.

One unique character of DNA is its unity and yet infinitismal variety. The DNA structure mentioned above is uniform in all organisms, whether it is man or virus. At the same time it is also true that DNA is the genetic material that decides the characters of individual organisms. Then how does DNA decide the individuality? The answer lies in the base pair sequence. The combination of A/T, G/C is uniform, but their sequence is not; for instance in one individual it may be AT, G/C, GC, AT, GC< AT etc. There is no limit to the variety in which base pair sequences are arranged.

This is the secret of individual differences. In no two individuals, the base pair sequence is same.

Unusual bases in DNA

The normal bases in DNA are A, T, G and C. In some viruses like PBS1 and PBS2, uracil replaces thymine. In some bacteriophages, C is replaced by 5 hydroxy methyl cytosine.

Fig. 3.3 Molecular basis of heredity
F.H.C. Crick (left and J.D. Watson (right)

Single strand DNA

Although DNA in most organisms is double stranded, in some phage viruses (ϕ x 174), single stranded DNA is found. The following table summaries the differences between double stranded DNA and single stranded DNA.

Character	Double Strand DNA	Single Strand DNA
1. U.V. Absorption	Constant from 10-80°C and then rises rapidly	Steady increase from 20° to 90°C
2. Action of formaldehyde	Resistant	Not Resistant

3.	Ratio of	equal, A=T and 1.33:0.98:0.75	A:T:G:C is
4.	Shape	Linear	Circular
5.	Respiration	Two strands separate and replication begins	Becomes double (replicativeform) temporarily.

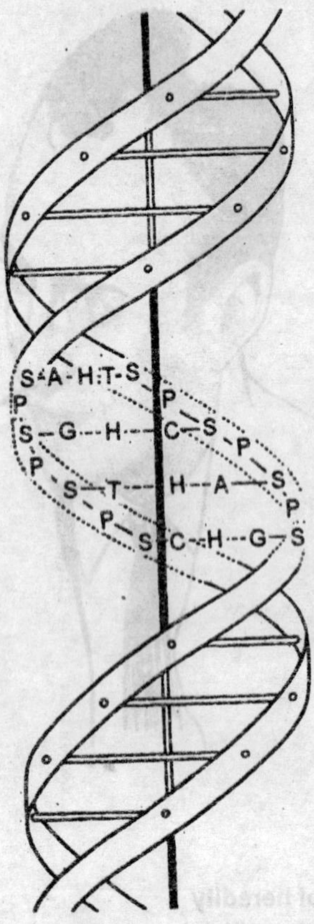

Alternative forms of DNA double helices

The structure of DNA molecule described above as visualized by crick and Watson assumes that the helical coiling of the molecule is right handed and this form of DNA has been called the B-form. Subsequent researches, however, have indicated that other types of coiling are also possible for the DNA helex. These alternative forms of DNA differ from one another in the following respects.

1. Number of residues per turn (n), and
2. Spacing of residues along the helix (h)

The existence of DNA as double helix has also been confirmed by experiments used for measuring the number of base pairs per turn of the helix and this works out to be 10.4 instead of 10 in the B-DNA. This leads to the modification of angle of rotation from 36 to 34.6. Thus, it is possible that there could be structural variants of DNA with varying values for n and h. Three structural variants of DNA A,B and C are known since a long time and transitions between them are also possible. The following table gives a brief summary of the characters of these forms of DNA.

Fig. 3.4 Molecular basis of heredity
The double helix model od DNA

Helix type bryophytes	Condition	Basepair	Rotation per
A	75%	11	+32.7 (right handed)
B	92%	10	+36.0 (right handed)
C	66%	9.33	+38.6 (right handed)
Z	very high salt	12	-30.0 (left handed)

The forms of DNA are assumed to be found in all molecules depending on the surrounding conditions. In addition to the above, D and E forms of DNA are also found very rarely with only 8 and 7.5 base pair per turn. These are found only in some molecules of DNA lacking in Guanine. In contrast A,B and C forms are found in all molecules irrespective of DNA sequence.

Z-DNA

This type of DNA was first synthesized artificially, but subsequently has been reported from living cells also. As this molecule follows a zig-zag course it is called Z-DNA. The resemblances and differences between B-DNA and Z-DNA are as follows –

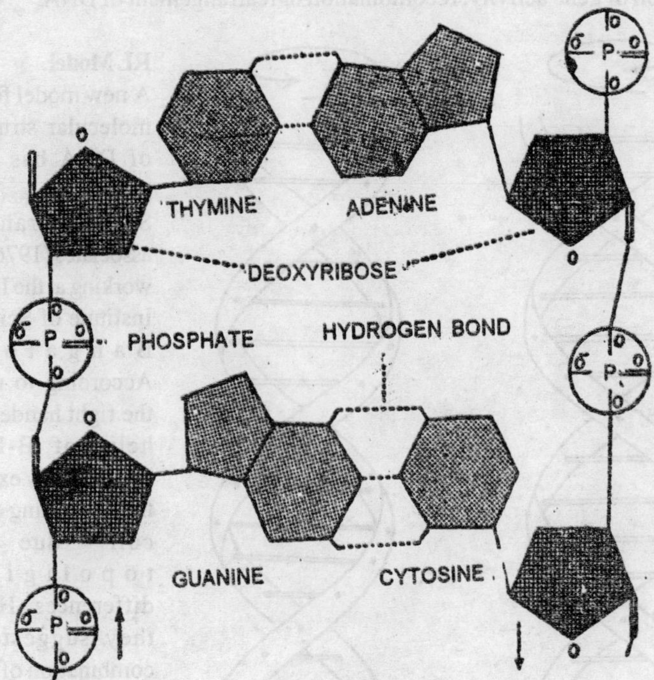

Fig. 3.5 Molecular basis of heredity
Arrangement of nucleotides in a segment of the DNA model

Resemblances between B-form and Z-form

1. Both are double helical.
2. The two strands are antiparallel in both forms.
3. Both forms exhibit A=T and G=C pairing.

Differences between Z-DNA and B-DNA

1. Z-DNA is left handed, while B-DNA is right handed.
2. The phosphate back bone is regular in B-DNA, while it is zig in Z-DNA.
3. In Z-DNA, sugar residues have alternating orientations as a result of which repeating, units are dinucleotide, whereas in B-DNA they are monocleotides.
4. Z-DNA has 12 bp for every twist of 360^0 while B-DNA has only 10bp.
5. As more base pairs are accommodated in one helix in Z-DNA, the angle of twist is 60^0 as against 36^0 in B-DNA.
6. One complete helix is $45A^0$ in Z-DNA, while it is $34A^0$ in B-DNA.
7. The diameter of the helix is $18A^0$ in Z-DNA, while it is $20A^0$ n B-DNA.

The role of Z-DNA is not certain, even though it is thought to play a role in regulation of gene activity, recombination or rearrangement of DNA.

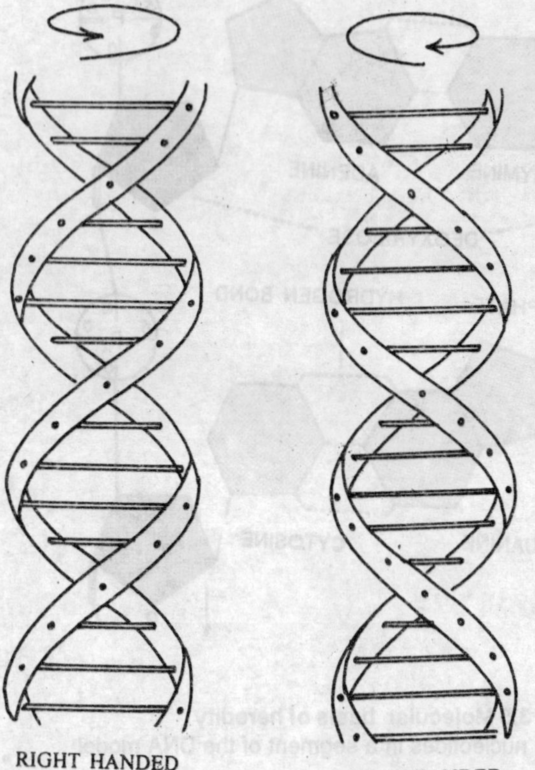

Fig. 3.6 Molecular basis of heredity
Two helices, one showing right handed coil as found in B-DNA and the other showing left handed sense as found in Z-DNA.

RL Model

A new model for the molecular structure of DNA has been proposed by Sasisekharan and associates (1976-89), working at the Indian Institute of Science, Bangalore. According to them, the right handed (R) helix of B-DNA cannot easily explain the unwinding of the coil due to topoligical differences. Hence they suggested a combination of right handed (R) and left handed (L) helices, which they called the RL model. In this model, a B-DNA molecule will have alternating left handed and right handed segments of approximately 5 base pairs in a repeat of 10 base pairs. This model provides conformational flexibility. Many molecular biologists opine that there is a strong experimental evidence for the RL model although one is not sure of the continuous length of right and left handed segments in a molecule of DNA.

Super Coiling of DNA

The double helix structure of DNA represents in linear molecule, but in living conditions the molecule is a closed structure without free ends. In bacteria, the DNA is circular while in eukaryotes also large loops appear to represent part of a circle. As a result of this, the DNA becomes supercoiled. One supercoil is introduced everytime when the duplex is twisted about its axis. The super coil may be negative (winding opposite to the helix coiling), or positive (winding in the same direction as the helix coiling).

RIBONUCLEIC ACID (RNA)

The RNA is also a polymer of nucleotides. These are called ribonucleotides. Each ribonucleotide has three parts viz., a pentose sugar, a nitrogen base and a phosphate group.

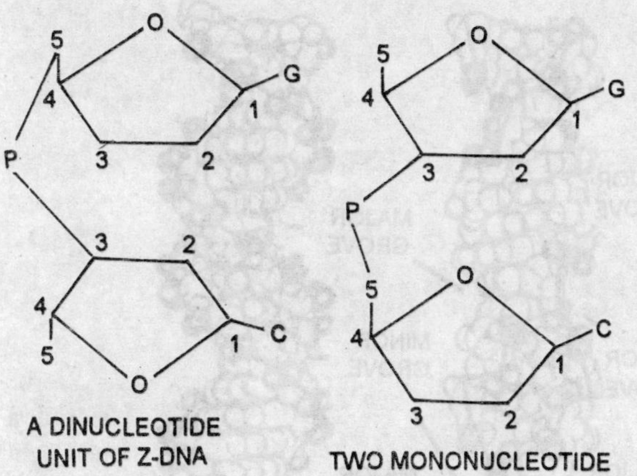

A DINUCLEOTIDE UNIT OF Z-DNA

TWO MONONUCLEOTIDE UNITS IF B-DNA

Fig. 3.7 Molecular basis of heredity
Orientation of adjacent sugar residues in Z-DNA and B-DNA showing opposite orientations in Z-DNA and same orientation in B-DNA. This results in dinucleotide units in Z-DNA as against mononucleotide units in B-DNA

The pentose sugar is a ribose sugar much like that of DNA, but with oxygen (present) attached to the 2' carbon position of the ring.

The nitrogen bases are – Adenine and guanine (purines), and cytosine and uracil (pyrimidines).

Molecular structure of RNA

Here also, the back bone is made up of alternating sugar and phosphate. The single chain folds to enable the base pairs to meet here and there to form the steps. The pairing is between A-U and C-G. Occasionally, the RNA may be double stranded but does not form a helix like DNA.

Single stranded RNA is the genetic material in plant viruses, poliomyletis viruses etc. Double stranded (non helical) RNA occurs as the genetic material in reoviruses.

Types of RNA

The following are the types of RNA found :

(a) Genetic RNA
(b) Ribosomal RNA
(c) Messenger RNA
(d) Soluble or Transfer RNA.

Genetic RNA

In certain viruses, RNA is the only genetic material. DNA being absent. This RNA is self- replicating; it is called RNA dependent RNA synthesis. The viral RNA directly functions as a messenger and produces RNA polymerase enzyme as well as proteins.

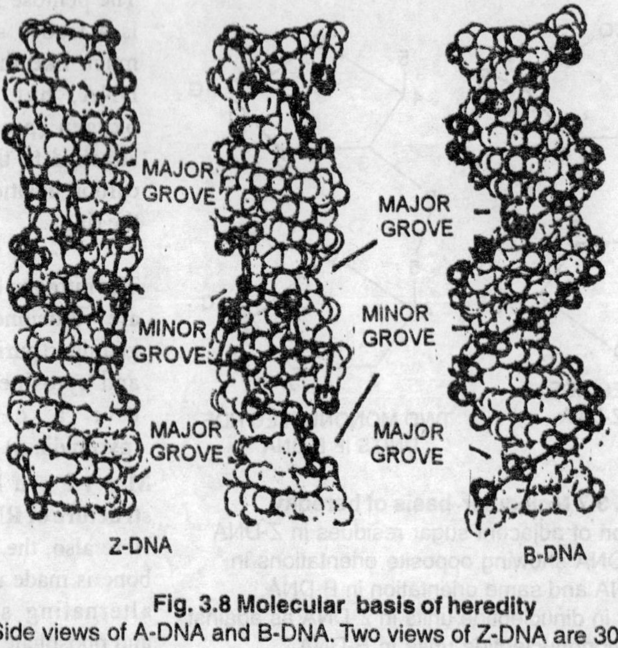

Fig. 3.8 Molecular basis of heredity
Side views of A-DNA and B-DNA. Two views of Z-DNA are 300 apart. Irregularity of backbone in Z-DNA is shown by heavy lines showing the path of phosphate residues, which is quite regular and smooth in B-DNA

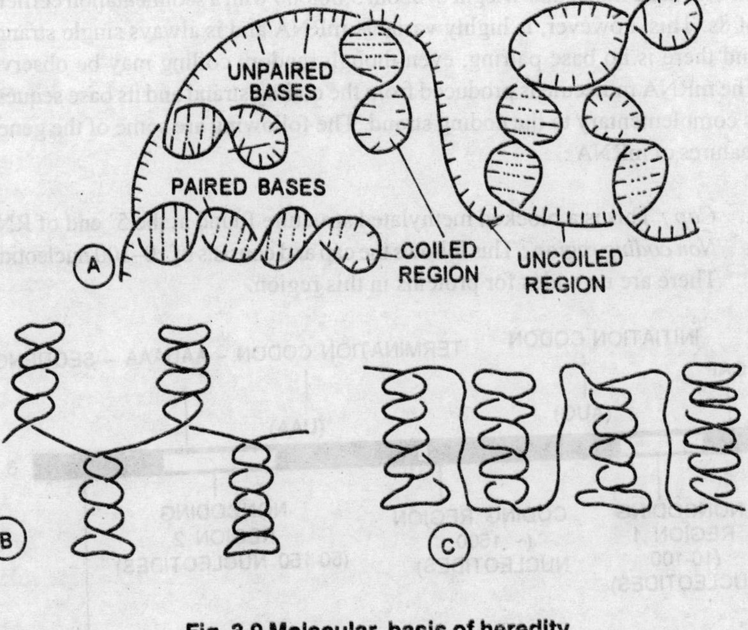

Fig. 3.9 Molecular basis of heredity

Secondary structure of ribosomal RNA (diagrammatic) **A.** Extended strand, **B.** Negative interaction showing repulsion, **C.** Positive interaction showing association

Ribosomal RNA (rRNA)

This is found in the ribosomes. It constitutes about 80% of total RNA of the cell. The base pairs of rRNA are complementary to that part of the nuclear DNA from where they are produced. rRNA consists of a single strand folding upon itself at some regions. The folded regions have paired bases, while the other regions have unpaired bases. Due to this in rRNA there is no purine pyrimidine equality.

There are three types of rRNA bases on sedimentation and molecular weight. These are (a) rRNA with molecular weight over a million, e.g. 21s-29s rRNA, (b) rRNA with molecular weight less than a million, e.g. 12s-18s rRNA, (c) rRNA with very low molecular weight (40,000) e.g. 5srRNA.

Messenger RNA (mRNA)

mRNA carries the message from DNA to ribosomes during protein synthesis. It is about 3-5% of total RNA content of the cell.

mRNA has a molecular wieght of about 5,00,000 with a sedimentation cefficient of 8s. This, however, is highly variable mRNA and is always single stranded and there is no base pairing, even though random coiling may be observed. The mRNA molecule is produced from the coding strand and its base sequence is complementary to the coding strand. The following are some of the general features of mRNA :

1. *Cap :* This is a blocked methylated structure found at the 5' end of RNA.
2. *Non coding region :* This follows the cap and consists of 10–100 nucleotides. There are no codes for proteins in this region.

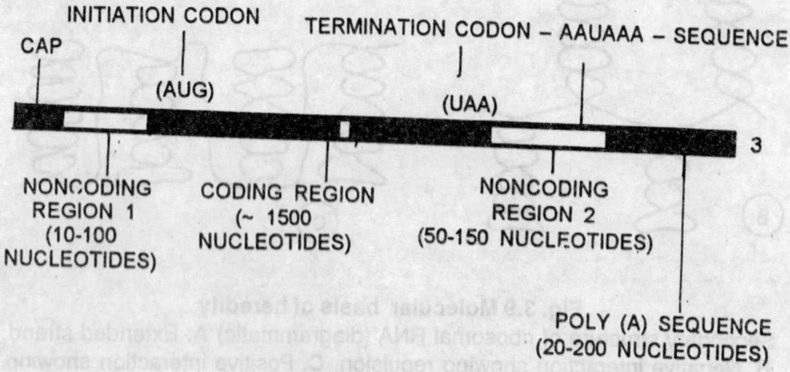

Fig. 3.10 Molecular basis of heredity
General features of eukaryotic mRNA

3. *Initiating codon :* This region begins the coding.
4. *Coding region :* Codes for proteins; and has about 1500 nucleotides.
5. *Termination codon :* The coding region is followed by a terminating codon, which stops translation.

Transfer RNA (tRNA or sRNA)
The transfer RNA also called soluble RNA, is next in quantity to rRNA. It has a low molecular weight of about 25,000–30,000 and the sedimentation co-efficient is 3.8s. It is made up of about 70–90 nucleotides.

tRNA is synthesised in the nucleus on a DNA template. Only a minute fraction of DNA (0.025%) codes for tRNA. The tRNA molecule consists of a single strand looped around itself with the 3' end always terminating in a C-C-A sequence. Some bases are paired.

The function of tRNA is to carry aminoacids to mRNA during protein synthesis. Each aminoacid couples with a specific tRNA. Since the protein molecule consists of about 20 types of amino acids, there must at least be 20 types of tRNAs. But there could be many more, sinçed there are two tRNAs for every amino acid in several cases.

Secondary structure of RNA

This refers to the arrangement of the chain of the molecule to enable it to function in a proper way. Many models have been proposed for the secondary structure of tRNA. Of these, clover leaf model proposed by Holley (1966), is most widely accepted.

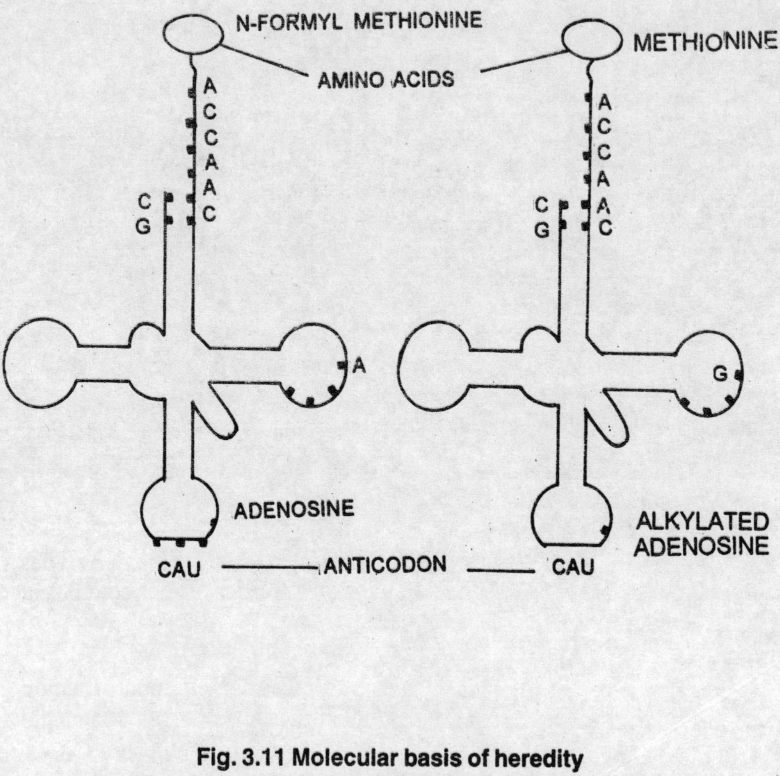

Fig. 3.11 Molecular basis of heredity
Secondary structure of tRNA (Clover leaf model)s

According to the clover leaf model, the single chain of tRNA is folded on itself to form five arms. As a result of this folding, the 3' end and 5' end lie adjacent to each other. Each arm consists of a basal stalk, and a loop. The double stranded stalks show the typical base pairing. There could be an exceptional mismatch. The loops do not have base pairing. One of their arms which does not have a loop is called the acceptor as this is the binding site for amino acids. The second arm is called the DNA arrm. This has a synthetase site which recognises the amino acid activating enzyme. The fourth is a variable arm; it has a stem, but the loop may or may not be present. The Tψ arm has a ribosome recognition site.

There is also a tertiary structure to the rRNA. This is brought about by H bonding between bases, and between bases and sugar phosphate backbone.

4

GENOME IN PROKARYOTES

The hereditary material in all organisms is remarkably similar. Having been evolved in the remote past, life has been able to maintain itself because of the ability of the hereditary materials to maintain a basic uniformity in structure and function. The hereditary material in all living beings (whether it is a border line entity like a virus, or an evolved human being) is Nucleic acids without exception, even though, there may be some stray instances of proteins (as prions) believed to carry out inheritance. The structure of the nucleic acids is unique; it has a basic uniformity, yet sufficient plasticity to allow for its manifestation from most primitive to the advanced organism.

In eukaryotic organisms, the genetic material is organised into large entities called chromosomes which allow for proper replication and distribution of genetic material during cell division and reproduction.

In prokaryotic organisms also,the nuclei acids constitute the chromosome, but the chromosome of prokaryotes is a naked strand of DNA (not associated with proteins) lying free in cytoplasm. The entire genetic system of each species is encoded in a single chromosome. Even in viruses, the genome is either DNA or RNA.

The prokaryotic genome is refered to as the chromosome, centrosome, nucleoid, chromatin body or the gneophore. Some traditional cytogeneticists object to the word chromosome and prefer to use the term genophore or nucleoid. However, as Swanson (1981), points out "this is more a matter of semantics than of cytogenetics, since the prokaryotic chromosome not only has the same fundamental structure based on nucleic acid sequences, but it also carries out, the same three basic functions of its eukaryotic counter part. That is, it is the organelle which stores, replicates and transmits the coded information of biological inheritance; which is so regulated in its action that its primary products appear at appropriate times and in appropriate amounts during the life-cycle of the organism; and which is capable of recombining its genetic content with other similar individuals in such a manner that the seggregating progeny of a hybrid union can have different genetic properties". Swanson (1981), therefore, opines that the term "chromosome" can be applied to both eukaryotic and prokaryotic species with distnctions in structure being made at appropriate times.

The chromosome of *Escherichia coli,* and *Salmonella typhimurium,* are probably the most thoroughly mapped genomes of any cellular organism. Hundreds of identifiable genes are located in a single circular linkage group. The entire genome of a bacterium is made up of a single chromosome. Occasionally, a single bacterial cell may have more than one chromosome at a given time. This should not be taken as the cell having two chromosomes. This represents a stage wherein the chromosome has replicated and formed two units, but the cell is yet to divide.

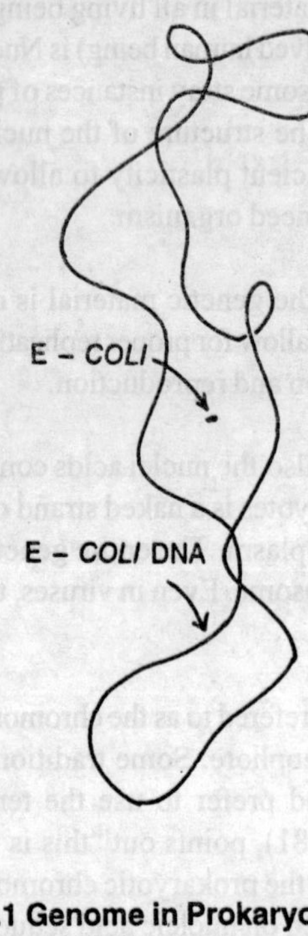

Fig. 4.1 Genome in Prokaryotes
Schematic diagram showing the relative sizes of E.coli and its DNA molecule

The chromosome of *E.coli* is a single, large double helix which is circular in outline. The DNA molecule is supercolied. It (DNA) has no association of histones as in eykaryotic chromosomes. However, small amounts of basic proteins have been found to be associated with the isolated DNA. It is quite likely that these might prevent some sections of the chromosome from transcription thus functioning like repressor proteins. Thus, it can be argued that the absence of DNA-histone association may not be such a fundamental point of difference between prokaryotic and eukaryotic DNA.

In *E.coli,* the total length of the circular DNA is 1300µm, whereas cylindrical bacteria have a maximum diameter of 1 µm only. This clearly indicates that the chromosome must be highly folded in the cell.

When the DNA of *E.coli* is isolated by techniques that avoid both DNA breakage and protein denaturation, a highly compact structure called *nucleoid* can be seen. This structure contains a DNA molecule, a fixed amount of protein and variable amounts of RNA. As has been pointed out already, the proteins might play a part in repression of transcription.

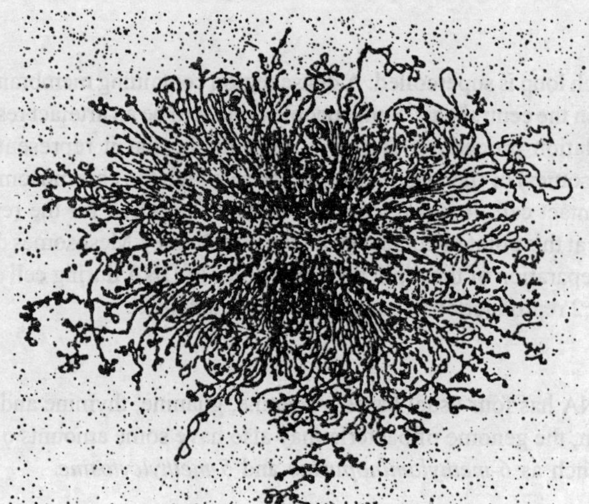

Fig. 4.2 Genome in Prokaryotes
An electron micrograph of an E.coli chromosome showing the multiple loops merging from a central region

The DNA molecule is about 1000 to 1100μm in length with a diameter of 20 Å to 40 Å. Such a molecule has more than 3.235×10^6 nucleotide pairs (each pairs is 3.4 Å long and there are 2941 base pairs per micrometer). The molecule has a molecular weight of 2.5×10^9 daltons (one dalton is equivalent to the weight of one hydrogen atom). This entire molecule is limited in the nuclear area of the cell which is no more than one micron and has no limiting boundary (such as a nuclear envelop), hence the molecule must be enormously folded.

It has been pointed above, that a small amount of RNA and basic proteins are associated with the DNA molecule. But the fact that the action of RNAse or protease does not alter the basic structure of DNA, indicates that neither RNA nor proteins are necessary for the basic circular natrue of the chromosome.

Detailed Structure of the Chromosome

The chromosome (DNA) of *E.coli* on isolation appears to be a highly folded compact structure. These folds appear to be highly regular and not random. If they were to be random, replication and proper seggregation would not be possible. The chromosome obviously has a high level of organization. Two features of the chromosome are noteworthy. These are – (1) The DNA is arranged in a series of loops and

(2) Each loop is supercoiled. A dense region containing membranous material is seen in the central part. Some people think it to be an artefact resulting out of the isolation procedure. But it is quite likely that it represents fragments (measosomes) of the plasma membrane to which the chromosome is attached in the intact cell. It is possible that enzymes required for the replication are located at this point of contact (with the membrane). Mesosomes do play a role in the separation of the replicated molecules of DNA during cell division. See chapter 2 reproduction in bacteria for more details).

The DNA has four usual bases – adenine, guanine, thymine and cytosine. In addition, the genome of bacteria may also have some amounts of methylated bases such as *6-methylaminopurine* and *5-methylcytosine.*

Introduction of a single strand break into the supercoil by a DNAse enzyme results in an abrupt change from supercoiled to non-supercoiled condition. This is because of free rotation about the opposing sugar phosphate bond. If the nucleoid, however, is nicked (cut), it does not become an open circle. But with each nick it passes through 40–50 intermediate stages in which the supercoiled loops turn into open loops one by one. This is a clear pointer to the fact that a single strand break does not cause free rotation of the entire molecule. The DNA is assumed to be fixed to the proteins in such a way that does not allow overlapping of the individual loops.

The enzyme (DNA gyrase) that brings about DNA replication is also responsible for supercoiling. This can be proved by the addition of *coumermycin,* an inhibitor of DNA gyrase, to the culture medium containing *E.coli* cells. The chromosomes of *E.coli* quickly lose supercoiling (on addition of coumerycin). In addition to DNAgyrase, another enzyme *topoisomerase* I, has the ability of unwinding the supercoils. Hence, it is believed that the two enzymes *–DNA gyrase* and *topoisomerase.* I act in a coordinated fashion and regulate the degree coiling. *In vitro* experiments have shown that purified topoisomerase I enzyme relases supercoiling to a non-supercoiled convalent circle. Mutants of *E.coli* lacking *topiosomerase* I have been seen to possess increased supercoiling to the extent of 32 percent more than the normal.

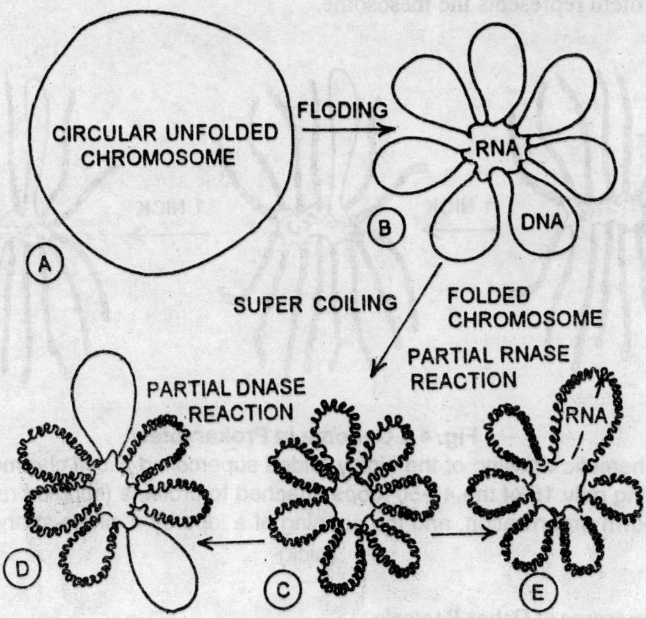

Fig. 4.3 Genome in Prokaryotes
Model of bacterial chromosome

A. Circular unfolded chromosome
B. Folded chromosome
C. Folded and supercoiled chromosome
D. Supercoiling relaxed into loops by DNAse nicks
E. RNAse cuts one RNA molecule

Some interesting observations have been made on these mutants lacking in topoisomerase. I. In *in vitro* conditions, the mutants undergo a further secondary mutation in or near the gyrase gene which seems to reduce the gyrase activity and return the supercoiling to near normal state.

Other Associated Structures of Chromosomes

The isolated genome of *E.coli* has about 30 percent by weight RNA and 10 percent protein. The RNA is largely a newly transcribed mRNA, tRNA or ribosomal RNA which is yet to be released into the cell machinery. A small amount of additional RNA in this is responsible for the folded character of the genome. RNAse activity can partially unfold the genome.

Most of the proteins have been identified as RNA polymerase engaged in transcription. This is quite understandable as a metabolically active bacterial cell would be virtually transcribing throughout its life cycle. A small portion of the protein represents the mesosome.

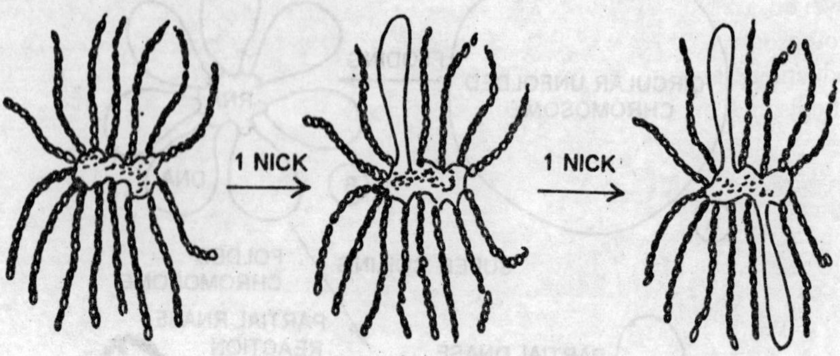

Fig. 4.4 Genome in Prokaryotes

A schematic drawing of the highly folded supercoiled *E.coli* chromosome, showing only 15 of the 40-50 loops attached to proteins (stippled region) of unknown organization, and the opening of a loop by a single-strand break (nick)

Chromosome of Other Bacteria

The thermophilic bacterium *Thermolplasma acidophilum,* possesses a chromosome that is quite unusual for a prokaryote. The genome is associated with a histone like protein which can complex with about 20% of the chromosome. The purpose of such an association is to maintain the stability of DNA during unfavourable (denaturing) environmental conditions.

The chromosome of *Bacillus subtilis,* is circular and has approximately the same dimensions and molecular weight like that of *E.coli.* There are suggestions that the genome of *B.subtils* replicates as a linear molecule rather than as a circular molecule. *Streptomyces coelicolor* and *Salmonella typhimurium* are also known to possess circular genome with the latter acomodating 323 identified genes. Species of *M.mycoplasma* have a small circular chromosome of about 265μ in length. *Haemophilus infleuzae,* the whooping cought causing bacterium, has a chromosome as long as 865μm which often appears linear with two free ends. Whether these ends really represent the points of a linear molecule or represent the ruptures of a circle during extraction, it is difficult to say.

PLASMIDS

The bacterial cell has an additional fragment of nucleic acids in addition to its normal chromosome. This additional genetic material was given the name episome (Jacob, Sccaeffer and Woolmann, 1960). Episome can exist either independently or it can integrate itself with the bacterial chromosome. Plasmids, however, are always independent and never integrate with the bacterial chromosome. Plasmids control a number of traits in bacteria such as –toxin production, antibiotic resistance, pilus production etc. Some people regard that episomes and plasmids are synonymus.

The word plasmid was first used by J.Lederberg (1952). However, the term is restricted only to bacterial cells where circular DNA molecules exist independent of the main genome.

Plasmids are circular double stranded molecules of DNA, and exist independently of chromosomal DNA in bacterial cells. They undergo replication during cell division and are carried to both the daughter cells. The plasmid has a right handed superhelical coil during the resting stage. There is one super helical turn for every 400–600 base pairs. The twisted conformation is known as **covalently closed DNA (CCD).** Cleavage results in the conversion of CCD to open circular form.

The following are some of the properties of the plasmids.

1. They are made up of DNA (double stranded).
2. They are capable of independent replication.
3. Some plasmids control bacterial conjugation.
4. Capable of reversible insertion into bacterial chromosome (only some plasmids).
5. They have the ability to get transfered from one bacterial cell to the other

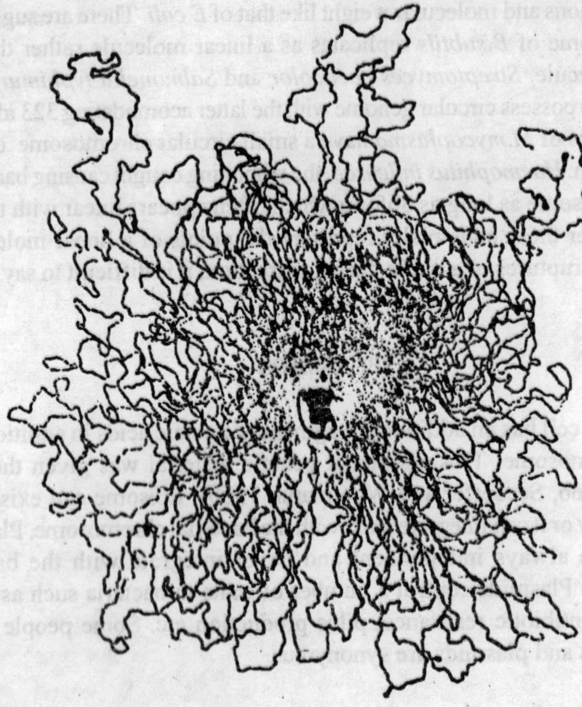

Fig. 4.5 Genome in Prokaryotes

An electron micrograph of the chromosome of the bacterium *Hemophilus infleunzae*

Plasmids are refered to as *conjugative* and *non-conjugative* based on the ability/ inability to promote their transfer during conjugation.

Classification of Plasmids

Basically plasmids are classified into three categories –

1. The F factor or sex factor.
2. The R Plasmid or resistance transfer factor (RTF), and
3. Col factor. There are some plasmids which may share the properties of more than one type of plasmid. Some bacteriologists recognise more than the three types based on some other properties.

F factor

The F factor plasmid confers the ability to the bacterial chromosome to get transferred to another cell (recepient). Hence, the donor is said to be F^+. F factor integrates with the bacterial genome facilitating its transfer. Because of this integration, it is often referred to as an episome.

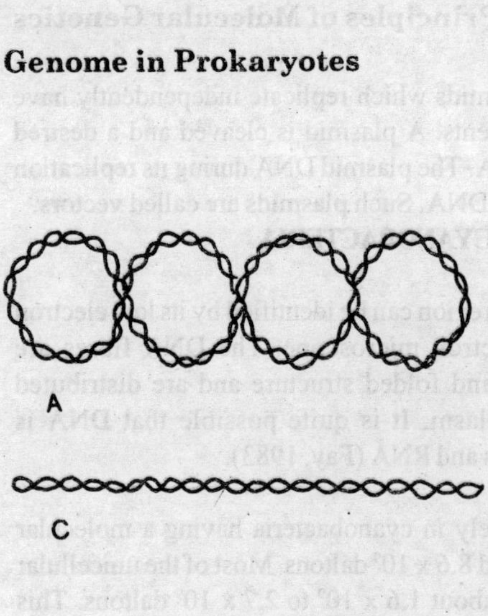

Fig. 4.6 Genome in Prokaryotes
Plasmid DNA, A. Supercoiled form, B. Open nicked circle, C. Linear duplex

R Plasmid

These plasmids, when present, confer on the bacterial cell resistance to several antibiotics such as streptomycin, tetracycline, chloramphenicol, sulphonamid etc. R plasmid mediated antibiotic resistance was first discovered in Japan in 1956, when *Shigella* became resistant to several antibiotics at one step. This was due to the transfer of R plasmid from one strain to the other resulting in not only antibiotic resistance but also the ability of a donor (it can donate R factor to other sensitive strains making them resistant).

R plasmid is a small extrachromosomal DNA ring. Sometimes, the R plasmid has two parts – (a) Part havin drug resistant genes and (b) Part having plasmid transfer genes.

The col factor (Plasmid)

Certain bacteria like *Escherichia coli, Shigella, Salmonella* etc., produce toxins called *colicins.* These are toxic proteins and can kill other bacteria. The ability to produce *colicin* rests in the plasmid called col plasmid or *colicinogenic factor.* Col factors resemble phages since they cause the death of bacteria, but they differ from phages as they are not released out during the lysis of the bacterial cell. Like the F factors, colplasmids also can be transferred from one bacterial cell to the other.

The number of plasmids per bacterial cell varies. It may be *a single copy* or *multiple copies*. Multiple copy plasmids which replicate independently have been used in gene cloning experiments. A plasmid is cleaved and a desired gene is inserted into the plasmid DNA. The plasmid DNA during its replication produces multiple copies of foreign DNA. Such plasmids are called vectors.

GENOME OF CYANOBACTERIA

In cyanobacteria, the nucleoplasmic region can be identified by its low electron density when observed under electron microscope. The DNA fibres are organized into a complex helical and folded structure and are distributed uniformly throughout the centroplasm. It is quite possible that DNA is associated with histione like proteins and RNA (Fay, 1983).

The size of the genome varies widely in cyanobacteria having a molecular weight ranging between 1.6×10^9 and 8.6×10^9 daltons. Most of the unicellular cyanobacteria possess genomes of about 1.6×10^9 to 2.7×10^9 daltons. This compares favourably with similar genomes found in other prokaryotes. Some filamentous cyanobacteria, however, have larger genomes. The distribution of genome sizes in cyanobacteria falls into four categories – 2.2, 3.6, 5.0 and 7.4×10^9 daltons. Members with a more complex morphology obviously possess larger genomes than those with a simple structure. There is a theory that the modern day prokaryotes evolved from a primitive bacterium with a small genome size of about 1.2×10^9 daltons. Replication of this genome during the course of evolution is supposed to have given rise to simple integer multiples seen in modern bacterial genome.

THE VIRAL GENOME

The core of the virion is made up of nucleic acids. A virus has only DNA or RNA, never both together. Within these parameters, however, considerable diversity exists. The nucleic acids may be double stranded, single stranded, linear or circular. Some have plus polarity; others may have minus polarity. Four types of nucleic acids are found in viruses with reference to the number of strands. These are –

1. Single stranded DNA —— (ss DNA) e.g., Colliphage virus
2. Double stranded DNA (ds DNA) e.g., Herpes virus
3. Single stranded RNA (ssRNA) e.g., TMV
4. Double stranded RNA (dsRNA) e.g., Reovirus

Single stranded DNA is found in ϕX174 virus. It was first discovered by Sinscheimer *et al,* in 1950's. ssDNA may be linear (Parovirus), or circular (ϕX174 virus). The ssDNA becomes double stranded during replication and at that time it is known as the replicative form.

Double stranded DNA is found in a number of animal viruses and bacteriophages. It has a variety of forms – linear (bacteriophages), cross linked (vaccinia virus) or closed circular duplex as in papova Viruses.

Single stranded RNA is found in a variety of animal viruses and icosahedral plant viruses. The strand may be plus (infectious), as in RNA bacteriophages, togaviruses etc. or minus (non infectious), as in rhabdoviruses and paramyxoviruses. Plus ssRNA acts directly as mRNA and translates proteins on the ribosomes of bacteria, whereas minus ssRNA first transcribes an mRNA and then it (mRNA) translates proteins (hence called non-infectious).

Double stranded RNA is found in animal viruses like reovirus, blue tongue virus etc. The dsRNA has 10 or more segments.

The nuclei acid core of Viruses may be summarised as follows –

1. Plant Viruses have only RNA (ss or ds), with the exception of cauliflower mosaic Viruses (DNA virus).
2. Animal Viruses have RNA (ss or ds), and DNA (only ds no ss).
3. Bacteriophages have DNA (ss or ds) or RNA (ss or ds). However, most phages have DNA Viruses.

5

REPLICATION OF DNA
(*In Prokaryotes*)

The DNA molecule regulates all the activities in the cell which includes both metabolism as well as reproduction. Because of its unique structure, it is able to control the synthesis of other macromolecules of the cell. At the same time when the cell reproduces, the DNA also should duplicate itself to be able to get distributed equally to the daughter cells. These two unique properties of the DNA molecule can be called *Heterocatalytic* and *Auto catalytic* functions. In the heterocatalytic function, DNA regulates the synthesis of other molecules like proteins. In the autocatalytic function which is perhaps very unique to DNA, it is able to make copies of itself in a precise and accurate manner. It is this ability of DNA that is responsible for the maintenance of life and inheritance of characters from one generation to another generation.

In this chapter, the various aspects of DNA replication in prokaryotes is discussed together with a brief account of the enzyme responsible for replication.

Types of Replication

Three different types of replication have been proposed at the molecular level for DNA. These are –
Semiconservative, Conservative and Dispersive.

1. Semiconservative

In this model of replication, the DNA helix unwinds but there is no rupture of a separated polynucleotide strands. Thus, the primary helices of DNA are not completely changed. According to this mechanism, the unwinding of the original strand would be followed by the synthesis of complementary strands for each of the two separated original polynucleotide strands. This mechanism was proposed by Crick and Watson, the discoverers of the molecular structure of DNA. In this method of replication, after replication of the original molecule, two molecules of DNA are formed. Each one will have one parental strand and one newly synthesised strand. Thus, in the semiconservative process, after replication the molecules of the next generation can be regarded as hybrids. The mechanism is called semiconservative because of the two strands of the parental molecule, only one strand remains in the molecule whereas the other goes to another molecule.

2. Conservative

In this mechanism, the original molecule is completely conserved and thus forms a part of the second generation molecule. Both the strands of the parental helix will synthesise an entirely new molecule, the new molecule synthesized would have both of its strands produced newly. After one generation of application, 50% of the molecules will be parental and 50% newly produced molecules. This is called conservative, because the original combination of both the parental strands will remain together and will not be separated.

3. Dispersive

In this method of replication, the parental DNA helix breaks up at several points forming many fragments. Each fragment of the helix would replicate and recombine at random to produce two daughter molecules of DNA. This is called dispersive because the original combination of the strands is completely disturbed as they get fragmented into bits.

Evidence for the Semiconservative Mode of Replication

Of the three types of replication mentioned above, there are overwhelming evidence to support the semiconservative mode of replication. Taylor *et al* (1957), demonstrated by autoradiography that the two chromatids during prohpase have one strand old and one strand new material. This clearly indicates that the replication is semiconservative. The following three experiments have clearly given experimental evidence for the semiconservative mode of replication.

1. Meselson and Sthal's Experiment

Meselson and Stahl (1958), conducted experiments on *E.coli* and demonstrated that the semiconservative mode of replication is correct. In their experiments, they used the following two basic principles.

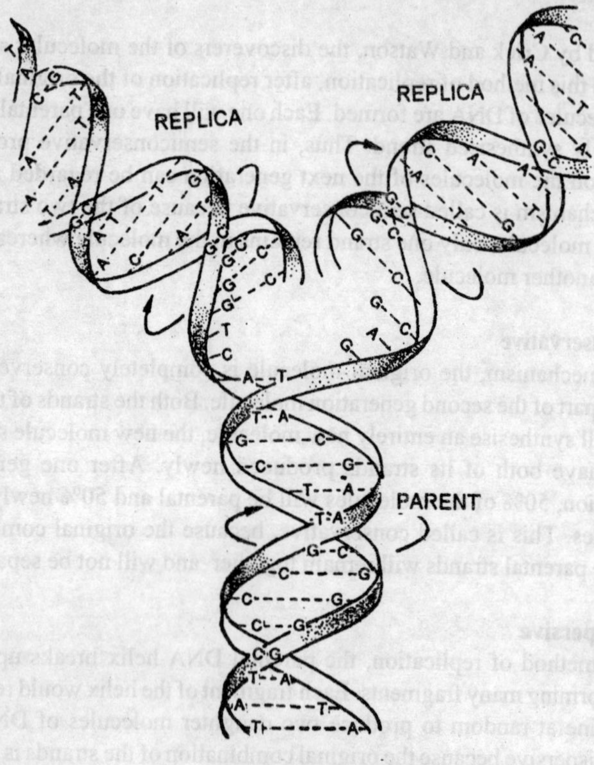

Fig. 5.1 Replication of DNA
Model of replication of DNA

(i) They cultured *E.coli* on a culture medium containing N^{15} (N^{15} is an isotope while normal nitrogen is N^{14}). The bacteria replicated in this medium for a number of generations and as a result, both the strands of the DNA molecule now contain N^{15}. In other words, the DNA of the bacterial populaton was now labelled with heavy nitrogen (N^{15}). The labelled bacteria were then transferred to a culture medium containing N^{14} and are allowed to replicate. After the first generation of replication, of the molecules of DNA, each one possesed one strand with N^{15} and the other strand with N^{14} (this should be the composition of the molecules if the mode of replication is semiconservative). In these molecules, the N^{15} strand represents a parental strand and the lighter N^{14} strand presents a newly synthesised strand. Thus, each molecule can be regarded as a hybrid DNA (having both N^{15} and N^{14}).

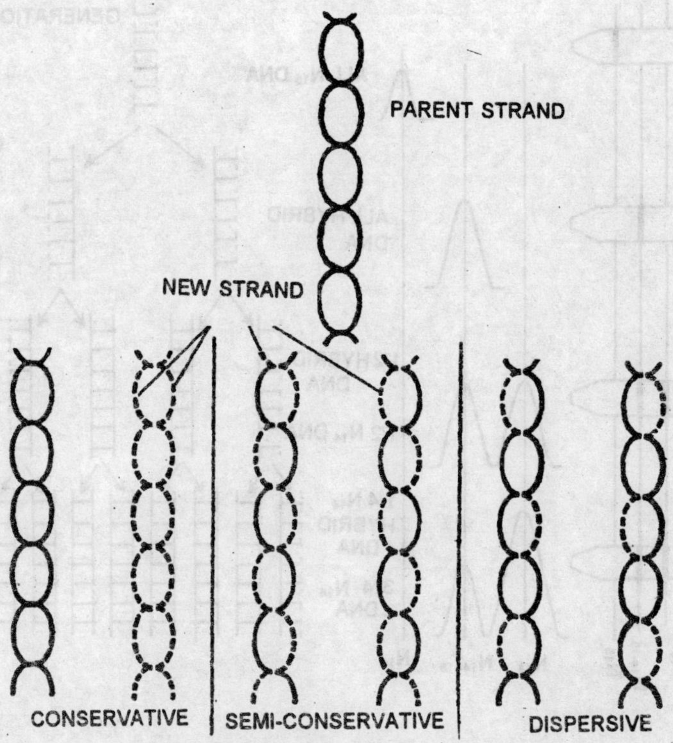

Fig. 5.2 Replication of DNA

Models of DNA replication - Conservative, semiconservative and dispersive

(ii) Meselson and Stahl used the principal of density gradient centrifugation and prepared gradients of Caesium chloride. This density gradient is based on the fact that when heavy salt solutions are subjected to ultra centrifugation, a continuous density gradient is set up (heavy would be at the bottom and lighter ones will be at the top with a range in between in the centrifuge tube). When a substance having a density within the range of this gradient is dissolved in salt solution, the substance will find place at its own level of density. Thus, with this technique even very slight differences in density can be detected.

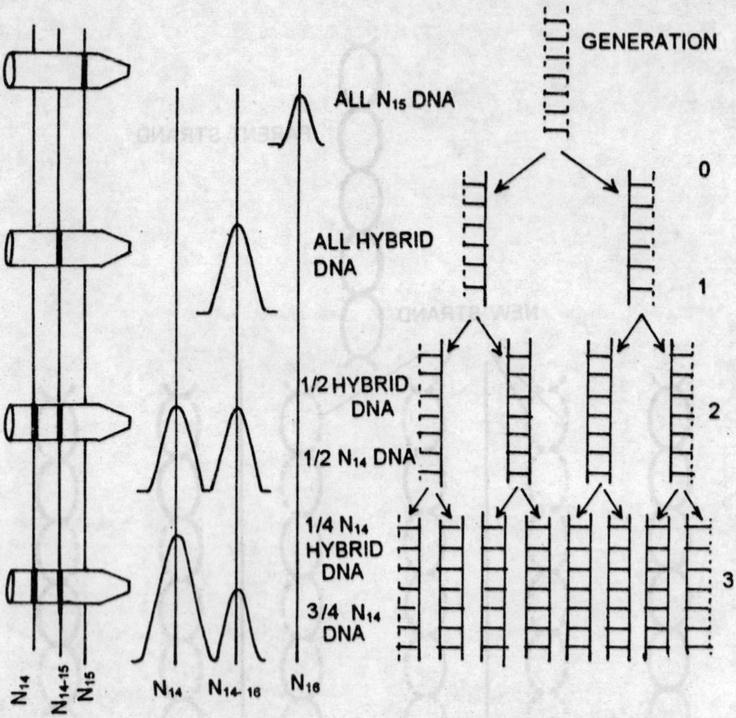

Fig. 5.3 Replication of DNA

Meselson and Stahl experiment in support of the semiconservative model of DNA replication

Based on the principles menthioned above, Messelson and Stahl subjected the bacterial cultures for density gradient centrifugation. At the first instance, N^{15} labelled bacteria were allowed to replicate on a medium containing N^{14}. After the first division, the centifugation showed the hybrid DNA had an intermediate density and formed a band inbetween N^{14} and N^{15} indicating clearly that each molecule contains one N^{14} and N^{15}. In the second generation two bands were found, one having normal N^{14} and the other having a hybrid (N^{14} and N^{15}). This is expected because the labelled nitrogen - N^{15} is getting reduced as the medium has only N^{14} and the original labelled N^{15} gradually gets decreased in the molecules. In the third generation, 3/4ths of the DNA and N^{14} and 1/4th of the DNA was hybrid. If the replication continues for some more generations N^{15} would disappear.

In the above experiment, if we assume that the replication is conservative there

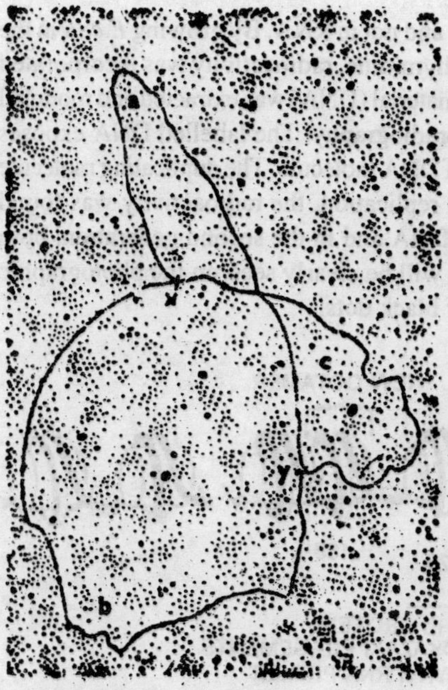

Fig. 5.4 Replication of DNA
Autoradiography of the single circular chromosome of *E. coli* labelled with tritiated thymidine-Cairans experiment

should be two bands in the centrifuge one with N^{15} and the other with N^{14} (newly formed molecule). On the other hand, if the mechanism is dispersive the molecule should spread all over the medium and should form only one band as both N^{14} and N^{15} in different combinations would be there in all the molecules. If the mechanism is semiconservative, the first generation would show only one layer having a mixture of N^{14} and N^{15}, and in the second generation there would, be two layers one N^{14} and the other N^{15} and N^{14}. Gradually, N^{14} would show greater band width and N^{15} would decrease. (Note. In the dispersive mode also only one band is formed in the first generation and it continues to be so for the second generation also. Where as in semiconservative mode, two bands aré formed in the second generation).

The above explanation clearly demonstrates that the semiconservative mode of replication proposed by Watson and Crick is the actual method of replication.

2. Cairn's autoradiographic experiment

The semiconservative mode of DNA replication has been proved in the chromosomes of *E.coli* by autoradiographic experiments conducted by Cairns.

Autoradiography is based on the fact that decaying radioactive elements emit electrons which collide with the silver grains of a photographic plate and turn it black. Thus in a autoradiogram, black dots represent positions of decaying radioactive elements. DNA of an organism can be labelled selectively by growing cells in a medium containing radioactive thymidine. Radioactive thymidine is obtained by using H_3– a heavy isotope of hydrogen known as tritium. Cairns used tritiated thymidine as it selectively labels only DNA and not RNA, because thymidine base is found only in DNA and not in RNA. By growing *E.coli* in the culture medium containing tritiated thymidine, radioactivity was incorporated in the daughter DNA molecules. The material (DNA) was extracted and mounted on a photographic emulsion for autoradiography. The labelled DNA exposes the film producing a diffused profile of DNA molecule. The duplicating molecule showed a replication fork. After the replication, the radioactivity was found incorporated in only one strands of DNA and in the second generation, both the strands become labelled and this will be clearly seen on the photographic emulsion because of the presence of black dots.

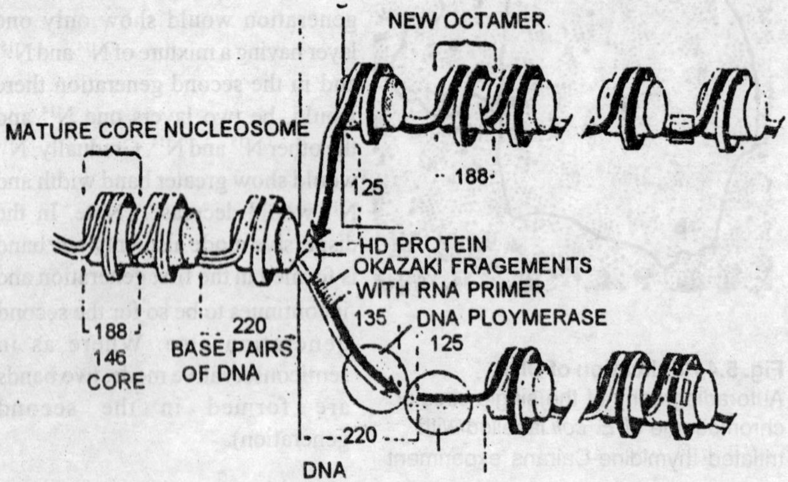

Fig. 5.5 Replication of DNA
Summary of the major steps in DNA replication

This experiment also clearly shows the mode of replication is semiconservative because after the first replication, each molecule will have only one labelledand the other non-tritiated strand. In the second generation, both the strands are tritiated in some molecules and in the other hybrid – one tritiated other non tritiated strands occur. and the other non-tritiated strand. In the second generation, both the strands are tritiated in some molecules and in the other hybrid – one tritiated other non tritiated strands occur.

Mechanism of replication
The mechanism of DNA replication in prokaryotes is as follows –

1. *Stage of replication during cell division* – It has been observed that

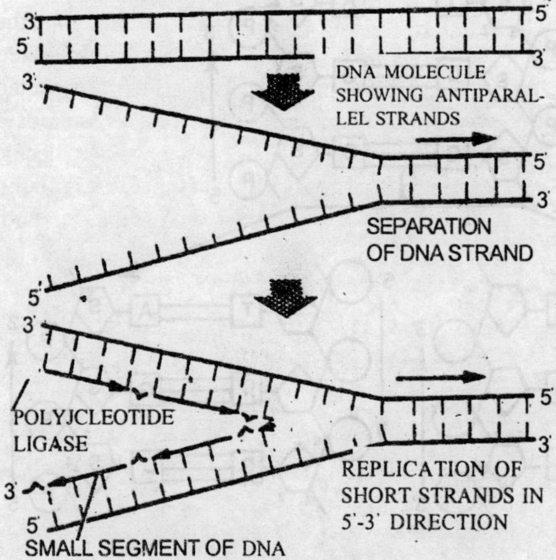

Fig. 5.6 Replication of DNA
Diagrammatic representation of discontinuous synthesis of DNA

replication takes place during interphase between two mitotic cycles. During interphase, the amount of DNA doubles due to synthesis. However, DNA synthesis does not take place during the entire interphase. It is confined to the S phase.

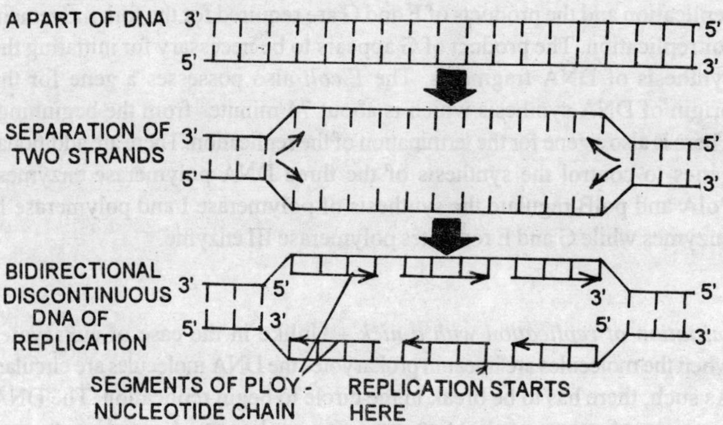

Fig. 5.7 Replication of DNA
Diagrammatic showing bidirectional DNA replication

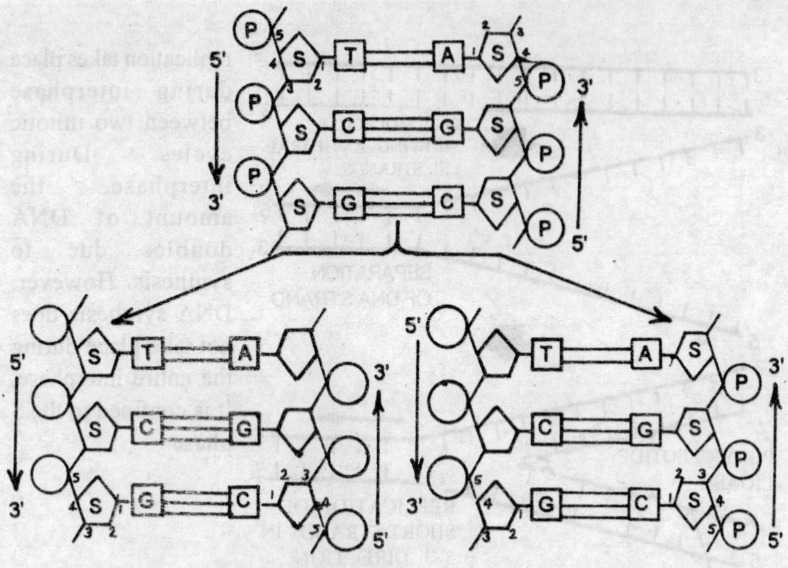

Fig. 5.8 Replication of DNA

Replication of DNA molecule. The two strands of DNA separate and each of them synthesizes its counterpart, thereby forming two similar molecules of DNA

2. *DNA Synthesising Genes* – In *E.coli* the overall process of replication requires the products of the genes *dnaA, dnaB, dnaC-D, dnaE,* and *dnaG.* The gene products of AC and D are required for the initiation of fresh replication and the products of E and G are required for the fork movement for replication. The product of G appears to be necessary for initiating the synthesis of DNA fragments. The *E.coli* also possesses a gene for the origin of DNA synthesis which is about 74 minutes from the beginning. There is also a gene for the termination of the replication. There are additional genes to control the synthesis of the three DNA polymerase enzymes. PolA and polB regulate the synthesis of polymerase I and polymerase II enzymes while G and E regulates polymerase III enzyme.

3. *Initiation of replication with a nick* – Unlike in the case of eukaryotes when the molecules are linear in prokaryotes the DNA molecules are circular. As such, there has to be break in the circle to begin replication. The DNA consists of many replicating units or replicons. According to one hypothesis, replication is initiated only after a break or nick in one of the two parental strands. The cut is presumed to take place in the helix at a

specific initiation point. This cut or incision is made by a specific endonuclease enzyme.

4. *Origin of replication* – Replication does not begin randomly, it always begins with a specific site called ori site and is regulated by a gene and its product. It has been discovered that in *E.coli*, the origin of replication point as about 74 minutes on the map.

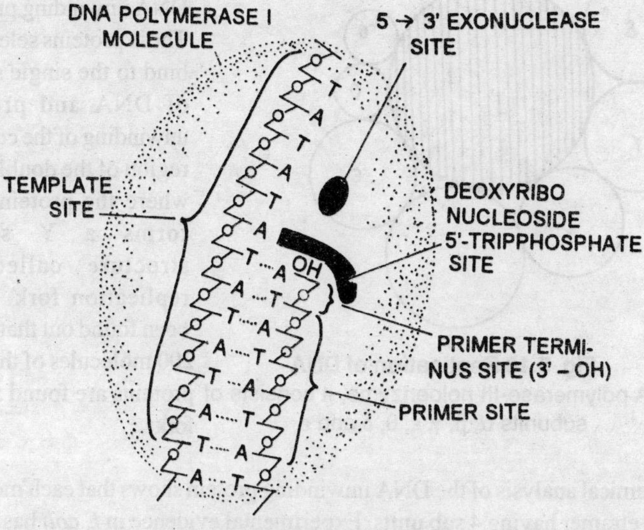

Fig. 5.9 Replication of DNA
A molecule of DNA polymerase-I showing specific binding sites

5. *Role of plasma membrane* – There are evidences to indicate that *E.coli* consists of an enzyme complex at the point of attachment of DNA to the membrane. This point of attachment is known as the replicating point. The DNA thread moves through the replicating point and duplicates itself in the process. In this process, the replication is exactly opposite of what is seen in eukaryotes where it is the enzyme that moves along the strand. The exact relationship between the plasma membrane and the DNA is not known. Quite probably, the membrane lipids may have some role to play in the process of initiation. Polymerase II and polymerase III enzymes are seen at the point of attachment.

6. *Strand unwinding* – The DNA helix has to first unwind in order to separate the two strands for beginning the replication. A study of the unwinding process in T_2 bacteriophages shows that each DNA molecule has to undergo 20,000 rotations in unwinding during replication. As the molecule itself

consists of 2 lakh nucleotide pairs, in these phages, replication must go on simultaneously along with unwinding.

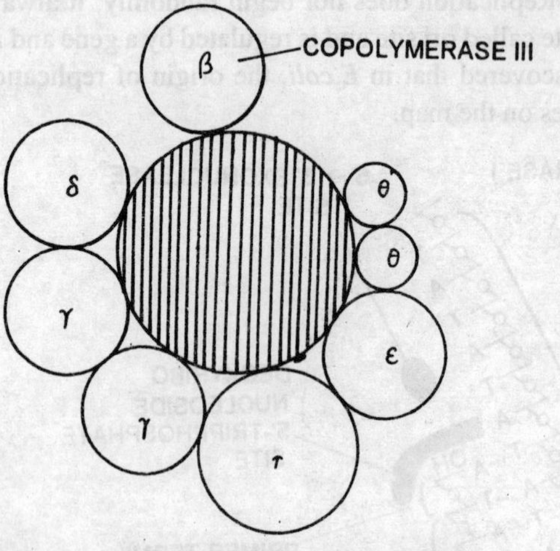

Fig. 5.10 Replication of DNA
DNA polymerase-III holoenzyme, it consists of subunits α, β, γ, ε, θ, δ and ε

The unwinding of the helix is controlled by enzyme proteins commonly referred to as DNA unwinding proteins. These proteins selectively bind to the single strands of DNA and promote unwinding of the coil. The region of the double helix where the proteins bind forms a Y shaped structure called the replication fork. It has been found out that nearly 200 molecules of the proteins are found at each fork.

Chemical analysis of the DNA unwinding protein shows that each molecule is a etramer having 4 sub units. Experimental evidence in *E.coli* has shown that the addition of DNA unwinding proteins to the medium increases the phase of the unwinding of the DNA molecule.

7. *Role of Swiveling Protein* – As the replication fork moves downwards along the parental strands, unwinding begins. For every ten nucleotides, there will be one rotation of unwinding. As the original configuration of the molecules is in the form of a wound, helix in circular DNA molecules present in prokaryotes untwisting or unwinding imposes a strain. To some extent, this strain can be received by additional twisting called super twisting in the unreplicated part of the DNA. However, this super twisting has a limit and ultimately even the supertwisting should be unwound for the replication fork to continue. The relieving of the strain imposed by unwinding in brought about by proteins called superhelix relaxing proteins. These proteins introduce small cuts in the non-replicating region of DNA thereby making one strand to rotate upon the other. Thus, relieving the strain. The breaks can then be healed by ligation.

8. *DNA Template* – To synthesise a fresh strand of DNAS apart from the

enzymes, the most important requirement are the nucleotides and a DNA template. The 4 nucleotides are adenine, guanine, cytosine and thymine. The previously existing DNA can serve as a template on which can be assembled fresh nucleotides.

The template DNA can be either single stranded or double stranded. Single stranded DNA is found naturally in some bacteriophages. Template DNA acts as a mould on which new strand of DNA is synthesised by complementary base pairing.

9. *RNA Primer* – DNA synthesis cannot be initiated on a DNA primer. DNA polymerase enzymes can initiate DNA synthesis only if there is a RNA primer. In other words, synthesis of a fresh strand can begin only if there is a small piece of a strand already. This small piece which is RNA is called RNA primer. The RNA primer is a short polynucleotide chain synthesised by the DNA template close to the point of replication. In all probability, this is catalysed by a modified RNA polymerase enzyme. It has been noticed that in *E.coli*, RNA polymerase is necessary to begin the replication.

10. *Elongation of chain* – Synthesis of new DNA strand takes place by the addition of fresh nucleotides to the 3'-OH group of the last nucleotide in the RNA primer. This synthesis takes place in 5'→3' direction and the enzyme that catalyses this is DNA polymerase III.

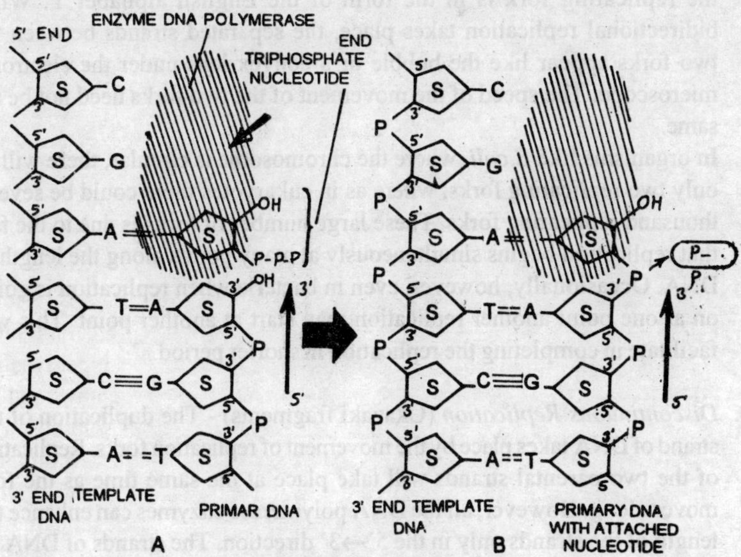

Fig. 5.11 Replication of DNA
Diagram showing action of DNA polymerase in DNA replication

The newly synthesised DNA chain has an RNA primer attached to its 5' end. After the synthesis of the chain, the RNA primer is hydrolysed by the exonuclease activity of DNA polymerase I. The resulting gap left by the RNA primer is filled by DNA nucleotides by the catalytic activity of DNA polymerase I. The freshly formed nucleotide chain is joined to the existing chain by the lygase enzyme.

Several opinions have been put forward to account for the presence of an RNA primer to initiate the synthesis of DNA chain. According to one opinion, if any errors occur during the initial synthesis, it can be eliminated when the RNA primer is removed.

11. *Direction of Replication* - It has been seen that in prokaryotes the replication may proceed in one or both directions from the point of origin. Thus, replication may be unidirectional or bidirectional. Bidirectional replication has been seen in *E.coli, Bacillus subtilis, Salmolella, Typhimirum, E.coli* phages etc. Unidirectional replication has been reported is bacteriophages P2 and 186, and also in the mit DNA of mouse LD cells. Bidirectional replication has an advantage in that, the origin is completely duplicated before the complication of replication.

12. *Formation of Replicating Forks* – At the point where the two strands get separated to initiate replication, a replicative fork is found. The shape of the replicating fork is in the form of the English alphabet Y. When bidirectional replication takes place, the separated strands between the two forks appear like the bubble or a convex lens under the electronic microscope. The speed of the movement of the two forks need not be the same.

    In organisms like *E.coli,* where the chromosome is circular, there will be only two replicating forks, where as in eukaryotes there could be several thousand replicating forks. These large number of forks is due to the fact that replication begins simultaneously at many points along the length of DNA. Occasionally, however, even in bacteria when replication is going on at one point another replication may start at another point. This will facilitate in completing the replication in shorter period.

13. *Discontinuous Replication* (Okazaki fragments) - The duplication of the strand of DNA takes place by the movement of replication forks. Replication of the two parental strands will take place at the same time as the fork moves along. However, all the DNA polymerase enzymes can enhance the length of the strands only in the 5'→3' direction. The strands of DNA on the other hand, polarise in opposite directions. This naturally poses a question as to how can DNA polymerase bring about simultaneous

replication. While the enzyme moves forward on one DNA strand, simultaneously it cannot move backward on the other.

Okazaki et al (1968), have provided an answer to the above problem. According to them only one strand 3'→5' is continuously replicated, while the other 5'→3' replicates in a discontinuous manner and synthesis, only short fragments of DNA. These short fragments have been called Okazaki fragments. The direction of synthesis of Okazaki fragments in opposite to the movement of the replicating fork. Eventually, Okazaki fragments are joined together by the enzyme polynucleotidfe ligase. In prokaryotes, the Okazaki fragments are as long as 1,000–7,000 nucleotides.

According to another suggestion (Okazaki et al), both the strands may replicate discontinuously and the fragments formed may join up later.
Evidence obtained from *in vitro* culture of *E.coli,* have shown that both the strands are synthesisd discontinuously.

Enzymes involved in DNA Replication

1. **DNA Polymerases** – Kornberg (1957), isolated an enzyme from *E.coli* which had the ability to polymerise the nucleotides dATP, dTTP, dCTP and dGTP to form a polynucleotide in the presence of a primer and DNA template. This enzyme was capable of synthesizing new DNA strands by proceeding in 5'→3' direction. Subsequent works of DeLucia and Carirns (1969), Knippers (1970), and Gefter et al (1969), showed that in prokaryotic cells there are at least three DNA polymerases – *poly I, pol II,* and *poly III.* Kornberg's DNA polymerase is equivalent to *Pol I.*

Polymerase I – This enzyme does not seem to be necessary as bacterial mutants lacking this are able to synthesize DNA. *Poly I* seems to be taking part in DNA repair. Structurally, poll is a single polypeptide chain with a molecular weight of 1,09,000. Electron microscopic studies of the enzyme have shown that the enzyme has a number of functional sites –

(i) A template site for binding template DNA.
(ii) Primer site for binding primer RNA segment.
(iii) A primer terminus site for 3' - hydroxyl terminus of primer.
(iv) A triphosphate site - locus for incoming deoxy ribonucleotide 5' group.
(v) Exonuclease site for 5'→3' exonuclease activity.

Polymerase II - This is a single polypeptide chain with a molecular weight of 90,000. *E.coli* cell has about 40 molecules of *pol III.* It helps in 5'→3' polymerization.

Polymerase III - Discovered by Kornberg and Gefter (1972), it is the most active among all the three enzymes. The enzyme is chiefly responsible for DNA chain elongation. It has a molecular weight of 5,5000. The enzyme also acts as 5'→3' and 3'→5' exonuclease..

DNA ligase– Also known as polynucleotide ligase, this enzyme joins the two fragments by catalysing the synthesis of a phosphodiester bond between a 3'-OH group at the end of one chain and 5'→3' - OH group at the other end of the other chain. The enzyme is necessary to join Okazaki fragments which are produced after discontinuous replication.

The enzyme has been isolated from *E.coli*. It is made up of a single polypeptide chain and has a molecular weight of 70,000. Each cell of *E.coli* is known to posses 2000–4000 molecules of this enzyme. DNA ligase has the following functions.

1. Joins Okazaki fragments during DNA replication.
2. Helps in sealing single strand nicks in DNA duplex.
3. Joins free ends of DNA (linear molecule) to form circular chromosome.
4. Joins the segments of DNA during recombination during meiosis.

MODELS FOR DNA REPLICATION (IN PROKARYOTES)

Jaboc and Brenner's Model

Studies with phase contrast microscope have shown that in the dividing cell, the sister chromosomes get oriented equidistantly inside the cell. In eukaryotic cell, this separation is achieved by the mitotic spindle. In prokaryotes like bacteria such a mechanism is lacking. Jacob and Brenner have proposed a model to explain the replication and separation of daughter chromosomes. According to this model, the chromosome gets attached to the mesosome at the point of origin of replication. After the completion of replication, a new membrane attachment site is formed adjacent to the old one. The strand which has acted as a template remains a circle, while the newly formed one is broken. One of the free ends of the newly formed chromosome attaches to the new binding site. During replication, the replicating fork and the enzyme system remains stationary while the DNA molecule moves along and gets replicated. Simultaneously a new membrane is synthesized between the two and the two genomes get separated. The newly synthesized DNA becomes circular eventually.

Carin's Model

At a specific point (origin), the DNA breaks and replication begins and proceeds in two opposite directions. As the replication proceeds, the DNA strand unwinds. This unilateral unwinding (the other cannot unwind freely) creates a pressure and it is transfered to the unreplicated part resulting in supertwisting. This results in the stoppage of replication. To begin replication once again, a 'neck' is made in one of the strands by a swivelling protein. This facilitates rotation of the strands and the break is closed subsequently.

Yoshikawa Model

This is a modification of the Cairn's model. According to Yoshikawa, the newly formed DNA strands covalently join to the ends of original chromosome.

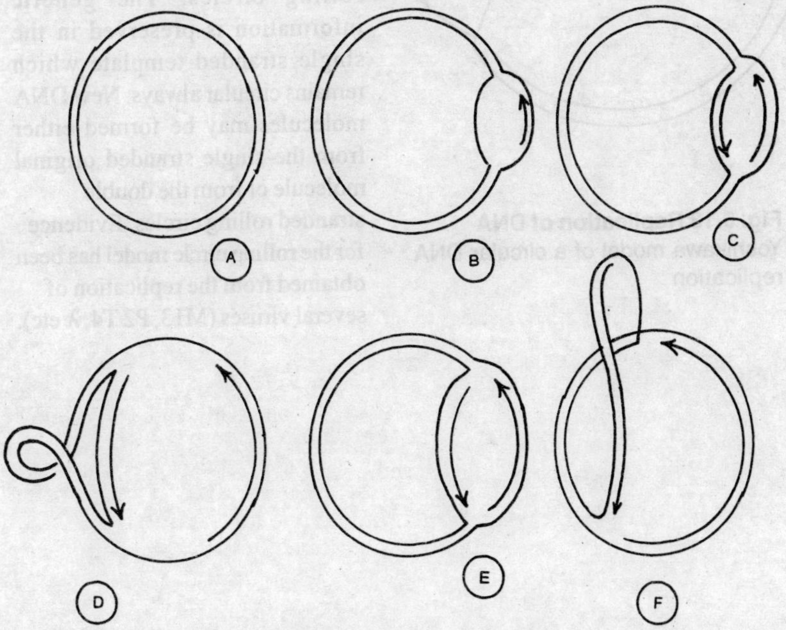

Fig. 5.12 Replication of DNA
Cairn's model for circular DNA replication

Rolling Circle Model

This is the model accepted for the replication of single stranded DNA in viruses (ϕ x174). The single stranded circular DNA becomes double stranded by the synthesis of another strand starting at the origin. One strand of this duplex ring is now cut at a point by an endonuclease. As a result of this, a linear strand with 3' and 5' end is created. The 3' end serves as a primer for the synthesis of a new DNA strand. The unbroken strand is used for this purpose; with the parental strand itself acting as a template. The 5' end of the strand becomes attached to the membrane. Meanwhile the original, unbroken single strand unwinds and

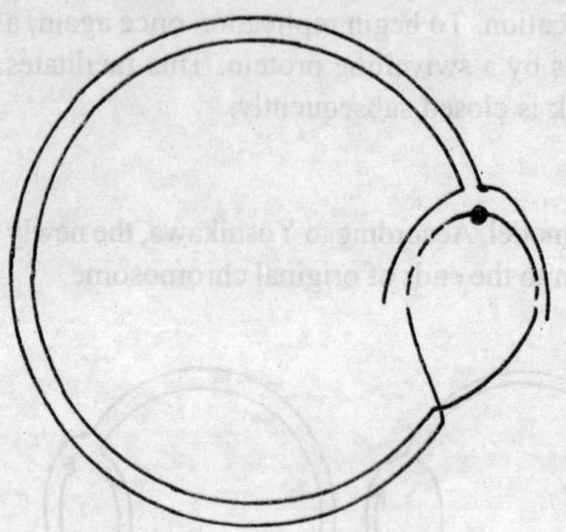

Fig. 5.13 Replication of DNA
Yoshikawa model of a circular DNA replication

separates leaving the tail attached to the membrane. The tail is cut off into a specific length – *Progeny rod.* The rod rolls and forms a circular molecule, with the enzyme ligase binding the two ends. The newly formed circular molecule which is double stranded can in turn become rolling circles. The genetic information is preserved in the single stranded template which remains circular always. New DNA molecules may be formed either from the single stranded original molecule or from the double stranded rolling circles. Evidence for the rolling circle model has been obtained from the replication of several viruses (M13, P2 T4, λ etc).

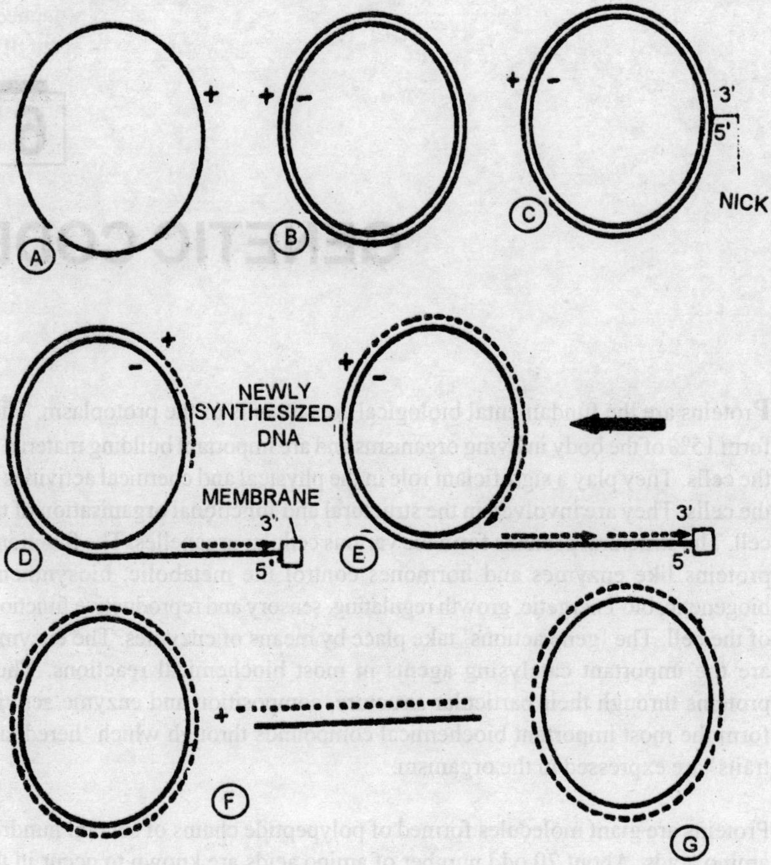

Fig. 5.14 Replication of DNA
Rolling circle model of a circular DNA replication

6

GENETIC CODE

Proteins are the fundamental biological constituents of the protoplasm. They form 15% of the body in living organisms and are important building material of the cells. They play a significiant role in the physical and chemical activities of the cells. They are involved in the structural and functional organisation of the cell. The structural proteins form the various cellular organelles. The functional proteins like enzymes and hormones control the metabolic, biosynthetic, biogenetic, bio-energetic, growth regulating, sensory and reproductive functions of the cell. The 'gene actions' take place by means of enzymes. The enzymes are the important catalysing agents in most biochemical reactions. Thus, proteins through their particular structure, composition and enzyme activity form the most important biochemical compounds through which 'hereditary traits' are expressed in the organism.

Proteins are giant molecules formed of polypeptide chains of several hundred amino acids. About 20 odd number of amino acids are known to occur in the cells. Amino acids are formed of a basic amino (NH_2), and an acidic carboxyl (COOH) groups. The variety of cells in an organism represent myriads of proteins. Different types of proteins are synthesised by different cells of the body. Different combinations of amino acids produce different proteins.

Synthesis of Protein

The process of protein synthesis is the most complicated biological activity. Basically, it is governed by the *genetic information* carried in the genes (DNA) of the chromosomes. Specific amino acids are linked in particular sequence or order to form a polypeptide chain of protein. The various cell organelles and chemical compounds involved in the synthetic proces are –

1. Deoxyribonucleic acid .... DNA
2. Three types of ribonucleic acids .... RNA
3. Amino acids

4. Enzymes and
5. Ribosomes

The discussion above will clearly reveal that proteins are the representative of the master molecule of the cell DNA to carry out and regulate all the metabolic activities. Proteins are important in another way also. They are important for the existence of DNA itself. A number of enzymes (proteins) are necessary for the replication of DNA itself. When proteins are of such fundamental biological importance, no wonder then the cell machinery to synthesize them (proteins) is not only complicated but precisely regulated.
Which molecule(s) present in the cell can regulate the synthesis of proteins? Obviously there cannot be another compound than DNA. The structure and properties of the DNA molecule are eminently suited to exercise such a regulation over the synthesis of proteins.

Having conceded that DNA directly is involved in the regulation of protein synthesis, we may now try to answer the two following questions – which aspect of protein structure is controlled by DNA? And how and by what mechanism DNA regulates the assembly of amino acids to synthesize the proteins?

Proteins are extremely complicated biological macromolecules. There are different dimensions for its structural complexity and integrity. These are theprimary, secondary, teritary and quaternary structures (for details – see the chapter on protein synthesis). DNA molecule specifies only the primary structure, while the other dimensions of protein structure like folding etc. is controlled by proeins themselves.

The primary structure of proteins involves the following aspects – quality, quantity and sequence of amino acids and it is this that is regulated by DNA.

What is the mechanism adopted by DNA to regulate the synthesis of proteins? Obviously, proteins cannot be synthesized *in situie* within the nucleus, because all the components needed for proteins–amino acids, ribosomes etc., are not present in the nucleus. It is not possible also for the DNA itself to migrate to cytoplasm to synthesize the proteins. The only alternative is for DNA to send a messenger or representative which carries the message to specify the quality, quantity and the sequence of amino acids to be asembled to produce the protein. This message is called the genetic code, because it specifies in genetic language the amino acids to be assembled.

The language of genetic code

A code is said to be a cryptic or hidden language, carrying a message and upon deciphering or translation, the message is understood. In the genetic code, the message is translated as the type of aminoacid to be assembled. What is the composition of the genetic code that comes out of DNA? To answer this, we must first understand the manner in which genetic information exists in the DNA molecule.

It is well known that eventhough the DNA molecule has three components namely sugar, phosphate and nitrogen bases, it is the last one that is unique to each DNA molecule in the sequence of their arrangement. The nitrogen bases vary from one DNA segment to another in arrangement, so the information is based upon the sequence of their arrangement. The sequence of nucleotides is then the code in which the DNA sends the message. How is the message sent? The DNA molecule replicates in a particular segment and the replicated segment in the form of RNA now called mRNA carries the genetic message in the form of nucleotide sequence (for details see protein synthesis). The sequence of nitrogen bases have been found to be identical to the linear sequence of amino acids in a peptide chain. In otherwords, there is colinearity between mRNA and polypeptide chain (see later in the chapter for details of colinearity).

It is thus clear that the genetic code is in the language of nitrogen bases and the reading of the code is in their sequence of arrangement.

Amino acids involved in protein synthesis

A large number of amino acids – almost upto 150 are found in nature. Of these, however, quite a number of them remain as free amino acids in the sense they do not join together to form a molecule of protein. It has been found out that only about a little more than 20 amino acids are protein amino acids in that all the known proteins are made up of only these amino acids. Some amino acids, however are found in almost all protein chains while a few others are found only in very few proteins. For example, cystine is present only in proteins which have a disulphide linkage. Hydroxylproline is another amino acid found only in few proteins.

The code is triplet (basis of crypto analysis)

A DNA molecule from every source contains only 4 types of nucleotides– adenine, guanine, cytosine and thymine. On the other hand the basic building blocks of proteins namely amino acids are more than 20 types. A basic problem that one has to analyse before finding about a genetic code is how many bases specify a unit of amino acis. Sevral opinions have been put forward in this regard.

Single code. According to one of the opinions, each nucleotide will code for one amino acid. But this would be insufficient to code for all the 20 amino acids as there are only 4 nucleotides. With 4 nucleotides only 4 amino acids can be coded. What about the rest?

Double code. This asumes that each code is formed of two letters i.e., 2 nucleotides. In other words, 2 nucleotides will specify one amino acid. In this type of a code, all the 20 amino acids cannot be coded but 16 codes (4 x 4) are possible. But there will be a number of ambiguous determination of amino acids.

Possible singlet, doublet and triplet codes of mRNA

Single Code (4 words)	Double (16 words)				Code Triple Code (64 words)			
					AAA	AAG	AAC	AAU
					AGA	AGG	AGC	AGU
					ACA	ACG	ACC	ACU
					AUA	AUG	AUC	AUU
					GAA	GAG	GAC	GAU
					GGA	GGG	GGC	GGU
					GCA	GCG	GCC	GCU
A	AA	AG	AC	AU	GUA	GUG	GUC	GUU
G	GA	GG	GC	GU	CAA	CAG	CAC	CAU
C	CA	CG	CC	CU	CGA	CGG	CGC	CGU
U	UA	UG	UC	UU	CCA	CCG	CCC	CCU
					CUA	CUG	CUC	CUU
					UAA	UAG	UAC	UAU
					UGA	UGG	UGC	UGU
					UCA	UCG	UCC	UCC
					UUA	UUG	UUC	UUC

Triple code. This was first suggested by the physicist, George Gamow (1954). According to this code, 3 bases (nucleotides) will code for one amino acid. In other words, each code will have 3 nucleotides which will specify the particular amino acid. These triplets of codes are given the name codons. With triplet codes, thepossibilities would be 64 (4x4x4) codons. This would be more than the number of amino acids (which are only 20). Obviously then each amino acid will have more than one code. Experimental evidence shows that the code is a triplet and about 61 of the 64 codes are used to specify amino acids during protein synthesis.

The maximum possible number of codons in the singlet, doublet and triplet codes

Type of code	Number of bases in codon	Number of codons	Ambiguous/ degenerate
Single code	1	4	Amgiguous
Double code	2	4 x 4 = 16	Amgiguous
Triplet code	3	4 x 4 x 4 = 65	Degenerate (more than required number)

Properties of code

1. The code is non-overlapping

The DNA molecule is a very long chain and has series of nucleotides. There is a gap between the nucleotides. Naturally a question arises that if three nucleotides constitute a codon will there be any gap between two successive codes or do the codes overlap. In otherwords, is the code overlapping or non overlapping? If the code is over lapping from where will the reading begin. Because the beginning of the code reading will alter the nucleotide combination and hence will alter amino acid which is specified by the code. When Gamow proposed the code he assumed it to be overlapping. Many people attempted to rationalise the excess number of codes i.e., 64 for 20 amino acids by assuming that the codes are overlapping and the codons can be reduced to about 20. But this involves a lot of problems. If we take for instance, the codes for lysine (AGG and AAA), and codes for phenyl alanine (UUU and UUC). The code letters of these two amino acids are mutual exclusive i.e., lysine forbidden combination of amino acids and proteins. But, on the other hand these amino acids are found in proteins proving that the genetic code is non-overlapping. Further, the overlapping code assumes that a nucleotide is read twice in different combinations to code for amino acids. If this is true, the entire precision of code would be disturbed. Further, any mutational change in an overlapping code would affect more than one code thus disturbing a number of amino acids.

Mutational studies have shown that the code is non-overlapping. In TMV (Tobacco Mosaic Virus), the change of one base in the DNA due to mutation would alter the composition of protein by a single amino acid proving that the code is non overlapping. If the code is overlapping change of one base should have disturbed more than one amino acid. Studies on normal and sickle cell haemoglobins have shown that single base mutations result in the replacement of only one amino acid.

2. The code is degenerate

(Synonym codons). As it has been mentioned above, there are 64 possible codons considering the code to be a triple one. Of these 64 codes about 61 code for amino acids. As they are only amino acids for exceed the number of amino acids. This leads to the conclusion that each amino acid must have more than one code. Of the 20 amino acids, only tryptophan and methionine have a single codon each, while all the others have more than one codons. *Phenylalanine, tryosine, histidine, glutamine, asparagine, lysine, aspartic acid* and *glutamic acid* have two codons each (see the codon table). *Leucine, argenine* and *serine* have 6 codons each. Thus, there is no specificity as to the number of codons that decide an amino acid. The number of codons (whether one or many) specifying the amino acid possibly have been evolved over a long period of time during the process of evolution. The variability in the number of codons perhaps may explain the differential distribution of amino acids in proteins (those with multiple codons occur more frequently).

When there are multiple codons for an amino acid, it must be noted that these codons are not totally different. Synonym codons specifying an amino acid have the first two bases of the triplet being constant. Whereas only the third will vary. For instance, proline is coded by the following codons – CCU, CCC, CCA and CCG. It is obvious here in these synonym codons, the first two bases are common.

The multiplicity or degeneracy of the code may pose a question as to why nature has provided alternate codons for an amino acid where a single codon would have been more precise and specific. One possible reason for the degeneracy of the codes is that multiple codons protect the organism from mutations which might eliminate an amino acid altogether if there were to be only one code.

Number of codons coding for different amino acids

	Amino acids	Number of codons
1.	Tryptophan, methoionine	1
2.	Phenylalanine, tyrosine, histidine, glutamine, asparagine Lysine, aspartic acid, glutamic acid, cysteine	2
3.	Isoleucine	3
4.	Valine, proline, threonine, alanine, glycine	4
5.	Leucine, agrinine, serine	5

3. Polarity of the code

It is very essential that a particular gene should specify an amino acid always and in all situations unless it undergoes mutational changes. It is necessary therefore, that the code must be read between a fixed beginning point and a fixed end point. These points are referred to as the initiation and termination codons respectively. Another necessity is the reading of the direction of the code because any change in the direction would alter the amino acid. All these features clearly point out that the code must have a polarity i.e., a fixed start, a fixed end point and a predetermined direction. For instance, if the codon direction as given below varies, the composition of the amino acids would completely vary.

Leucine	Isoleucine	Valine	Serine →
UUG	AUG	GUG	UCG
← Valine	Leucine	Leucine	Alanine

It is obvious from the above that the direction of reading the code would alter the sequence of the amino acids and obviously changes the composition of the proteins. Available molecular biological evidences indicate that the message in mRNA is read in 5' →3' direction. Based on this message, the polypeptide chain is synthesised from the direction of amino group to carboxyl group.

4. The code is comaless

One of the most interesting questions with reference to the non-overlapping code is, are there any gaps between the codes in the form of punctuation marks so that one code is separate from the other. In genetic terms this means that there could be bases between codes indicating where one code ends and the other begins. For instance, codes with intervening punctuation marks can be represented as follows –

UUG	X	AUG	X	GUG	X	UCG
Valine		Leucine		Leucine		Alanine

In the above, X can be regarded as a base separating the two codons. If a mutation occurs in any one of the codes, it will not affect other codes and consequently, the disturbance in the protein synthesised would be to the extent of only one amino acid. On the other hand, if the code is commaless (no punctuation mark-base between codes), mutation in one base would result in a drastic change of the genetic message. Researches of Dr.Khorana and others had clearly indicated that the code is commaless i.e., there is no delimiting point between one codon and the other. Khorana and his associates used artifically synthesised long polynucleotide chains which had repeating sequences to produce polypeptide chains. For example, the repeating sequences CUCUCU contains the codons for leucine (CUC), and serine (UCU). When this sequence was used for producing the proteins neither amino acid was incorporated with the protein unless the other one is also present. This clearly explains that the triplet code has to be commaless, and only then there would be alternate translation of the codes into leucine and serine.

5. Universality of the code

The code was first deciphered and worked out in microorganisms but subsequent researches have clearly shown that these codes are same in their amino acid specification from bacteria to man. Thus, the code is said to be universal. Marshall Caskey and Nirenbrg (1967), have shown by their experiments that diverse organisms such as *E.coli* (bacterium), *Xenopus laevis* (amphibian), and Guinea pig which is a mammal used the same code for their amino acyl tRNAs.

Gene mutations provide additional evidence for the universality of codes. These mutations result in the substitution of one amino acid for other. Studies have indicated, gene mutations bringing about amino acid substitutions effect the same amino acid in TMB, *E.coli* and also in man. For instance, a single point mutation, changes the amino acid in the coat protein of TMV, alpha chain of tryptophan synthetase in *E.coli,* and also in the haemoglobin of man.

Another remarkable feature of the code is not only that it universally occurs in all organisms, but it has also remained probably without any change ever since it evolved in some bacteria a few billion years ago. Any mutation that occurs would change the composition of mRNA, and consequently the amino acid composition in the protein. Changes such as these apparently have proved to be harmful to the organisms and as such there will be a strong selection pressure against the occurrence of such mutations. Hence, natural selection has persisted with the codes perhaps ever since they originated.

Ambiguity in the genetic code

There is no evidence to indicate that the genetic code is ambiguous *in vivo*. Ambiguity means that a single codon may code for more than one amino acid. The only exception to the non-ambiguity is the AUG codon in prokaryotes–it codes for formyl methionine at the initiation site while at other positions it specifies methionine. It is not clear as to how this distinction is always met precisely but it is believed that the secondary structure of the mRNA at the initiation site may have got something to do with the coding for formyl methionine. In *in vitro* conditions, however, genetic codes sometimes appear to be ambiguous. For example poly-U codon is known to incorporate some amounts of leucine in addition to phenylalanine. Treatment of ribosomes with streptomycin is known to introduce ambiguity to codons. For example in addition to coding for phenylalanine UUU is also known to code isoleucine, serine and leucine.

Codons and anticodons

During the process of translation, the message carried in the mRNA would be converted into the sequence of amino acids which are picked up by tRNA. This is achieved by the process of binding. For this binding to take place, obviously then there should be a specific relationship between the nucleotide. Sequence of mRNA is called the codon and that of tRNA is called anticodon. The reading of mRNS is in the 5'→ 3' direction and the codons are also written in the same direction. For example, codon AUG is written as 5' AUG 3'. The corresponding anticodon and tRNA should therefore be represented as 5' CAU 3'. In such a configuration, the first nucleotide of both codon and anticodon would be the ones at the 5' end and third nucleotides would be at the 3' end.

Very often, the anticodon is written in the 3' → 5' direction so that it will be easy to correlate between the bases of codon and anticodon.

Initiation codons

In most of the polypeptide chains, the first amino acid which begins the chain is methionine in eukaryotes and N-formlymethionine in prokaryotes. Methionine or N-formylmethionine tRNA complex specifically binds to the initiation sites in mRNA containing the codon AUG which codes for methionine. The AUG code, therefore, is referred to as the initiation codon. Less frequently, however, GUG (code for valine) may also serve as the initiation codon in bacterial protein synthesis. In the bacteriophage MS2, GUG is the initiation codon for the protein referred to as A protein. Researches have indicated that the GUG may replace AUG as initiation codon when the latter is lost due to chromosomal abbreation. It has also been found out that chain initiation by GUG is not efficient as AUG possibly because the former has less affinity for formyl methionyl tRNA.

Termination codons (nonsense codons)

It has been mentioned above that AUG (occasionally GUG) initiates the synthesis of a protein chain. The natural question would be which codon specifies or signals the end of termination of the synthesis of the protein chain. It has been found out that three of the 64 codons do not specify any tRNA and as such they cannot assemble any amino acid. As such, these codons whenever they are found in mRNA would signal the end of the chain. They are called termination codons as they terminate the synthesis of the polypeptide chain.

They are also called *nonsense codons* as they donot specify any amino acid. The nonsense codons are UAG (*amber*), UAA (*ochre*), and UGA (*opal*). Of the three termination codons, UAG was the first termination codon to be discovered. It was named amber after a graduate student in Germany, named Bernstein (means amber in German). This graduate student helped in the discovery of the class of mutations. The other two termination codons were obviously given the name of the colours only to maintain uniformity.

Termination codons found in mRNA will not specify any amino acid and as such stop the process of translation and bring about the release of the peptide chain from ribosomes. No tRNA molecules have anticodons complementary to the termination codons found in mRNA. Termination codons seem to have some affinity to proteins called release factors which bring about the release of the protein chain. In prokaryotes, there are three release factors RF-1, RF-2, and RF-3. RF-1 recognises the amber and ochre codons, while RF-2 recognises ochre and opal codons, RF-3 stimulates RF-1 and RF-2 ultimately releasing the newly synthesised peptide chain. In eukaryotes there is a single factor (RF protein) which is capable of recognising all the three termination codons.

Evidence for genetic code (deciphering the code)

Earlier to the pathbreaking experiments of Nirenberg, Matthaei and H.G. Khorama, work on the genetic code was largely speculative and restricted to development of theoritical concepts. The great physict George Gamow, who proposed the triplet code gave the comprehensive theoretical concept for various aspects of genetic code. The main features of his proposals are –

1. The code is a triplet.
2. Translation is a direct procedure between codon and amino acid.
3. Translation of codes is of overlapping type.
4. Code is degenerate.
5. Nuclei acid and the peptide chain are colinear.
6. Code is universal.

The works of Brenner (1957), showed that the triplet code is of non-overlapping type. Gamow's idea of the direct translation process became unacceptable when Crick proposed adoptor hypothesis. As per this hypothesis, during the translation process some mediator molecules intervene between amino acids and nucleic acids. It is now known that the intervening molecules are tRNA.

Experimental evidence for the existence of genetic code may be obtained by two basic approaches – *in vitro* approach and *in vivo* approach. In *in vitro* method, nucleotide sequence are compared with the amino acid residues of the proteins in a culture medium. In *in vivo* method, proteins and nucleic acids are obtained from intact cells and then compared.

The *in vitro* approach has the following four methods :

1. Genetic analysis of deletion mutatants of a cistron.
2. Incorporation of amino acids by synthetic polyribonucleotides.
3. Binding of known triplets of aminoacyl tRNA.
4. Amino aicd sequence generated by mRNA transcribed from synthetic tRNA of known nucleotides sequences.

1. Genetic analysis

Genetic analysis of deletion and insertion mutants (spontaneous as well as induced) in phage T4 was reported by Crick *et al* in 1961. This deletion or insertion of a nucleotide in a DNA chain leads to a frame shift mutation after the point of deletion or insertion. As a result of this, the reading of the codon subsequent to the mutation gets altered. Obviously then, the amino acid composition of the resultant protein would be different. As a result, the protein may be functionally inactive. All the mutations tested by Crick *et al,* produced a phenotype called mutant II. It was further observed by Crick *et al,* that a series of three mutations was necessary to supress the effects of a previous mutation. In other words, three supressor mutations would nullify the effect of three frame shift mutations and bring back the original phenotype. This is a clear evidence to show that the altered codon reading would return to the normal state after three point mutations thus indicating that the codon must have three nucleotides. The study also indicates that the code is commaless. If there were to be a punctuation mark between the two codes, a deletion or insertion of a base would change the number of bases between two commas adjacent to the point of insertion or deletion.

2. Incorporation of amino acids by synthetic polyribonucleotides

Synthetic polyribo nucleotides were used as mRNA by Nirenberg and Matthai to determine the meaning of codons. They produced a synthetic RNA by using the enzyme polynucleotide phosphorylase discovered in 1955 by Grunberg. Manago and Ochoa. This enzyme does not use DNA as a template and can incorporate ribonucleotides in a random order. The first synthetic RNA used for protein in a self-free system (the *in vitro* system consists of ribosomes, tRNA's and other necessary factors for protein synthesis) was poly U i.e., the mRNA consists of only one nucleotide namely, U in a series. This can be called a homopolymer as it is a polymer of only one nucleotide. The poly U mRNA was able to synthesise a homopolymer of a polypeptide chain containing only phenylalanine residues. This proved that the triplet code UUU specifies phenylaline. Subsequently, many such homopolymer species of mRNA were synthesized and were directed to produce polypeptide chains. For example, AAA was shown to code for lysine and CCC specified proline.

Nirenberg and Ochoa subsequently used heteropolymeres of mRNA to produce polypeptide chains. A heteropolymer includes a mixture of two, three or all the four nucleotides in mRNA. This yielded further information on the possible codons for certain amino acids. For example, poly CG directs the incorporation of arginine and alanine in addition to the proline coded by poly C. Some additional information on the assignment of codons to individual amino acids was obtained by varying the ratios of different nucleotides in the synthetic mRNA. In this experiment, the code assignment was at best ambiguous since all the four bases were not required for the incorporation of any of the amino acids.

3. Binding of known triplets to aminoacyl tRNA

A triribonucleotide cannot direct the synthesis of a polypeptide but it leads to binding of specific aminoacyl tRNA to the ribosomes. Nirenberg *et al,* used this property for assigning specific codons of specific amino acids. Their technique is known as Nirenberg – Leder technique. Conventionally, a trinucleotide with 5' terminal phosphate are more active and those lacking this phosphates in inducing the binding of aminoacyl tRNA to ribosomes. On the otherhand, those with 3' terminal phosphate are inactive. This shows that the codons must be triplet. In the Nirenberg - Leder technique, trinucleotides with 5' terminal phosphate are added to a cell free system containing a mixture of 20 amino acids bound to their respective tRNA. One of the 20 amino acids had radioactive carbon. For different experiments, different labelled amino acids were used but with the same trinucleotide after the synthesis of protein, the cell free mixture containing ribosomes were analyed to detect radioactivity confirming the binding of aminoacyl tRNA to the ribosomes. These studies showed that UUU, CCC and AAA induced the binding of phenylalnine, proline and lycine respectively to ribosomes. Using this technique Nirenberg *et al,* assigned most of the codons to their respective amino acids.

4. Amino acid sequence generated by mRNA transcribed from synthetic tRNA of known nucleotides sequences

H.G. Khorana and his associates used the above technique for codon assignment in *in vitro* conditions. Known as Khorana technique, in this method long specific DNA sequences, as first synthesized. These are then transcribed to mRNA molecules using the DNA directed RNA polymerase enzyme. These mRNA molecules which have known sequences of nucleotide are directed to synthesize polypeptides. The polypeptides were then subjected to sequential degradation to know the amino acid sequences. The nucleotides sequence is then compared with the amino acid sequences. The nucleotides sequence is to deduce specific codons for different amino acids. Using this technique, almost all the codons for amino acids were designated. In this technique, Khorana and commaless nature, non overlapping and triplet nature also.

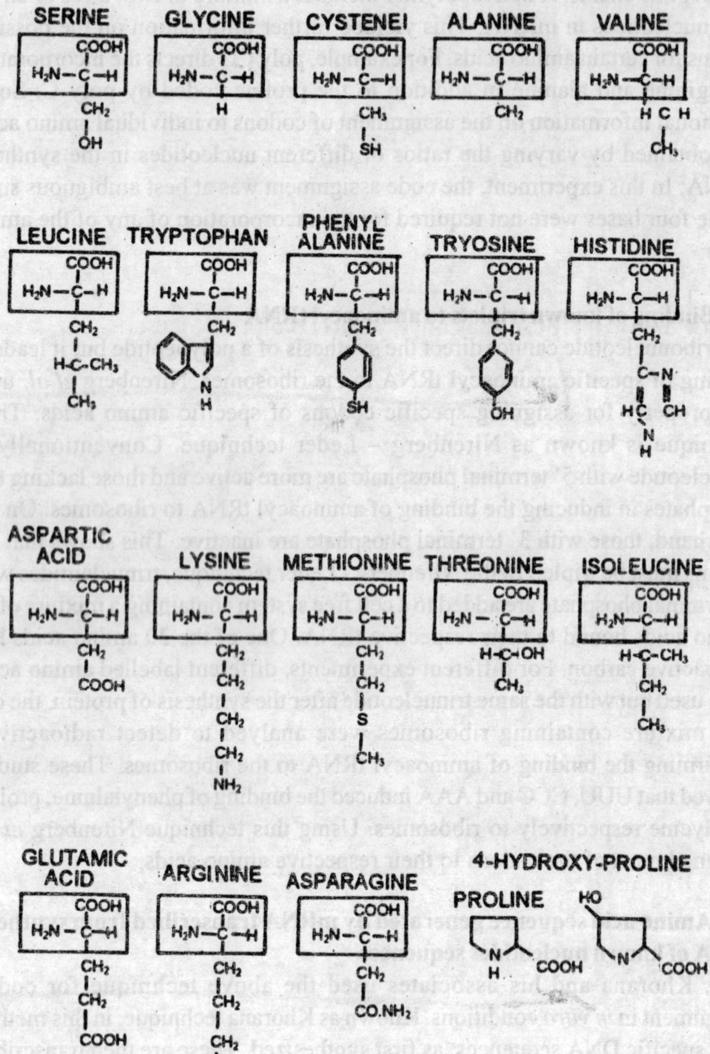

Fig. 6.1 Genetic code
The codon dictionary

In Vivo codon assignment

In the cell free systems, genetic codes have been deciphered with reference to their amino acid affinity but a doubt may arise whether these codes are the same in living systems also. In otherwords, whether the living systems use the same codes for protein synthesis. In order to prove the similarity of codes in both in *vitro* and in *vivo* three different techniques were used by three different molecular biologists.

(i) Amino acid replacement studies
(ii) Frame shift mutation studies
(iii) Comparison of a nucleotide sequence of a cistron and the amino acid sequence of the polypeptide produced by it

Amino acids replacement studies on the synthesis of tryptophan synthetase in *E.coli* (Yanofsky, 1963) have clearly indicated that the codons for the replaced amino acids, and for the amino acid replacing it prove the specificity of a codon for a particular amino acid. Frame shift mutations on lysozol enzyme of T4 bacteriophages have also shown the affinity between a specific codon and a particular amino acid.

The direct in *vivo* evidence for the genetic code comes from a comparison of the base sequence of a cistron and the amino acid sequence in a polypeptide chain produced by it. Such a study has been done with RNA bacteriophage MS2 which has only three cistrons. One of these cistrons produces the coat protein having 129 amino acids. The cistron responsible for it has 387 nucleotides. There is a total agreement between amino acid sequence expected from the base sequence of the cistron and the actual amino acid sequence of the head protein. In othewords, each codon of the cistron precisely designates the same amino acids as expected from the coding dictionary.

Wobble hypothesis - for details see protein synthesis

Colinearity of the gene structure and its product

The gene is an unbranched chain of nucleotide pairs in a particular sequence. The polypeptide chain is also unbranched and is made up of amino acid residues attached in a linear series. Thus, a linear relationship between the gene and the polypeptide chain produced by it can be indicated. Yanofsky *et al* (1964), established this colinearity in their study of tryptophan synthetase enzyme of *E coli.*

The enzyme tryptophan synthease is made up of two chains α and β. A genetic map of the α chain gene was established by determining the recombination percentage of mutations for the amino acid tryptophan. The complete amino acid residues of the α chain of the enzyme tryptophan synthetase was thus established. Subsequently, the location of amino acid substitutions in the chain for 10 mutations was determined. It was then found out that the order of mutations corresponded with the order of amino acid substitutions in the α chain of the polypeptide.

Sarabhai et al (1964), came to the same conclusion in their study of nonsense mutations in bacteriophage T4.

Pattern of the genetic code

A distinct pattern emerges when one analyses the amino acid structure and the type of codons that specify them (fig 13.1). Amino acids with similar structural properties tend to have related codons. For example, codons for aspartic acid (GAU and GAC) are similar to the codons for glutanic acid (GAA and GAG). Same relationship is found between the codons of phenylalanine, tyrosine and tryptophan – all these begin with uracil. It is believed that this feature of the code is evolved to minimise the consequences of mistakes that may arise during translation and base substitutions due to mutation. In otherwords, if an amino acidin a protein is replaced by another with almost similar properties, the protein may still remain functional.

Another pattern that is seen in the code is that for synonym codons (multiple codes for thesame amino acid), the first two bases are always constant whereas only the third one varies. The flexibility of the third nucleotide of the codon is another method to help minimise the harm posed by mutations. Crick (1966) has called this wobbling and proposed a hypothesis called Wobble hypothesis (see protein synthesis for details).

New genetic codes in mitochondria and ciliate protozoa

It has been mentioned above that the code is universal and ever since it originated it has remained static. However, during 1980s it was observed that the genetic codes of mitochondrial DNA in yeasts may be different from the others. In yeast, mitochondria UGA codes for tryptophan, while in nuclear genes in the same organism UGA is the termination codon. This, however, is not to say that in all mitochondria of all organisms, the UGA codes for tryptophan. For example in maize cytochrome, oxidase tryptophan is coded by CGG. This code in nuclear genes codes for arginine. There are also indications that AUA which codes for isoleucine in nuclear genes codes for methionine in *mit* DNA of mammals, yeast, Drosophila etc.

A different genetic code in DNA is perhaps necessiated because of the small *mit* genome size. T.H. Jukes (1966), presented a concept called *archetypal code* which will allow one anticodon (of tRNA) to pair with 4 different codons due to wobbling at the first base of anticodon and the third base of codon.

In 1986 a different genetic code were shown to be working in a ciliate member of Mycoplasma (*mycoplasma capricolum*). In the genetic code of this organism, UAA and UAG are not nonsense codons but code for glutamine.

Evolution of genetic code

Many of the features of the genetic code like degenerase, non-overlapping commaless nature etc. have led to a suggestion that the present version of the genetic code has evolved from a more primitive version which are perhaps a double than a triplet. It is believed that this shifting from doublet to triplet occurred very earily in the origin of cellular life. It is not certain, however whether all the codes for all the amino acids were evolved simultaneously or the codon dictionary gradually arrived at the present stage of 64 is not certain. One thing however, is certain eversince it came into existence in the present form, the biological necessity of it imposed such a selection pressure on it that it has remained unchanged upto the present day.

7

PROTEIN SYNTHESIS

Protein Structure And Classification

Of the two important macromolecules that mastermind all the activities of the living systems, proteins are one, the nucleic acids being the other. Proteins are present all over the cell and are important both structurally and functionally.

The name protein was first suggested by Berzelius (1838). *Protein* in Greek means 'First rank'. Proteins are highly complex organic molecules with molecular weight ranging upto many millions. Chemically, proteins are madeup of C, H, O and N. In addition to these major elements, some proteins may also haave sulphur, phosphorous, iron zinc, iodine etc. as components. The ratio of major elements in proteins are C= 50-55%, O=20.23%, N=15-18% and S=0-4%.

Proteins are made up of building blocks called amino acids (i.e. on hydrolysis, proteins yield these units). The idea that proteins are build up of amino acids was first proposed by Emil Fischer.

Amino acids are carboxylic acids which have atleast one COOH group and one NH_2 group. Other functional groups may also be present. There are atleast 22 different types of amino acids encountered in plants.

The amino acids are bound to one another in a protein chain by means of linkages or bonds called peptide linkages. Peptide linkages are formed between the COOH group of one amino acid and the NH_2 group of the other, with the elimination of a molecule of water. Several amino acids combine together with the peptide bond and form a peptide chain. Several of these peptides constitute a polypeptide chain. The polypeptide chain folds and refolds in a characteristic fashion to form a protein molecule.

Peptides

These are made up of amino acids linked with CONH (peptide) bonds. Peptides are intermediate compounds produced during the synthesis or degradation of proteins.

Amino acids ——— peptides ——— proteins

Proteins

These are the final productsin the N2 metabolism, and are made up of folded peptide chains which have many additional linkages (depends upon the point of folding), in addition to the CONH bond. Proteins have a three dimensional configuration, and have a very high molecular weight reaching upto 10^4 - 10. Although proteins contain not more than 20 amino acids, they are of immense

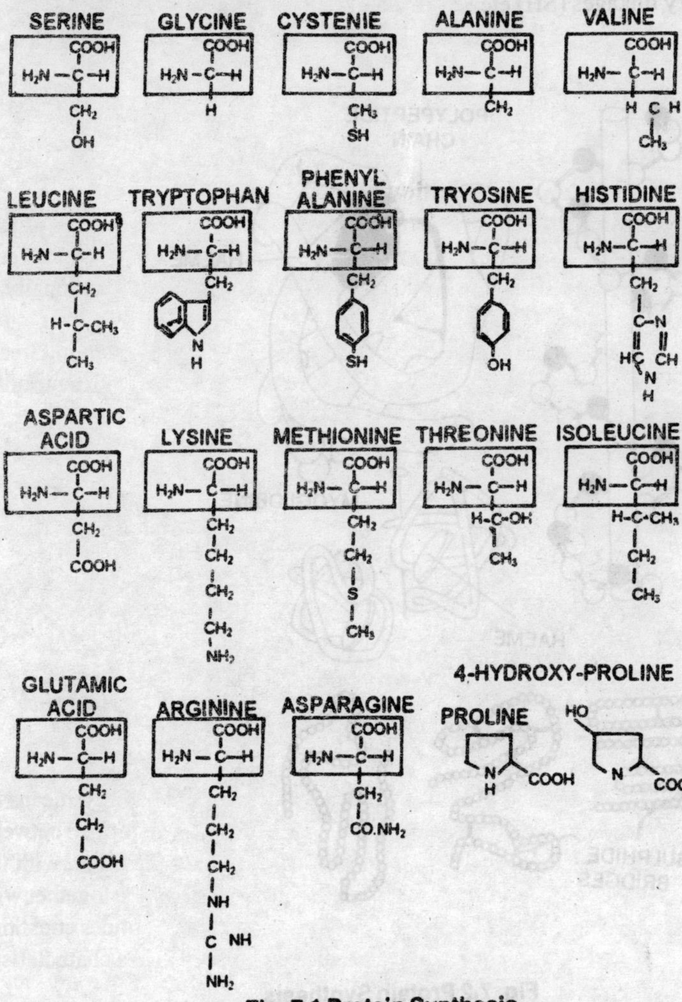

Fig. 7.1 Protein Synthesis
Structure of the amino acids

variety and complexity due to variations in combination of amino acids. In general, the three following aspects pertaining to the amino acids decide the type of protein. These are – the quality, quantity and sequence of amino acids in the peptide chains.

In other words, the type of amino acid, its quantity and the sequence in which these acids are arranged forms the basis for the categorization of proteins. Besides, proteins may differ in the number of peptide chains per molecule, type of secondary linkages (SH) etc.

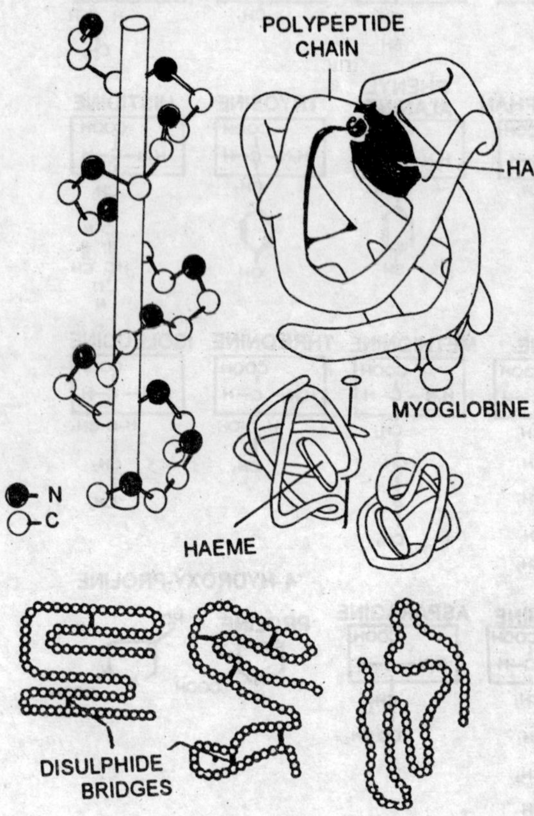

Fig. 7.2 Protein Synthesis

Protein structure. **A.** Polypeptide chain helical configuration, **B.** Tertiary and quaternary structure of myoglobin, **C.** Denaturation of proteins

Structure of proteins

Modern biochemists regard four structural levels in the protein molecule which accounts for its complexity. These levels are – primary, secondary, tertiary and quaternary structure, in the order of increasing complexity.

Primary structure

This is the simplest and the most basic structure and refers to the linear peptide chain. Besides the CONH bond, disulphide bonds (-S-S-) are found among the peptide chain linkage and the sulphur containing amino acids.

Secondary structure

The primary polypeptide chain spirally coils forming the L helical chain. The characteristic bonds of the secondary structures are hydrogen bonds between the carbonyl and amino groups of peptide links, salt linkages and Vander Wall's forces which help maintain the helical structure.

Tertiary structure

Folding and superfolding of the helical chain in a three dimensional orientation constitutes the tertiary structure. Interaction between the amino acid residues situated far apart in the chain brings about the folding.

Quaternary structure

This consists of more than one identical sub units (poly peptide chain) which associate with one another to form a functional unit. Individually, each subunit is non-functional. Such complicated structures are usually seen in enzymes. For example, the enzyme phosphorylase has two sub units, which are inactive individually, but as dimers, they are active. A protein with identical sub units in the quaternary structure is known to be homogenous or if the sub units are dissimilar, it is known as heterogenous. A protein with more than one sub unit (monomer) is referred to as an oligomeric protein.

Classification of Proteins

As the proteins possess, a very complex strucuture, their classification also poses a problem as many categorizations are possible based on structure, function etc. The following is a generalized classification. The major groups are recognized in proteins based on their solubility and chemical properties.

I. Simple proteins

II. Conjugated proteins and

III. Derived proteins.

I. SIMPLE PROTEINS

Chemically, these have only amino acids. No additional compounds are present in the molecule. Simple proteins are of the folowing kinds –

1. Albumins

These are soluble in water, dilute salts, acids and alkalies. Heat denatures the protein structure by coagulation. Some of the commonly occuring albumins are Legumelin (from legume seeds), Amylase (from barely), Leucosin (from cereals) etc.

2. Globulins

Less soluble in water than albumins but soluble in less concentrated salt solutions. Heat coagulation is seen in globulins also. Eg. Legumin (pears), Tuberin (potato) etc.

3. Glutelins

Insoluble in water and salt solutions, but soluble in weak acids and alkalies. Glutelins are generally found in plants and are heat immune. Seed proteins found in cereals like Wheat (glutenin) and rice (oryzenin) are best examples of glutelins.

4. Prolamines

These proteins are water insoluble, but dissolve in 70–80% alcohol. Prolamines have a high concentration of the amino acid proline and other amides. E.g. Gliadin (Wheat), Hordein (barley) and Zein (maize).

5. Scleroproteins

These are fibrous proteins found in integral part of the structural framework of plants. These proteins are insoluble in most of the solvents.

6. Histones

Most of the nucleoproteins, haemoglobin and globins are histones. They are insoluble in water but soluble in dilute ammonia. Histones have a high concentration of lysine and arginine.

7. Protamines

These are soluble in water, dilute acids and ammonia and do not undergo heat coagulation. Basic amino acids are rich in protamines, while tryptophan and tyrosine are completely absent. Eg. Nucleoproteins of sperms in fish.

II. CONJUGATED PROTEINS

In these proteins, non-amino acid components are associated with the protein molecule. The non-amino components are called prosthetic groups. There are seven major categories of conjugated proteins.

1. Nucleoproteins

Proteins are found in combination with nucleic acids. Hydrolysis can separate nucleic acids and simple proteins. Nucleoproteins are water soluble.

2. Glycoproteins

Also called mucoids, these are conjugated with carbohydrates like macromolecular polysaccharides, sugars, sugar phosphate etc. Glycoproteins are found in cell membranes.

3. Chromoproteins

These constitute the pigments and are coloured, the coloured appearance is due to the presence of metals such as Cu, Fe, Mg etc. Chlorophylls, flavoproteins, carotenoids, cytochromes etc. are typical examples of chromoproteins.

4. Lipoproteins

Proteins in conjugation with lipids are called lipoproteins. Lecithin, Cephalin etc. are the lipids found in combination with proteins. Lipoproteins are an integral part of all cell membranes.

5. Metalloproteins

Certain types of enzymes which require metals as activators belong to this group. Many of the respiratory enzymes belong to this group.

6. Phosphoproteins

These are proteins with phosphoric acid as the prosthetic group. Phosphoproteins are insoluble in water, but soluble in alkalies. E.g. milk and egg protein (casein and viteline).

7. Leeithoprotein

Proteins with lecithin as the prosthetic group (E.g. white of egg) belong to this group.

III. DERIVED PROTEINS

These are degradation products of proteins. Primary proteins on hydrolysis with acids or alkalies yield derived proteins.

(A) Primary derived proteins and

(B) Secondary derived proteins

A. Primary derived Proteins

There are two categories of primary derived proteins. These are Metaproteins and Coagulated proteins.

(i) *Metaproteins* : These are insoluble in water but dissolve in dilute salt solutions, acids and alkalies. Hydrolysis of natural proteins on long treatment with alkalies or acids yield metaproteins.

(ii) Coagulated proteins: Treatment with heat or alcohol of natural proteins yields these proteins.

B. Secondary derived proteins

Three types of secondary derived proteins have been identified. These are proteoses, peptones and peptides.

(i) Proteoses : Prolonged hydrolysis of metaproteins yield proteoses (eg. Albuminoses from albumin). These are water soluble but immune to heat treatment.

(ii) Peptones : Peptones are obtained by continued hydrolysis of proteoses. Like proteoses, these are also water soluble and heat insensitive.

(iii) Peptides : Prolonged hydrolysis of natural proteins yields peptides. These are also water soluble and not coagulated by heat.

Functions of proteins

Most of the vital activities of the cell are controlled by the two master molecules, namely proteins and nucleic acids. Proteins are important both from the point of view of structure and function of the cell. Structurally, proteins constitute in integral part of membranes, pigments etc. Functionally, in the form of enzymes, proteins play such a vital role in cell physiology, that without proteins (enzymes), no metabolic reaction is possible. Even the synthesis of DNA is regulated by proteins. In the form of Chromoproteins (Chlorophyll, Cytochromes etc), proteins mediate all energy transformation reaction. It is because proteins are so important that the nucleic acid directly regulates the protein synthesis.

Protein structure is highly complex (as has been pointed out earlier), as well as specific. The primary structure is determined by the quality, quantity and the sequence of amino acids. Even an alteration of the position of one amino acid would alter the structure of the protein, and if it happens to be an enzyme regulating an important reaction, the entire reaction is disturbed leading to several abnormalities. Hence it is but natural that there is direct control of DNA in the synthesis of proteins. DNA, RNA and ribosomes of the cell are involved in the synthesis of proteins.

The central dogma : The basic mechanism of protein synthesis is that the nuclear DNA regulates the assembly of amino acids in the ribosomes to produce a polypeptide chain. Since DNA cannot move out of the nucleus it sends a messager – mRNA which carries the message in the form of sequence of nucleotides. This mRNA forms a template on the ribosomes. Another type of RNA, tRNA found in cytoplasm (whose nucleotide sequence is complementary to that of mRNA), picks up the amino acids and assembles them on the ribosomes to produce a peptide chain.

This may be briefly expressed as follows :
DNA Replication → DNA Transcription → RNA Translation → Proteins translation. Transcription is the copying of a complementary messenger RNA strand on a DNA strand. In translation, the genetic information present in mRNA (in the form of sequence of nucleotides) directs the sequence of amino acids to be assembled. We shall study these events in detail.

Transcription

Copying of a complementary strand of mRNA requires the following – a template, activated precursors, a divalent metal ion and RNA polymerase.

DNA molecules serves as a template for the production of mRNA. Of the two strands of DNA, only one strand called the 'sense' or the 'coding' strand transcribes mRNA. Single stranded DNA also can serve as a template.

The activated precursors required for mRNA synthesis are – Adenine triphosphate, Guanine triphosphate, Uracil triphosphate and Cytosine triphosphate (ATP, GTP, UTP and CTP)

Mg^{++} or Mn^{++} are the divalent metal ions required for mRNA synthesis.

The enzyme RNA polymerase is necessary for transcription. The nucleotides of mRNA are assembled together with the help of RNA polymerase. The enzyme consists of two parts - a core enzyme and a sigma factor. The sigma factor initiates transcription of mRNA on the DNA template and the core enzyme continues transcription. The DNA base sequence are thus transcribed into mRNA base sequence.

Genetic Code

This is the message in mRNA in the form of nucleotide sequence. Three nucleotides code for one amino acid. It is called a triplet code or codon. There are 64 codons for the 20 odd amino acids. i.e., there is more than one code for each amino acid. These alternative codes are helpful in the cae of random mutations which may alter a code (nucleotide).

The nucleotides of mRNA are complementary to the nucleotides of tRNA (present in cytoplasm). The triplet nucleotides of tRNA are called anticodons. Each tRNA has three nucleotides. The pairing between the codon and anticodon (mRNA and tRNA), follows the A-U, G-C combination. For example, if mRNA consists of the first code UUU, it pairs with a tRNA having an anticodon. AAA. Thus the series of codons of mRNA determine the series of anticodons of different molecules of tRNA, and hence the amino acids (Each tRNA molecule is specific to a particular amino acid). Since the triplets of mRNA are determined by the base sequence of DNA, it follows that the DNA molecule decides the sequence of the amino acids and hence the structure of the protein molecule.

Some triplets do not code for any amino acid (UAA and UAG). These are called nonsense codes. The so called nonsense codes are actually terminator codons and are now called ochre and amber (UAA and AUG). Code AUG is called initiator codon since it initiates the synthesis of a polypeptide chain.

The triplet codes for all amino acids collected together is called a codon dictionary. The dictionary was first proposed by Nirenberg (1965), and subsequently worked out in further detail by Khorana H.G.
The mRNA is released out of the nucleus into the cytoplasm where it is attached to the smaller sub unit of the ribosomes. Many ribosomes are connected by a single mRNA molecule, constituting a polysome or polyribosome. The mRNA attaches itself to the ribosomes in such a way, that its bases remain free (to be able to undergo pairing with the bases of tRNA). The whole structure (polyribosomes + mRNA), is called mRNA complex.

tRNA (transfer RNA) : This is present freely in cytoplasm. Also called sRNA (soluble RNA), it is the smallest among the various types of RNA molecules found in the cell. Although the tRNA is single stranded like other types of RNA, it appears double stranded because of coiling. Complementary base pairs of the same strand form pseudo bondings of hydrogen due to folding. There are two regions which are free without bonding – one at the acceptor and which always has CCA base pair sequence, and the other in the middle region of the molecule called the anticodon region, where there bases turn out of the folded portion and remain free. The types of tRNA is based on the types of these free bases. tRNA has the following functions.

(i) To attach to a particular amino acid at the acceptor, and (CCA) with adenine nucleotide. The choice of amino acid, however, depends on the type of free bases present at the anticodon end.
(ii) to attach itself at a specific site on the mRNA, and
(iii) to free the amino acid after it gets attached to another amino acid by means of peptide linkage.

The Wobble hypothesis
The triplet code or codon is degenerate i.e., there are many more codons than there are amino acids. Only 20 amino acids are involved in protein synthesis, while there are 64 possibilities or types of codons (43). Hence, more than one codon can code for an amino acid. It is interesting to find out as to why the

codes are degenerate? There could have been only so many codons as there are amino acids. But there is a problem here. If there were to be only one code for an amino acid, what happens if there is a mutation at that site in DNA? Obviously the code changes and the amino acid cannot be incorporated in the protein leading to abnormalities. On the other hand, if there are alternate codes, they act as a protection against mutation in a specific base. Hence, degeneracy of codes is a must.

Regarding tRNA present in cytoplasm, since they have to bind themselves to a specific codon, there should have been as many tRNA's as there are codons. But the number of tRNA is only as many as there are amino acids. This means, the anticodons of the tRNA's must be able to 'read' more than one codon of mRNA. How is this reading possible when bonding between base pairs is highly specific?

Crick (1966), proposed **'the Wobble hypthesis'** in order to solve this apparent dilemma. According to this hypothesis, only the first two bases of the codon have a precise pairing with the bases of the anticodon of mRNA, while the pairing between the third basis of the codn and anticodon may Wobble (non-specific). The pairing in the third base is ambigous. Thus a single tRNA can pair (bind) with more than one mRNA codon differing in only the third base. For example, the anticodon UCG of tRNA can read codons AGC and AGU of mRNA. The pairing between UGC and AGU is a perfect pairing as per the Crick and Watson pattern. But in the second case i.e., between UGC and AGU, the pairing between the first two bases is normal, while between G and U is against the normal pairing pattern. This unsual bonding G and U is called Wobble pairing.

The degeneracy of the code is not totally randon. Multiple codons for an amino acid always carry the same bases in the first two positions in a triplet with only the third 'Wobbling'. For example all the four codons for Valine have the first two bases - GU and for alanine it is GC.

Valine - GUU, GUC, GUA and GUG.
Alanine - GCU, GCC, GCA and GCG.

Translation

The process of translation consists of activation of amino acids, transfer of activated amino acid to tRNA, initiation of synthesis of polypeptide chain, chain elongation and chain termination.

Activation of amino acids

Before amino acids (20 of them) get themselves attached to the specific type of tRNA they have to be activated (energized). The activation is carried out by the ATP molecule in the presence of the specific enzyme called aminoacyl synthetase with Mg as the cofactor. The reaction produces aminoacyl adenylate (AAA) or aminoacyl AMP. Pyrophosphates are released in the process.

Amino acid + ATP + Enzyme ———— Enzyme AA> AMP complex + Ppi

Transfer of activated amino acids to tRNA : The activated amino acid (aminoacyl AMP) is transferred on to a specific tRNA, during which AMP and the enzyme (aminoacyl synthetase) are released. The amino acid gets attached to the CCA and tRNA.

Enzyme AA. AMP + tRNA ———— AA. tRNA + AMP + Enzyme

Initiation of Polypeptide chain synthesis

This involves two steps viz., preparation of the site of protein synthesis and initiation of the polypeptide chain.

(i) Preparation of the site of synthesis

Ribosomes in cytoplasm are the actual sites of protein synthesis. These are usually into their subunits (50s and 30s), when they are not participating in protein synthesis. The mRNA usually attaches to the 30s subunit through its 5' end carrying the AUG codon. In the process of attachment certain substances called initiation factors (IF) help. These factors are of three types in prokaryotes (IF1, IF2, IF3) and are of six types in eukaryotes (eIF2, eIF2al, eIF2a2, eIF2a3, eIFa4 and eIF3).

The attachment of the mRNA to the 30s submit requires the help of IF3 and Mg^{++}. The IF3 probably unwinds the rRNA (ribosomal RNA), so that it can bind to mRNA. After the attachment of mRNA, an mRNA - 30s complex is formed.

The starting amino acid in a polypeptide is methionine *(met)* in the case of eukaryotes, and N-formly methionine *(f-met)* in the case of prokaryotes. The amino acid (met or f-met) tRNA complex attaches to the initiation codon (AUG) on the mRNA 30-s complex through its anticodon (UAC). This forms the 30s initiation complex and requires the participation of IF2, IF3 and GTP (energy donor).

IF3

(i) 30s + mRNA ———— mRNA - 30s complex

(ii) mRNA - 30s complex + met tRNA ——— 70s - mRNA - met - tRNA

During the formation of the initiation complex , the 50s subunit combines with the 30s subunit to form the 70s unit. This union is facilitated by the participation of IF1. With the formation of the 70s ribosome), the initiation of the polypeptide chain begins.

(ii) Initiation of the polypeptide chain

Each ribosome has two cavities - P site (polypeptide site) and met-tRNA (aminoacyl sites). AA tRNA complex occupies the A site, while the met - tRNA (tRNA with the initiator aminoacid - methionine) will be at the P site. This occupation requires a molecule of GTP and the participation of IF_1. The energy of GTP is utilized for the formation of the peptide linkage.

Peptide linkage occurs when the carboxyl (COOH) group of peptidyl tRNA (met. tRNA) present at the P site reacts with the amino group (NH_2) of AA tRNA at the A site. The enzyme peptidyl transfers and catalyses the formation of peptide bond. Later the AA+tRNA moves from A site to the P site in the presence of GTP and IF_2

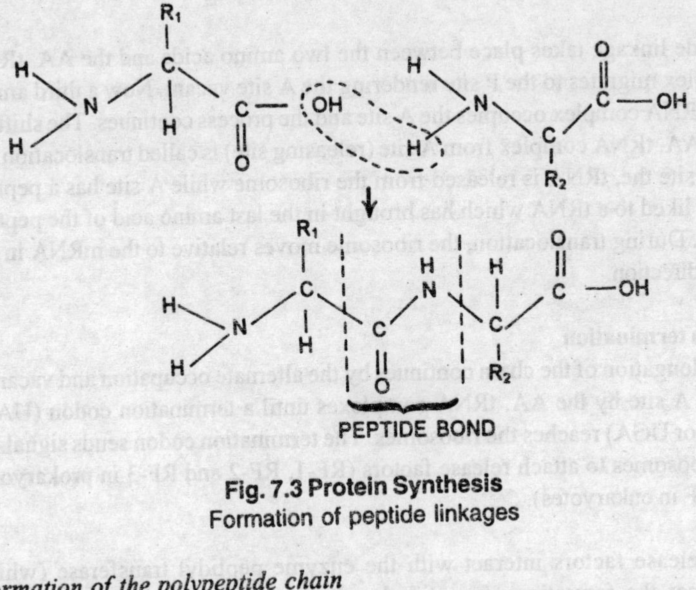

Fig. 7.3 Protein Synthesis
Formation of peptide linkages

Formation of the polypeptide chain

A specific type of tRNA (which has bases complementary to the codon of mRNA), linked to a particular amino acid gets attached to a specific site on mRNA when its anticodons can form a pairing with the first codon of the mRNA. Soon after, a second tRNA with its amino acid joins the second site (triplet) on mRNA. The first and the second amino acids attached to different tRNA form a peptide linkage as both are linked to mRNA *via* the anticodons of tRNA. A molecule of water is released during the formation of the peptide linkage.

$$COOH + NH_2 \xrightarrow[\text{Peptide bond}]{} CONH + H_2O$$

As soon as the peptide linkage is formed, the tRNA is made free and comes back to cytoplasm. This whole process in which a specific tRNA carries its amino acid to a specific site is called translation i.e. the code (triplet) is translated into an amino acid. Here the sequence of bases in tRNA will be identical with the base sequence of that part of DNA coding for mRNA.

Chain elongation

Elongation of the chain requires the participation of certain factors called elongation factors (EF). These are EF-Tu, EF-TS, and EF-G in prokaryotes, and EF-1 and EF-2 in eukaryotes. After the first amino acid tRNA complex has moved from the A site of the ribosome to the P site, a second AA. tRNA complex now occupies the A site. The codons are recognized enzymatically. This requires the participation of Ef-Tu, GTP and Mg^{++}.

Peptide linkage takes place between the two amino acids and the AA. tRNA complex migrates to the P site rendering the A site vacant. Now a third amino acid tRNA complex occupies the A site and the process continues. The shifting of th AA. tRNA complex from A site (releasing site) is called translocation. At the P site the, tRNA is released from the ribosome while A site has a peptide chain liked to a tRNA which has brought in the last amino acid of the peptide chain. During translocation, the ribosome moves relative to the mRNA in the 5'-3' direction.

Chain termination

The elongation of the chain continues by the alternate occupation and vacancy at the A site by the AA. tRNA complexes until a termination codon (UAA, UAG or UGA) reaches the ribosomes. The termination codon sends signals to the ribosomes to attach release factors (RF-1, RF-2 and RF-3 in prokaryotes and RF in eukaryotes).

The release factors interact with the enzyme peptidyl transferase (which catalyses the formation of peptide bond) resulting in the hydrolysis of the (i)bond between the polypeptide chain and the tRNA (linked to the last amino acid of the peptide chain at the site) and the former is released from the ribosome.

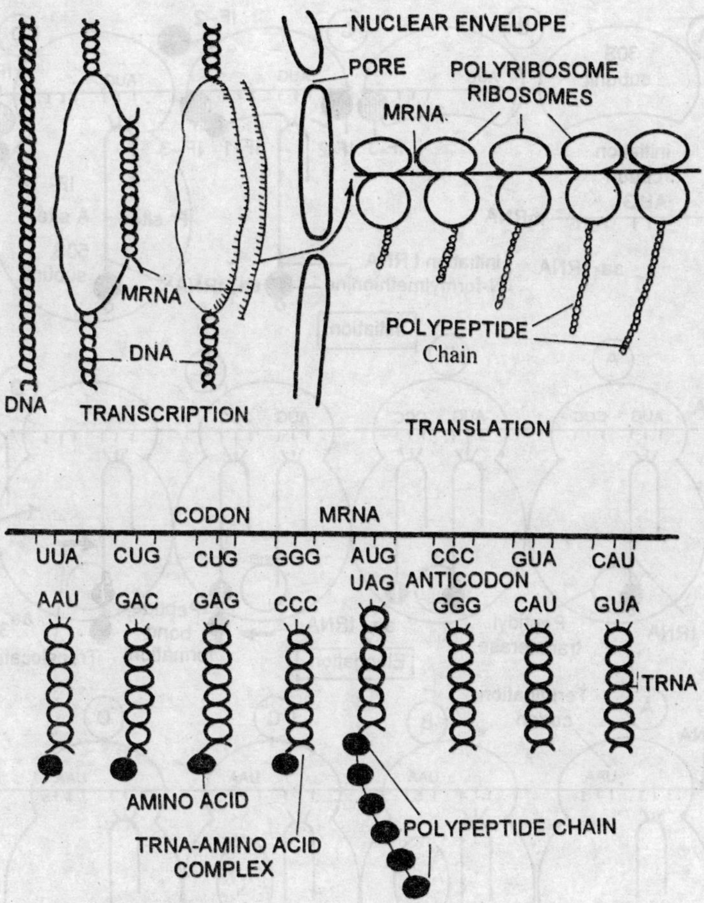

Fig. 7.4 Protein Synthesis
General scheme of protein synthesis

Other events that occur at the termination of the polypeptide chain are –

Release factors get disassociated from the ribosomes due to the hydrolysis of GTP.

(ii) tRNA is also released.

(iii) Ribosomal sub units (30s and 50s) get disunited.

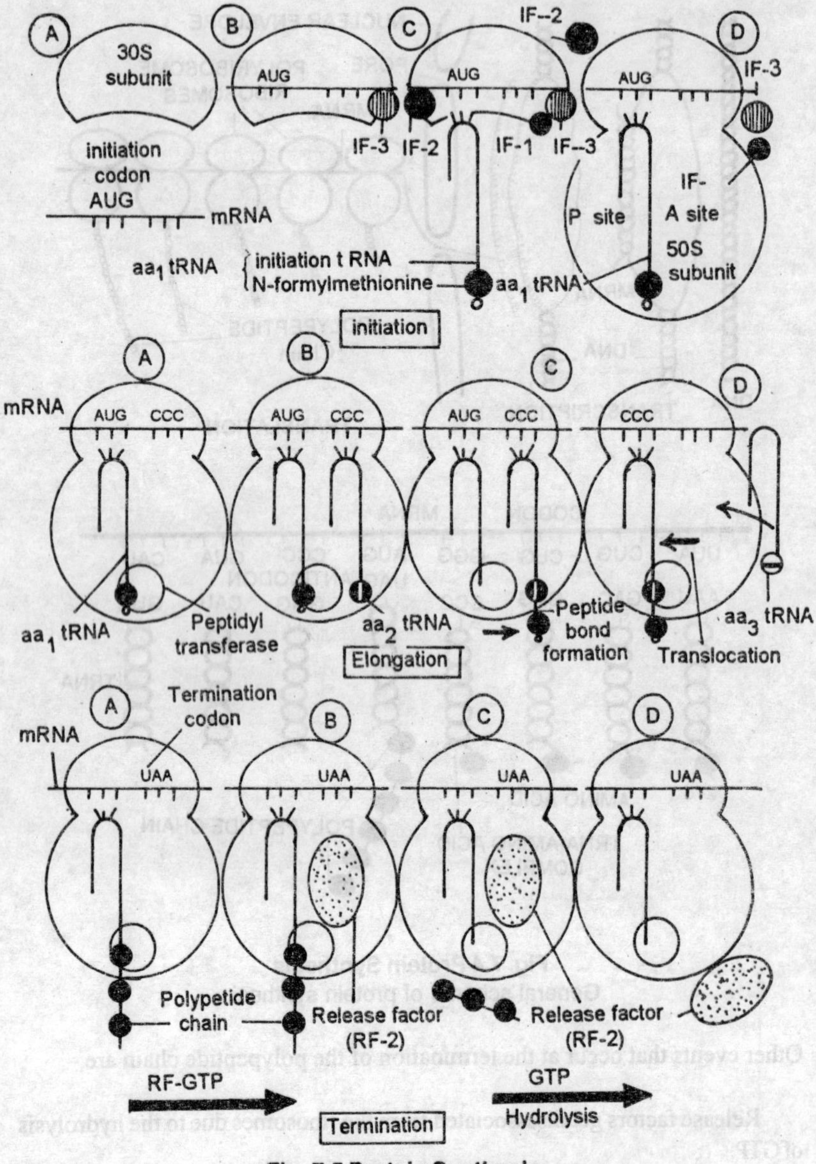

Fig. 7.5 Protein Synthesis
Different stages of translation

(iv) The polypeptide chain is processed by the removal of the methionine residue from the chain. Depending upon the circumstances, the chain may get cleaved into two or more pieces.

In eukaryotes, the polypeptide chain passes through a tunnel between the two subunits of the ribosome into the cavity of the cisterna of the endoplasmic reticulum.

Rate of Protein synthesis

The rate of transcription in *Escherichia coli* DNA for the enzyme controlling trypotophan synthesis is 28 nucleotides per second at 39°C. The speed of translation is 7 amino acids per second. For the haemoglobin molecule, the rate of translation is 4 aminoacids per second.

DIFFERENCE IN PROTEINS SYNTHESIS BETWEEN PROKARYOTES AND EUKARYOTES.

No.	Character	Prokaryotes	Eukaryotes
1.	Ribosomes involved in protein synthesis	70s, with 30s and 50s sub units.	80s with 40s and 60s sub units.
2.	Occurrence of transcription and translation	Translation begins even while transcription is transcription and still going on in mRNA	Translation begins after complete mRNA comes out of the nucleus.
3.	Migration of mRNA	Ribosomes pull out mRNA from the DNA	mRNA moves out of the nucleus through the pores in the nuclear membrane.
4.	Sites of initiation of protein synthesis in mRNA	many	one
5.	Initiator amino acid for protein synthesis.	N-formyul-methionine	Methionine
6.	Initiator factors	IF1, IF21 and IF3	eIF2, eIFwa1, eIF2a2 eIF2a3 and eIF3
7.	Elongation factors	EF-Tu, Ef-Ts and Ef-G	Ef-1, Ef-2
8.	Release factors	Rf-1, Rf-2 and RF-3	Rf (single factor)

8

GENETIC REGULATION

All the activities of an organism, be it unicellular or multicellular, are ultimately controlled by the genes. In multicellular organisms, where cells differ widely in structure and function all of them carry the same genes, since all of them are traceable to a single cell viz., zygote. This poses an interesting questioen as to the function of genes. How do similar genes function differently? The answer to active at all times. The genes are *switched* on and *switched off* at different times. This differential gene action is also responsible for the synthesis of different types of proteins (enzymes) at different times. Not all enzymes are required at all times. Enzymes required during early stages of development may not be required at later stages in the life cycle and *vice versa*. The metabolic products produced may induce or repress the enzyme synthesis.

Genetic control on protein synthesis is exercised at three stages – transcription stage, translation stage and at the release of polypeptide chain (post translation stage).

The control mechanism ensures that the cell is not unnecessarily flooded by the enzymes which are required only at particular times and only in particular quantity. Depending on the requirement the enzyme synthesis can either be induced or suppressed. Genetic regulation is seen both in prokaryotes and in eukaryotes. While the purpose of genetic regulation is same in both prokaryotes and eukaryotes, the mechanisms differ. In prokaryotes, the mechanism of genetic regulation is simple when compared with the eukaryotic organisms. The reasons for this are not far to understand. As the eukaryote organisms are far more complex they do require a very sophisticated mechanism of genetic regulation. In this chapter we will discuss genetic regulation in both prokaryotes and eukaryotes.

GENETIC REGULATION IN PROKARYOTES

Induction and Repression

In *E.coli,* the synthesis of the enzyme β *galactosidase* has been studied in considerable detail. This enzyme breaks up lactose into glucose and galactose. If lactose is not supplied to *E.coli* cells, the presence of the enzyme β *galactosidise* is hardly detectable, but as soon as the substrate (lactose) ·is added, the production of the enzyme β *galactosidase* increases by as much as 10,000 times. The production of enzyme again gets reduced if lactose is removed from the system. Such enzymes whose sythesis can be increased many times by the presence of the substrate are called **inducible enzymes.** The genetic system responsible for synthesis of such enzymes is called the *inducible system.* The inducible enzyme is an example of adaptive enzyme in the sense its (enzyme) synthesis gets increased due to the presence of the substrate.

In some other systems, the reverse of the inducible system occurs. For instance, in a culture of *E.coli,* if no amino acids are supplied from outside, the cells can produce all the enzymes required for the production of amino acids. On the otherhand, if a particular amino acid like histidine is added to the system, the production of histidine synthesising enzymes will fall down. In these systems the addition of a product of a synthetic pathway will check the production of enzymes responsible for the synthetic pathway. In other words increased presence of a substrate would reduce the synthesis of an enzyme which has the enhancing effect of the production of that substate. Systems such as these are called *repressible systems,* and the enzyme whose synthesis gets checked is known as the **repressible enzyme.**

Inducer and Co-repressor

It has been mentioned above that an enzyme gets induced or represented by the presence of a particular substrate. The substance which induces the synthesis of an enzyme is called the inducer and the substance which supresses the synthesis of the enzyme is called the repressor. In the synthesis of β *galactosidase,* lactose acts as the inducer and in the supression of histidine synthesising enzyme, histidine acts as the repressor.

Enzyme induction or repression is generally brought about by the substrate on which an enzyme can act. But in some instances, there will be molecules which resemble the substrate which also can act as the inducers. A typical example of this kind of induction is IPTG (isopropyl thiogalactoside). IPTG resembles lactose and can induce the production of the enzyme β *galactosidase.* Such inducers which are not substrates themselves but can bring about induction because of similarities with the substrate are known as **gratuitious inducers.**

In the above paragraphs, we have discussed as to how in the absence of lactose there is no synthesis of β *galacotosidase.* This means then in the absence of the inducer, the genes responsible for the production of the enzyme protein do not function. It will be interesting to find out as to why the enzymes do not function in the absence of lactose, and how in the presence of lactose the genes regulating the enzyme synthesis get activated. It has been found out that a group of molecules called *repressors* are present in the cell and they check the activity of the enzymes. In the presence of the inducers, the repressors get inactivated. In some instances, ordinarily the repressors are inactive but they become active in the presence of another group of molecules called co-repressors. The induction and repression is supposed to work in the following manner :

1. Inducible system : active repressor + inducer = inactive repressor
2. Repressible system : inactive repressor + corepressor = active repressor

The operon concept

F.Jacob and J.monod (1961), the Nobel laureates have studied in detail the inducible system in the synthesis of the enzyme β *galactosidase* in *E.coli.* In order to explain the induction and repression of enzyme synthesis, they have proposed the involvement of a group of genes along with the presence of cytoplasmic inducers and repressors. Their model of gene action is popularly called the operon model. This scheme of gene action proposed by Jacob and Monod as originally conceived by them consisted of an operator gene and a number of structural genes. As per the current understanding, however the operon system consists of structural gene, operator gene, promoter gene, regulator gene and repressor, inducer and co-repressor. The following is an account of the components of the operon model.

Structural genes

There are many structural genes associated in an operon system. These direct the synthesis of the mRNA and govern the sequence of amino acids in a protein molecule. Each structural gene might produce a particular kind of protein or all structural genes might regulate the production of a single protein. The activities of the structural gene(s) are controlled by the promoter and operator site of the operon system. The most well studied structural genes (z,y and *ac*), are those of the lactose operon system in *E.coli.*

Operator gene

The operator gene is situated adjacent to the first structural gene. It switches on or switches off the functioning of the structural gene (protein synthesis). In case a structural gene has to be suppressed, a repressor attaches itself to the operator to form an Operator-repressor complex. In the case of protein synthesis, the operator-repressor complex prevents the transcription by blocking the movement of RNA polymerase.

Promoter gene

The promoter gene is continuous with the operator gene and is believed to lie left to it. It is suggested that RNA polymerase binds to the promoter site during transcription. Three regions have been recognised in the promoter site. These are (i) recognition site, (ii) initial binding site and (iii) the mRNA initiation site (operator site).

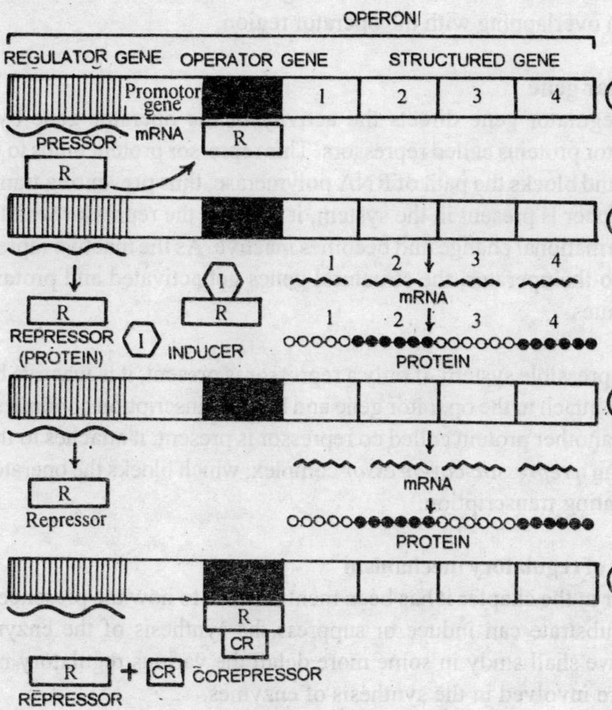

Fig. 8.1 Genetic regulation
Genetic regulation of protein synthesis in prokaryotes
(the Operon system of Jacob and Monod, 1961)

(i) Recognition site

Also called the *cga* site (catabolic gene activator site), it consists of certain palindromic sequence of DNA. These symmetrical sections of DNA are recognised by proteins having symmetrically placed sub units. This site, also called the CRP site (cyclic AMP receptor protein site) binds a Carotenoids protein to the promoter gene and thus facilitiates the binding of the enzyme

RNA polymerase. It has been found that in *E.coli*, CRP combines with cAMP (cyclic adenosine monophosphate), forming a CRP + cAMP complex which binds to the promoter enhancing the binding of RNA polymerase and activates transcription. This regulation is called positive control.

(ii) Initial binding site

This consists of seven bases (DNA) to which the RNA polymerase binds.

(iii)RNA initiation site

The site where transcription begins is called initiation site. This is the region overlapping with the operator region.

Regular gene

The regulator gene directs the activity of the operator gene by producing inhibitor proteins called repressors. This repressor protein binds to the operator gene and blocks the path of RNA polymerase, thus preventing transcription. If an inducer is present in the system, it binds to the repressor which undergoes conformational change and becomes inactive. As the inactive repressor cannot bind to the operator, the structural genes get activated and protein synthesis continues.

In a repressible system, if only a repressor is present, it is inactive by itself and cannot attach to the operator gene and hence transcription continues. However, when another protein called co repressor is present, it attaches to the repressor forming a *repressor-corepressor* complex, which blocks the operator gene thus preventing transcription.

Types of regulatory mechanism

Earlier in the chapter it has been mentioned as to how the presence or absence of a substrate can induce or suppress the synthesis of the enzyme protein. Here we shall study in some more detail the various regulatory mechanisms that are involved in the synthesis of enzymes.

Induction. It has already been mentioned that in inducible systems, the inducible enzymes are normally present in small quantities and their synthesis can be increased manifold by the presence of the inducer. Such as example of inducer in *E.coli* is lactose and the inducible enzyme is β *galactosidase.* In the absence of lactose, only one ore two molecules of the enzymes β *galactosidase* may be present in the system. When the inducer is added, within three minutes about 3,000 molecules of the enzyme are synthesised by the gene.

When the inducer is absent, the repressor protein will block the production of the enzyme by not allowing transcription of mRNA by the structural genes. The repressor protein actually blocks the operator gene. When the inducer is present the active repressor binds to the inducer to form a repressor inducer complex. As a result, the repressor gets inactivated. An inactivated repressor cannot bind to the operator and hence the operator will allow the structural genes to transcribe mRNA and the enzyme is synthesised.

Repression. The presence of a metabolite inhibiting the synthesis of an enzyme in the system is caled repression. It has been mentioned already that histidine acts as a repressor and blocks the production of the enzyme responsible for histidine synthesis. This is called enzyme repression. Another example of repression is the blocking of the production of the enzyme tryptophan synthetase by the substrate tryptophan. The enzyme tryptophan synthetase is produced only when the bacterium (*E.coli*) is cultured in a tryptophan free medium.

In the repressible system, the protein produced by the regulator gene is an inactive repressor (*aporepressor*). On combining with a corepressor (effector), the aporepressor molecules get activated due to a conformational change. As a result of this the aporepressor becomes active and binds to the operator gene. As a result of the operator genes getting blocked the structural genes cannot by itself bind to the operator gene. The operator gene being free will allow the structural genes to transcribe mRNA. The following is a summary of the regulation of enzyme synthesis in inducible and repressive systems.

I. Inducible system

1. ***Inducer present :*** Regulator gene → Active repressor + inducer → repressor inactive → operator gene free → structural genes transcribe mRNA → synthesis of enzyme protein.
2. ***Inducer absent :*** Regulator gene → active repressor → binds to the operator gene which becomes inactive → no mRNA transcription → inhibition of enzyme synthesis.

II. Repressible system

1. ***Corepressor present :*** Regulator gene → inactive repressor + corepressor → active repressor → binds and inactivates operator gene → structural genes do not transcribe mRNA → inhibition of enzyme synthesis.
2. ***Corepressor absent :*** Regulator gene → inactive repressor → operator gene not blocked → structural genes transcribe mRNA → enzyme protein synthesised.

Negative control

In the inducible and repressive systems, protein synthesis is allowed to take place when the operator gene is free and it (protein synthesis) stops if the operator gene is blocked. In other words, gene expression is possible only when the operator is free. Such a controlled mechanism of protein synthesis is said to be of a negative type. A typical example of negative control is the production of the enzyme protein in the presence of lactose and in the absence of glucose. The presence of glucose actually inhibits the production of the enzyme β *galactosidase* while the presence of lactose promotes the production of the eyzyme.

In negative control, the regulator protein acting as the repressor prevents the transcription of mRNA by structural genes. The main controlling factor is the operator gene.

Positive control

In this control, the regulator protein functions as an activator and increases the enzyme synthesis. The activator binds itself to DNA in a location called the initiator site. Here the initiator site need not necessarily be the operator gene. As the regulator protein enhances the transcription of mRNA by structural genes this kind of regulation is called a positive control. A typical example of a positive control is the arabinose operon seen in *E. coli* (see later for details).

Replication of DNA is also an example of positive control. The bacterial genome is made up of replicating units called replicons which are capable of independent duplication. One of the structural genes of a replicon is known to produce an activator protein which in turn is induced by some cytoplasmic factors. The activator binds to a specific site on the replicon called the replicator and initiates replication from this point.

Catabolic repression

The works of Magasaine and others have shown that the presence of glucose inhibits the synthesis of enzymes involved in the breakdown of sugars such as lactose, arabinose or galactose. This phenomenon of repression is known as catabolic repression. This has been in many microorganisms including *E.coli*, yeasts etc. For instance, when a culture of *E.coli* maintained on lactose is supplemented with glucose, it has been seen that the synthesis of β *galactosidase* enzyme becomes suppressed as long as glucose is present in the medium. The inhibitory effect of glucose is due to a marked drop in the intercellular levels of a nucleotide cyclic AMP. The studies of Zubay and Beckwith (1970), have shown that glucose suppresses the level of cyclic AMP which in turn inhibits the transcription of mRNA.

In vitro studies of Eron, et al (1971), have indicated that lactose operon transcription requires not only cyclic AMP but also a protein called CAP (catabolic activator protein). CAP binds to the cyclic AMP. Thus, catabolic repression has two steps at the first stage CAP binds to cyclic AMP and at the second stage, this complex (CAP cyclic AMP) exerts its effect on the transcription process.

Feed back inhibition

In some instances, the final product of a particular biosynthetic pathway gets accumulated if it is not utilised for metabolic activities. Such a kind of accumulation many a time will cause the stoppage of further synthesis of the substance. For instance, if the normally growing *E.coli* cells are supplied with isoleucine (an amino acid), it has the effect of blocking the isoleucine synthesising enzyme. This inhibition is due tothe binding of isoleucine to enzyme *threonine deaminase*. This kind of inhibition is called feed back inhibition. Feed back inhibition may be defined as the inhibitory effect of the final product on the enzyme involved in the first step of production when a synthetic pathway is made up of several steps and is controlled by the activity of specific enzymes. Feed back inhibition has an advantage in the sense if the enzyme of the first step is blocked, there is no necessity to block the intermediary enzymes as in any case they would not get the substrate.

Feed back inhibition is mostly through the allosteric transformation of the enzyme so that its active site gets inactivated. Feed back inhibition is different from repression in that in the latter there is no mRNA synthesis, and hence no enzyme synthesis. Feed back inhibition is actually a case of control of gene expression after translation.

Role of Leader sequences in genetic regulation

Gene expression in prokaryotes is controlled both at the level of transcription as well as after the transcription. In the mechanisms which operate after transcription, a prominent role is played by leader sequences and attenuators. For example, in the tryptophan operon of *E.coli,* the mRNA of gene trpE (gene for anthranilate synthetase) has a sequence of 162 bases between initiator codon and the promoter operator region of this gene. This sequence of 162 bases is called leader sequence and is thought to exercise post transcriptional control. This leader sequence of the gene trp E has a mechanism which allows transcription when synthesis is needed, and stop the transcription whenever synthesis is not needed.

Regulation by antisense RNA

Some of the recent researches have shown that regulation of protein synthesis can be controlled at the translational level by using RNA which has complementary bases to mRNA. When such an RNA is introduced, it binds to mRNA forming RNA mRNA hybrid and prevents the translation, it (RNA) which is this RNA is complementary to mRNA and interferes with the translation, it (RNA) is called antisense RNA. Antisense RNA can be used in experimental conditions to stop the translation of mRNA. In *in vivo* conditions also, antisense RNAs are known to regulate the translation of mRNA. In transposon Tn 10, the production of the enzyme *transposàse* is prohibited by antisense RNA. The production of antisense RNA is controlled by a promoter which directs transcription in a direction opposite to the transposase gene, thus producing antisense RNA.

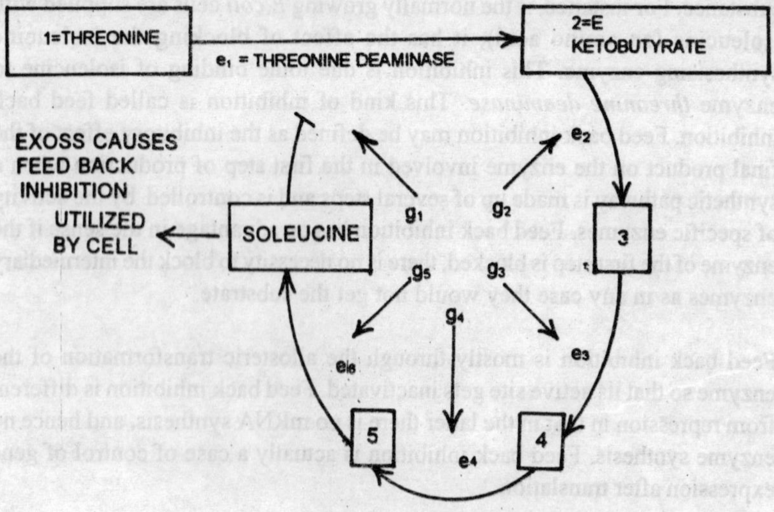

Fig. 8.2 Genetic regulation
Feedback inhibition in *E.coli* (after Stansfield, 1969)

Role of antisense RNA in regulating gene expression has also been demonstrated in the case of the two genes involved for the synthesis of outer membrane protein.

Examples of a few operon systems

The lac operon of E.coli. The lac operator consists of three structural genes.

1. Gene *z*. This is made up of 3,063 base pairs and codes for enzyme β *galactosidase*, which is active as a tetramer and breaks lactose into glucose and galactose.
2. Gene *y*. This is made up of 800 base pairs and codes for the enzyme β *galactoside permease* which is a membrane bound protein and helps in the transportation of metabolites.

3. Gene *ac*. This consists of 800 base pairs and codes for β *galactoside transacetylase*, an enzyme that transfers an acetyl group from acetyl coA to β *galactosides*.

In addition to the structural genes, the lac operon consists of a regulator gene *i*, an operator gene *o* and a promoter gene *p*. These six adjacent cistrons (genes) constitute the lac operon system. An operon can be defined as an integrated genetic unit.

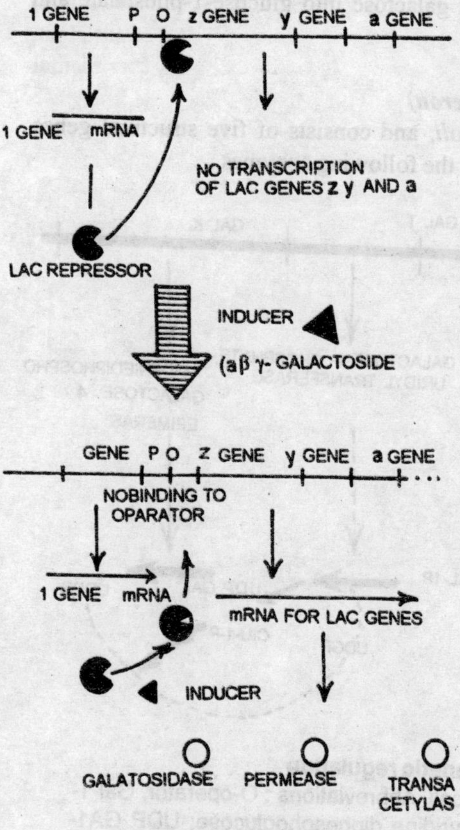

Fig. 8.3 Genetic regulation
Lac-operon model

In lac operon, the gene *i* produces a repressor (protein) which inactivates the operator by binding with it. Subsequently, an inducer such as lactose binds with the repressor molecule there by inactiviting it. As a result, the operon gene *o* is free from the repressor and becomes active. When the operator gene is active mRNA is produced from the structural genes *z*, *y* and *ac*. The production of mRNA starts from the promoter gene. It has been found that all the three structural genes produce a single mRNA molecule which codes for the sequence amino acids for all the three enzymes – β *galactosidase, permease* and *transacetylase.* Thus equal amounts of all the three enzymes are produced and the first message translated is for β *galactosidase;* this is followed by the synthesis of permease and then followed by transacetylase. Thus the operons show polarity. This has been proved by mutational studies. Mutations in the *z* region will reduce the quantities of all the three enzymes. While mutations in *y* locus will have no effect on β *galactosidase* but will reduce the quantities of other two enzymes. Such mutations are said to be **polar mutations.**

The gal operon system

In the gal operon system in *E.coli,* the sequence of three structural genes are gal E, gal T and gal K. These genes are directly responsible for the sequence of the enzymes produced namely galactokinase, galactose-1-phosphate uridyl transferase, and uridinediphospho-galactose-4- epimerase. These enzymes are responsible for the degradation of galactose into glucose-1-phosphate and UDP glucose.

Tryptophan operon system (trp operon)

The trp operon is also seen a *E.coli,* and consists of five structural genes ABCD and E. These genes code for the following enzymes.:

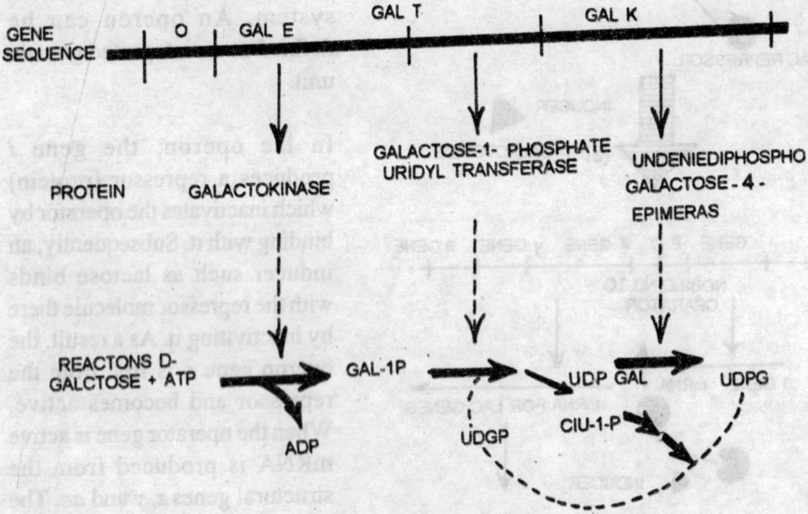

Fig. 8.4 Genetic regulation

The galctose (gal) operon of E.coli. Abbreviations : O-operator, Gal-1-galactose-1-phosphate, UDPG-uridine diphosphoglucose; UDP GA1-uridine diphosphogalactose; Glu-1-P=glucose-1-phosphage (after Goodenough and Levine, 1974)

1. Genes A and B code for tryptophan synthetase.
2. Gene C codes for PRA isomerase and IGP synthetase.
3. Gene D codes for phosphoribosyl pyrophosphate anthranilate transferase.
4. Gene E codes for anthranilate synthetase.

The levels of these enzymes is low when the amount of tryptophan is more and the level high when the amount of tryptophan is less. All these enzymes are regulated in a coordinated fashion. The operator and promoter are overlapping and are located next to *trp* E gene. The regulator gene (R) is located away from the operon and the repressor produced by it is inactive until it combines with tryptophan. All the enzymes are produced constitutively if the repressor is either absent or fails to combine with tryptophan. All the structural genes are translated together and form a polycistronic mRNA. This polycistronic mRNA besides the transcription of the bases of structural genes has a 160 nucleotide long leader sequence. This contains the attenuator site to which the substrate tryptophan can get attached and thus reduce the rate of transcription. Thus in *trp* operon, there are two regulation – the regulator gene and the attenuator site in the RNA itself.

The *trp* operon thus seems to have a dual control–a control by regulator and also by attenuator. In addition to this, feed back regulation is also seen in *trp* operon.

Arabinose operon

This is also seen in *E.coli* and is an example of positive control. Like the lac operon, the arabinose operon also consists of three strcutural genes. These are –

(i) *ara* A gene codes for arabinose isomerase

(ii) *ara* B gene codes for L-ribulokinase

(iii) *ara* B gene codes for L-ribulose-5-phosphate epimerase

These structural genes are located together on the *E.coli* chromosome. Their expression is under the regulation of a regulator gene C. This gene C called activator must be present for transcription of the arabinose operon. The product of C gene binds to the operator in the presence of arabinose and forms the activator protein. As a result of this, transcription takes place in the arabinose genes. If the activator is not bound to the arabinose or if the arabinose is absent, the operon is not expressed because the regulator protein binds to the operator gene (*ara* o) and functions as a repressor. In arabinose operon, the gene C acts both as an activator and repressor. In this way, arabinose operon is different from lac operon.

Gal genes in yeasts

In yeasts, galactose is fermented by the activity of three enzymes–kinase, transferase and epimerase. These are coded by the three structural genes GAL 1, GAL 7 and GAL 10. The regulator gene is located away from the structural genes. If a mutatkon occurs in a regulator gene, all the three enzymes are synthesised without control indicating thereby that the product of the regulator gene is a repressor. The regulator gene of gail in yeasts is very much similar to the gals in *E.coli*. A detailed genetic analysis has revealed that the product of the regulatory gene controls the functioning of an unlinked gene GAL 4. The product of GAL 4 is a positive regulator for the three linked genes. The GAL 4 gene has an operator gene *(c)* also. The regulator gene is *i,* the product of this gene exerts a negative control over gene c (operator). As a result, transcription of GAL 4 is reduced. When GAL 4 is free it produces a product which stimulates the transcription by the structural genes.

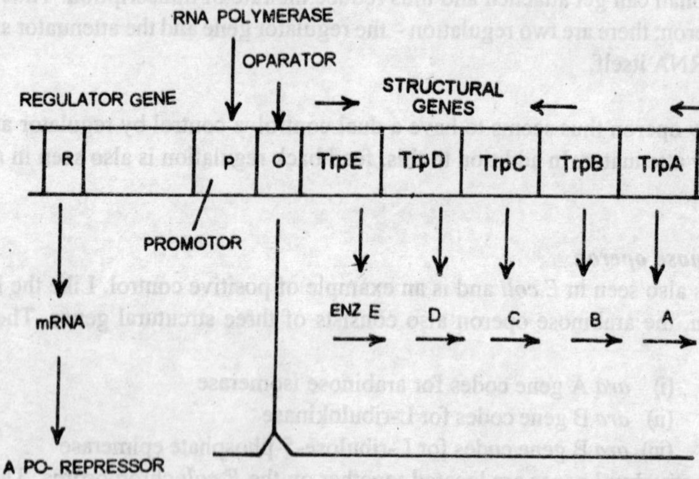

Fig. 8.5 Genetic regulation
Tryptophan operon

Regulation of transcription in λ phage DNA

Regulation of the λ phage DNA has both negative and positive control mechanisms. The phage exists in a lytic or lysogenic state and this is controlled by some regulatory proteins. Depending on when the genes get activated during infection, multiplication or maturation phase, the λ genes can be classified into *early genes* and *late genes.* The early genes as the name indicates are required at the beginning of the infection cycle, while the late genes are required for the assembly and maturation of phages. The following table gives a list of genes identified on the λ genome along with their functions.

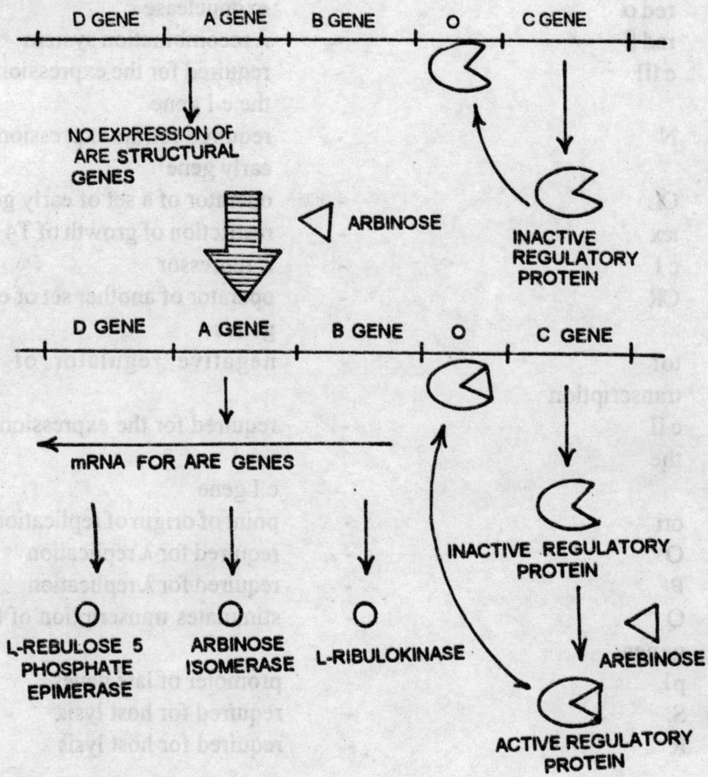

Fig. 8.6 Genetic regulation
The L-arabinose operon of E.coli

List of genes of the λ phage genome

A,W,B,C,D,E and F	**-**	**Head genes**
Z,U,V,G,T,H,L,K,Iand J	**-**	**Tail genes**
att λ	**-**	**site of attachment and insertion**
of the		**phage DNA into the host DNA**

Gene	Function
int	integration of λ DNA e
xis	excision of λ DNA
red α	exonuclease
red β	λ recombination system
c III	required for the expression of the c I gene
N	required for the expression of early gene
OL	operator of a set of early genes
rex	restriction of growth of T4 R II
c I	λ repressor
OR	operator of another set of early genes
tof	negative regulator of c I transcription
c II	required for the expressions of the c I gene
ori	point of origin of replication
O	required for λ replication
P	required for λ replication
Q	stimulates transcription of late genes
pl	promoter of late genes
S	required for host lysis
R	required for host lysis

The cI gene codes for the repressor which binds to the operator genes thereby blocking the transcription of both the early operons. The repressor can bind to the operons blocking its integration and replication resulting in immunity due to a negative control exerted by a repressor. The product of gene N, however, acts as a positive regulator and allows transcription. The product of the N gene is known as antiterminator because instead of stopping, it allows transcription. Another example of positive regulation is the influence of Q gene product on late genes. This product combines with the coat transcriptase enzyme of *E.coli* and acts favourably on the promoter of the late genes. This is the λ phage, there are both negative and positive regulatory mechanisms to control the initiation and completion of lytic or lysogenic cycle. This is very necessary for an organism which has to survive in a continuously changing environment such as the bacterial cell.

GENETIC REGULATION IN EUKARYOTES

The mechanism of gene regulation of protein synthesis explained above illustrates the phenomenon in prokaryotes, whereby a repressor protein transcribed by regulator genes may activate or prevent the protein synthesis depending on the presence of inducer or corepressor.

In eukaryotes, genetic regulation takes place by a slightly different mechanism. Two types of proteins associated with the DNA of the chromosomes possibly play a role in gene regulation. These are histones and non-histone chromosomal proteins (NHC proteins). Since histones of different organisms have similar amino acid composition it is unlikely that they can selectively repress gene action. At best, histones may probably bring about a general (non specific) repression. It is believed that histones under certain situations bring about supercoiling of DNA preventing the movement of RNA-polymerase and thus blocking transcription. This is known by the fact RNA precursors are incorporated only in those regions where chromatin is diffuse (no super coiling).

Three models have been proposed by various scientists to account for gene regulation in eukaryotes. These are - 1. Frenster's model, 2. Non-histone depressor model and Britten and Davidson's model.

Frenster's model

This is also called gene specific derepressor RNA model. As has been pointed out earlier, histones bring about non-specific repression of transcription. Derepression (initiation of transcription) is brought about when nuclear polyions separate histones from DNA at certain specific parts (of DNA), and form complexes with histones.

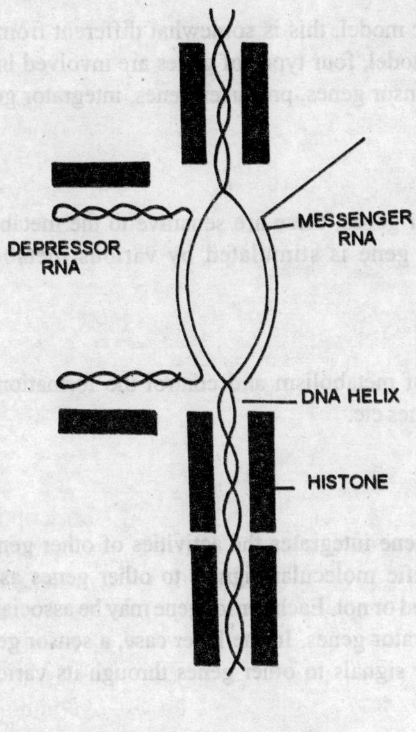

Fig. 8.7 Genetic regulation

Genetic regulation of protein synthesis in Eukaryotes-Frenster's model of gene regulation

Non-histones derepressors

The model proposed by Paul et al (1971), is similar to Frenster's model except the derepression is brought about by NHC acid proteins instead gene specific RNA. The fact that NHC proteins are of immense variety in different organisms, and also at different stages of growth even in the same individual goes to show that they can undertake selective derepression at specific gene loci. Derepression is brought about when NHC proteins bind themselves at specific loci to histones forming complexes. This complex separates itself from the DNA and allows it to transcribe, only at those specific loci.

Gene regulation by NHC proteins

(a) Non specific repression by histones.
(b) Specific binding of NHC proteins to histones.
(c) Removal of histone by NHC proteins.
(d) De repression of DNA, leading to transcription.

Britten Davidson Model

Also known as operon operator model, this is somewhat different from the other two models. As per this model, four types of genes are involved in the regulation process. These are sensor genes, producer genes, integrator genes and receptor genes.

1. *Sensor genes*

Comparable to the promoter genes, these are sensitive to the metabolic status of the cell. The sensor gene is stimulated by various metabolic substances.

2. *Producer genes*

These regulate the output of metabolism and control the formation of enzymes, cell organelle, membranes etc.

3. *Integrator gene*

As the name suggests, the gene integrates the activities of other genes. The integrator gene sends specific molecular signals to other genes as to whether the sensor gene is activated or not. Each sensor gene may be associated with a single or a group of integrator genes. In the latter case, a sensor gene may send a variety of molecular signals to other genes through its various integrator genes.

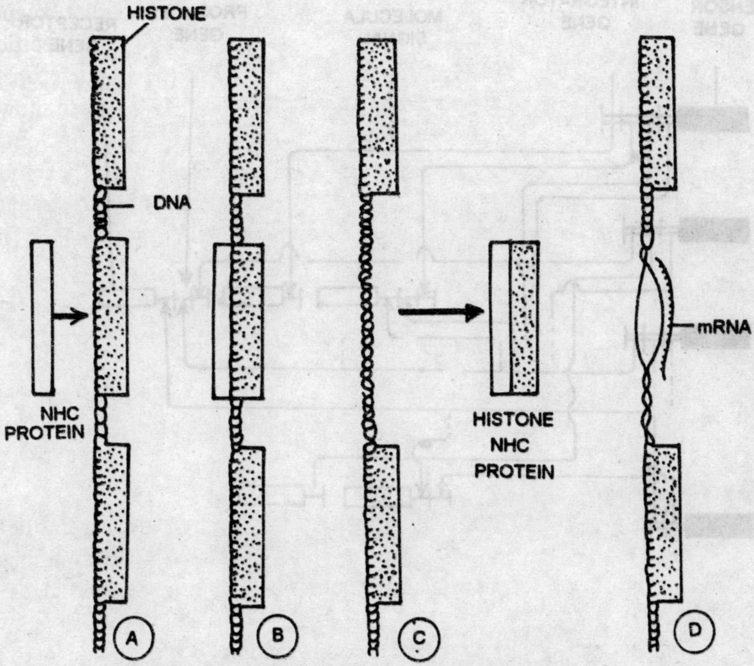

Fig. 8.8 Genetic regulation
Gene regulation by NHC histones

A. Non specific repression by histones
B. Specific bindng of NHC proteins to histones
C. Removal of histone by NHC proteins
D. De repression of DNA, leading to transcription

4. *Receptor gene*

This acts as an intermediate between the integrators and the producers. When the molecular signal reaches the receptor gene, it inturn activates the producer gene. A producer may have one or a group of receptor genes associated with it in the same way as sensors have seveal integrators. This helps, not only in the variety of origin of molecular signals, (from sensors) but also variety in reception of these signals (by receptors). Perhaps, multiintegrator gene complex and multireceptor gene complex may have a specific multichannel relationship.

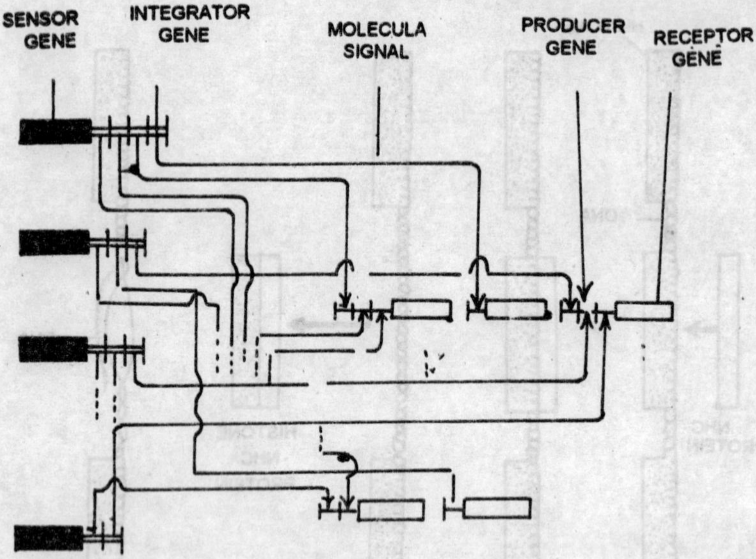

Fig. 8.9 Genetic regulation
Britten Davidson's Operon-Operator model

9

GENETIC RECOMBINATION

Genetical studies in microorganisms is extremely difficult, as they are very minute in size and do not possess any variability in their phenotype which can be subjected to genetic analysis. Another difficulty of microbial genetical studies is that microorganisms reproduce asexually to multiply their number and microbiologists thought that microbes like bacteria do not reproduce sexually. Extensive and repeated reseaches were made to explore the possibility of sexual reproduction in bacteria. Most of the earlier attempts were unsuccessful, because they were based purely on morphological observation of cell fusion. When Luria and Delburck established the assay procedure for obtaining the meaningful results, bacteriologists knew that their approach towards bacterial sexuality should change and a new technique is necessary for the analysis of sexuality in bacteria. The first successful experiment in bacterial genetics was performed by Joshua Lederberg and E.L. Tatum. These two collaborated in 1946, and using K_{12} strains of *E.coli* conducted what can now be called a classical experiment in bacterial crossing. They were able to demonstrate, that in *E.coli*, exchange of biochemical traits is possible. For their work on bacterial genetics, lederberg and Tatum shared with G.W. Beadle, the Nobel prize in medicine, for the year 1958.

With the establishment of the fact, that bacteria do exhibit sexuality, a vast area of microbial genetics opened up for geneticists. Prokaryotic organisms like bacteria were favoured for genetic experiments because –

1. Large amount of material could be handled as they are small.
2. Several generations could be obtained within a short period.
3. Pure cultures could be easily obtained for analysis.

With the advantages mentioned above, microbial genetics has made rapid strides in the last few decades.

Types of genetic material transfer

As in any other organism in bacteria also, sexuality involves transfer and recombination of genetic material from one bacterial cell to the other. Three methods of genetic recombination have been demonstrated in bacteria. These are conjugation, transformation and transduction.

CONJUGATION

Genetic recombination

Lederberg and Tatum (1944), demonstrated this in experimenting with the biochemical mutants of *Escherichia coli.* These bacteria require the following four nutrients 1. Biotin 2. Methionine 3. Leucine 4. Thereonine. But they can be grown on a minimal medium (a medium containing only agar and carbohydrates) because, by using carbohydrates they can synthesize all the above four nutrients. They are called 'Paratrophs' and can be genotypically designated as follows :

$$B^+ M^+ L^+ T^+$$

In this, letter stands for the nutrient (B = biotin etc) and the sign + stands for the capacity to synthesize that particular nutrient.

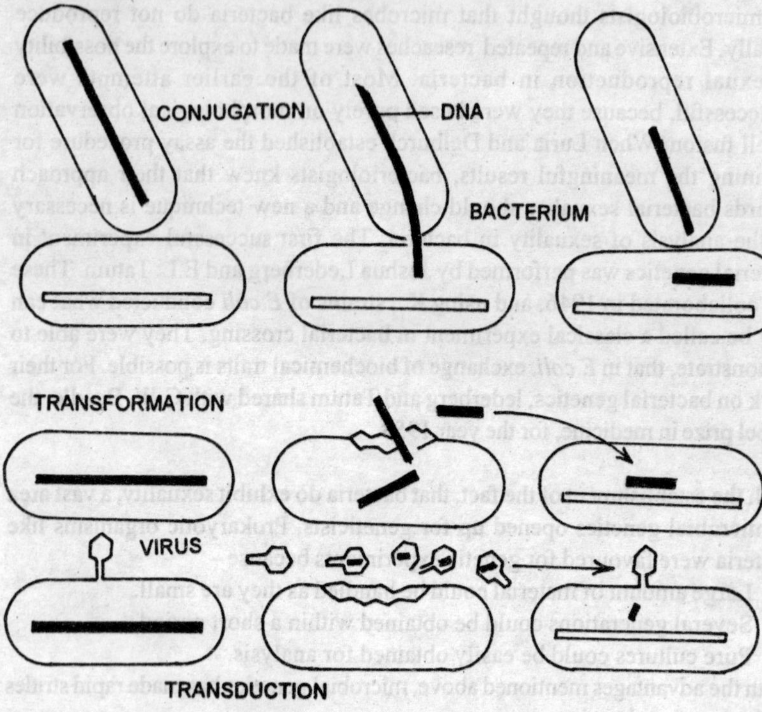

Fig. 9.1 Genetic Recombination

Genetic recombination in bacteria

By subculturing in the laboratory, the paratrophs undergo mutation and lose the capacity to synthesize one or more nutrients. The mutants are called *Auxotrophs.* By careful isolation, Lederberg and Tatum obtained two complementary polyauxotrophic varieties of *E.coli*. They can be designated as follows

1. $B^+ M^+ L^+ T^+$
2. $B^- M^- L^- T^-$

(The sign–stands for the loss of the synthesizing capacity. For ex : L means leucine cannot be synthesized, and hence it has to be added to the culture).

It is obvious that both strains 1 and 2 cannot be cultured on a minimal medium. Strain 1 cannot be cultured because it cannot synthesize leucine and theronine. Strain 2 cannot be cultured because it cannot synthesize biotin and methionine.

But when both of them are kept on a minimal medium, curiously they survive (normal expectation is, they should not live). On nutritional analysis, it was revealed that they were Paratrophs. The formation of paratrophs from two polyauxotrophic varieties can be explained only on the possibility of sexual recombination between strain 1 and 2.

$B^+M^+L^+T^+$ $B^-M^-L^-T^-$
1 2

Because of recombination of 1 and 2 the following combinations may be obtained.

$B^-M^-L^-T^-$ and $B^+M^+L^+T^+$
(a) (b)

Of the above, (a) will not survive because it has lost the capacity to synthesize all the nutrients, but (b) will survive because it has become a paratroph.

Mechanism of conjugation

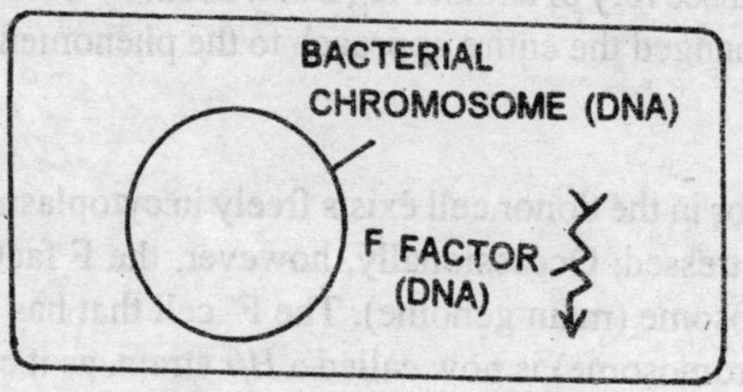

Fig. 9.2 Genetic Recombination
A bacterial cell showing circular DNA and F factor

The exact mechanism of genetic exchange was worked out by three groups of scientists led by Lederberg in Wisconsin USA, Hayes in London and Woolman in Paris. Genetic transformation was not involved in the process, was confirmed when cell free extracts of one bacterium failed to transformer other cells. It was demonstrated that the two bacterial cells physically come together and DNA from one passes into the other through the conjugation tube.

It was observed that in bacterial cultures, conjugation did not occur randomly between cells; rather it was selective. Genetic material seemed to migrate selectively only from particular cells to certain other specific cells. What is this specificity? What determines that only certain cells can conjugate while the others cannot. Careful studies indicated that the K_{12} *E.coli* has two strains – *Donor strain* and *recipient strain.* The donor strain possesses an episome (plasmid), while the recipient does not have this episome, as F+. The recipient strain does not have F factor and can be designated as F-. The donor strain may also be called male and the recipient female.

The fertility factor is free of the main genome of the bacterium and lies freely in cytoplasm. It is made up of a small DNA molecule having a molecular weight of 2,50,000. Normally, only one F factor is present in a cell and it replicates along with the main genome during cell division.

When conjugation occurs between F and F cells, the main genome is not involved. Consequently, chromosomal genes are not transferred from donor to recipient cell. In normal conjugation, the F factor of donor replicates and one copy of the F factor migrates along the conjugation tube to the recipient cell. As a result, both the cells now have a F factor. Consequent to this transfer, even the original recipient cell now becomes a donor as it has got an F factor. In other words, both the cells are now F^+. This means that the female cell has become male. The newly converted donor cell can now transfer its F factor to other recipient cells by *infecting* them with F factor.

If the entire conjugation process were to be limited to only transfer of the F factor with chromosome genes not being involved, it would have been perhaps of little genetic interest. But the discovery of another K_{12} *E.coli* strain by Cavalli-the *Hrf* (High fertility) strain changed the entire approach to the phenomenon of bacterial conjugation.

What is a *Hfr* strain? The F factor in the donor cell exists freely in cytoplasm – a point that has already been stressed. Occassionally, however, the F factor gets integrated with the chromosome (main genome). The F^+ cell that has an integrated F factor (with the chromosome) is now called a *Hfr* strain, as it can increase the rate of recombination many fold. So a F^+ male that brings about recombination has now been converted into a supermale (High fertility male). Integration of F factor into the chromosome involves a break (in the circular chromosome) and attachment of F factor. As a result of this integration, the chromosome along with its F factor becomes linear with two free ends. It is this chromosome. As the F factor becomes a part of the chromosome and as the chromosome has free ends, one end enters the other bacterium through the conjugation tube.

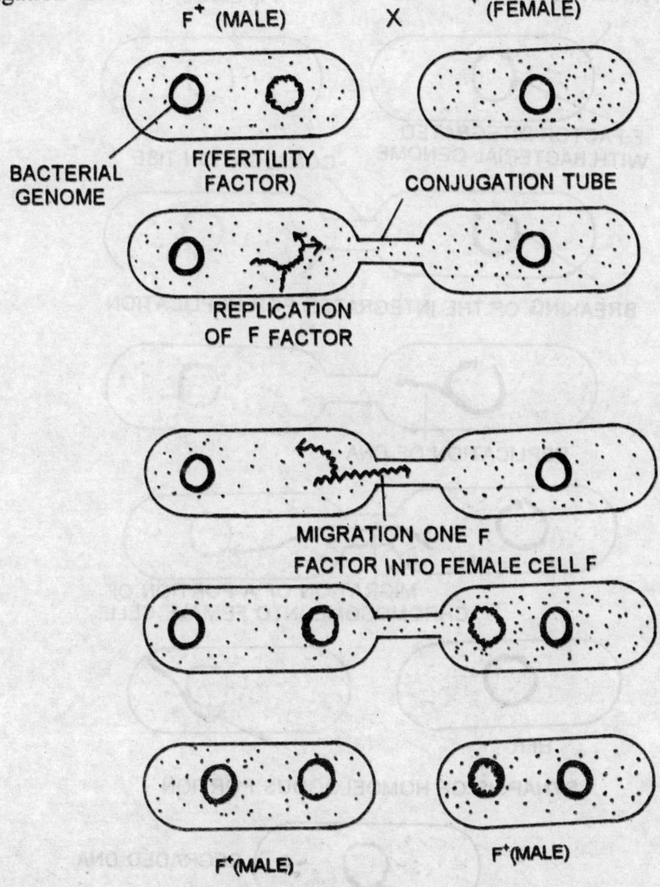

Fig. 9.3 Genetic Recombination
Conjugation between F+ (male) and F- (female) resulting in the conversion of female into male due to transfer of the fertility factor, F.

When a *Hfr* cell conjugates with a F cell, the latter receives the chromosomal genes. Which end of the chromosome of Hfr cell enters the F-cell? It has been found out that when F+ joins the chromosome with a single nick, it will be at one end. This end can be called posterior and the other end, anterior. Usually, chromosome movement from *Hfr* to F begins at the anterior end and the F factor enters the F strain only when conjugation is complete. This, however, rarely happens as conjugation and the F factor remains in the Hfr due to partial conjugation. Two possibilities exist here.

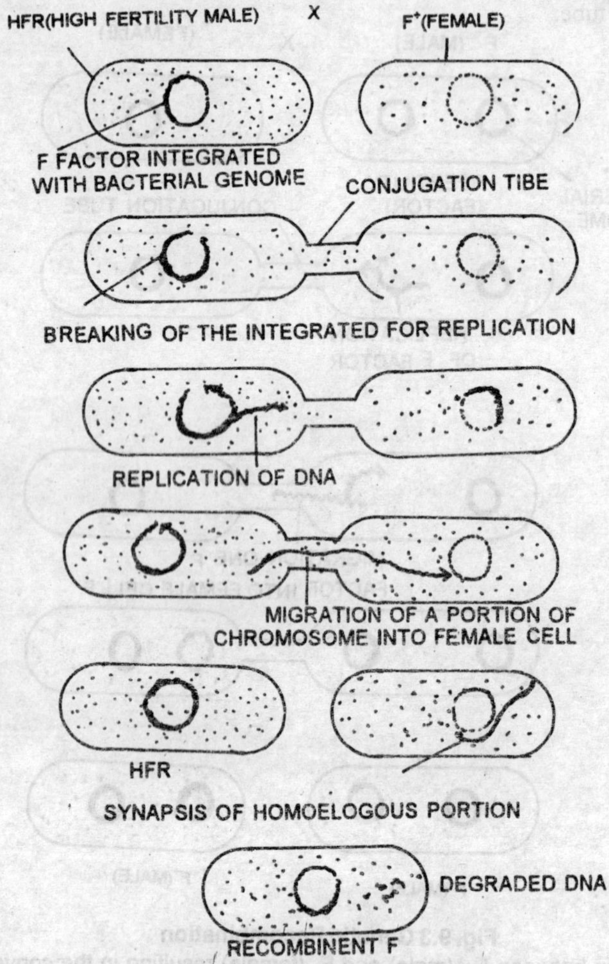

Fig. 9.4 Genetic Recombination

Conjugation between high fertility male (Hfr) and female (F-) cells, resulting in production of recombinant female

1. Complete conjugation

Total transfer of genome along with F factor from *Hfr* to F. As a result, not only the F acquires the genetic traits of *Hfr*, it becomes a Hfr. This is total recombination.

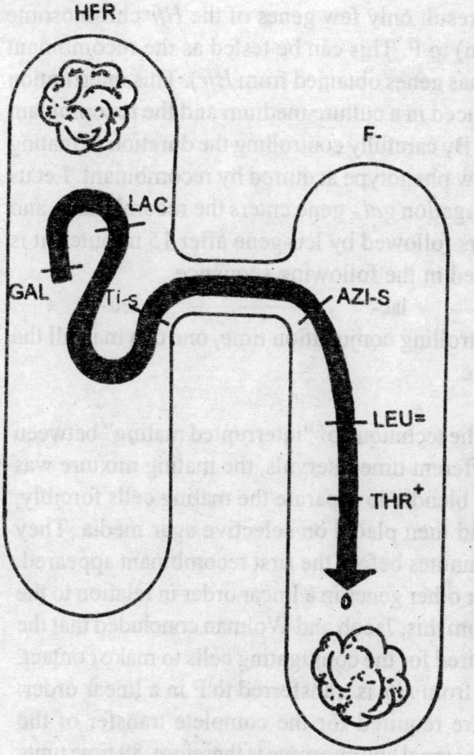

Fig. 9.5 Genetic Recombination

Transfer of genes from Hfr to F in a definite linear order

2. Interrupted or partial recombination

Here, chromosome entry from *Hfr* to F stops in the middle. As a result, F remains female (F) only, but acquires new chromosomal genes as a result of partial recombination. Whichever genes have entered from *Hfr* to F, they express themselves i.e., the F acquires new phenotype.

The F factor attachment to the chromosome is not at any fixed locus. There are several loci at which the chromosome can have a single break and the F factor can integrate to change the cell into a Hfr from F+. In otherwords, the integration of F factor into the chromosome may take place at any one of the possible loci. This interesting observation has led to the conclusion of the existence of several Hfr strains ategorized on the basis of locus of attachment of F factor to the chromosome. This can be further clarified as follows–let us suppose that the bacterial genome has genes A to Z on it. Theoretically, the F factor can attach between A and B or Z and A, C and A and soon. The point of attachment of F factor can be proved by the fact, that if gene A appears first in the recombinant followed by B, then the attachment is between A and B. The F factor always forms the rear end of the genome. For different Hfr strains, the recombination values of various genes in a population varies. But when once, the recombination frequency for a gene is known, the frequency of other genes is always fixed. In otherwords, in a chromosome from Hfr whichever gene may enter first, its distance from other genes will not vary. This aspect has been exploited for mapping the genes on bacterial chromosome.

Interrupted conjugation and chromosome mapping

It has been pointed above, that in a mating between *Hfr* and F, the process of conjugation is rarely completed. More often, it is interrupted by several environmental factors and as a result only few genes of the *Hfr* chromosome get transferred (after duplication) to F. This can be tested as the recombinant acquires a new phenotype (as it has genes obtained from *Hfr*). This interruption of mating can be artificially induced in a culture medium and the recombinant tested for acquiring the new trait. By carefully controlling the duration of mating one can observe, the series of new phenotype acquired by recombinant. Let us suppose, after 5 minutes of conjugation *gal* - gene enters the recombinant, and after 10 minutes *lac* - gene enters followed by leu-gene after 15 minutes, it is obvious that the genes are located in the following sequence.

gal- ———————— lac- ———————— leu-

By following the method of controlling conjugation time, one can map all the genes on a bacterial chromosome.

Jacob and Woman (1957), used the technique of "interrupted mating" between a *Hfr* strain and a F strain. At different time intervals, the mating mixture was diluted and agitated on a waring blender to separate the mating cells forcibly, thus interrupting conjugation and then plated on selective agar media. They demonstrated a lag period of 8 minutes before the first recombinant appeared. The recombinants then appear for other genes in a linear order in relation to the time allowed for conjugation. From this, Jacob and Wolman concluded that the latent period of 8 minutes is required for the conjugating cells to make contact, whereafter, the genetic material from *Hfr* is transferred to F in a linear order. They showed that 89 minutes are required for the complete transfer of the chromosome form *Hfr* to F. The bacterial chromosome is therefore, 89 time units long. Wolman and Jacob in 1956, produced the first genetic map of a bacterial chromosome in a *Hfr* strain.

The transfer of genes in linear order from male bacterium (tagged with virus) to female; the length of transferred segment is dependent the time taken for transfer, and the chromosome is, therefore, mapped in terms of time units.

Sex duction

A *Hfr* strain originates as a result of integration of F factor with the bacterial chromosome. After integration into the chromosome, the F factor might get separated again and remain as an independent entity in the cytoplasm. If this happens, the *Hfr* strain would change into a F^+ strain. In otherwords, *Hfr* and F^+ strains are interconvertible. (Fig. 9.7)

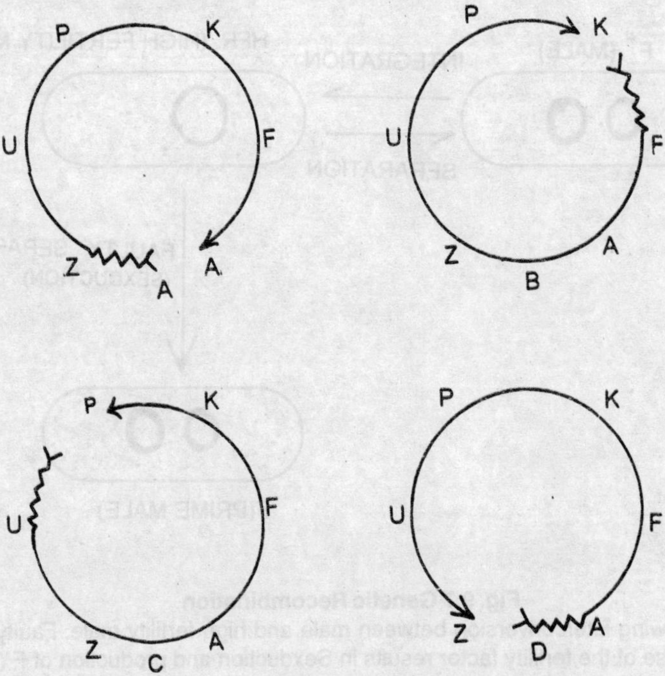

Fig. 9.6 Genetic Recombination
(A,B,C,D). Four hypothetical Hfr strains having different genes at their proximal ends

In some unusual cases, when an integrated F factor gets detached from the chromosome, it may carry along with it some of the chromosomal genes (some porition of the chromosome gets torn and remains attached to the F factor). Such an independent F^+ factor is given the name F^1 (F prime). An F^1 cell is male and like the F^+ except that the F factor has some genes of the chromosome.

When a F^1 (male) cell conjugates with a F cell, the recipient cell on getting the F^1 becomes a donor. In addition, it becomes diploid for certain genes of the chromosome which are present in the F^1. Because these genes are present on the chromosome as well as an F^1, this phenomenon is known as *sex duction* (Jacob and Alderberg, 1959). The recipient cells become *paritial diploids* for the genes introduced by the F^1.

GENETIC TRANSFORMATION

Every individual born out of sexual reproduction will have the mixture of traits of both the parents. In genetic terms, this means that there is genetic recombination in the off spring. Genetic recombination or transformation which is an index of genetic change has also been reported in bacteria.

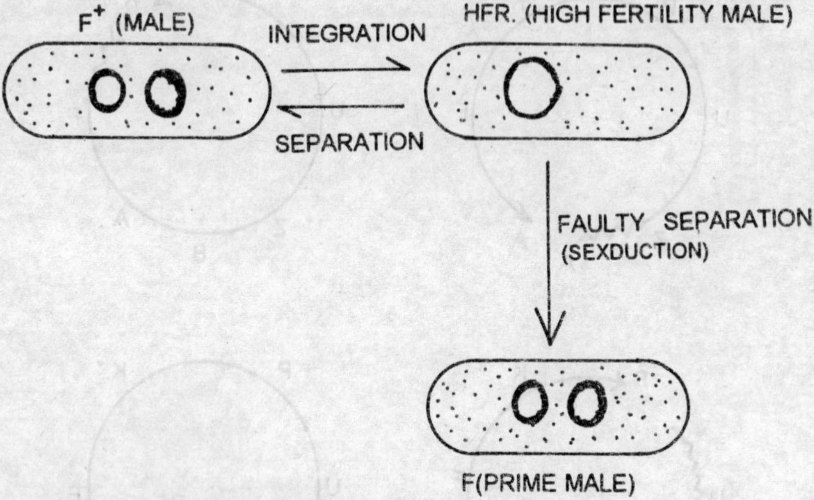

Fig. 9.7 Genetic Recombination

Showing interconversion between male and high fertility male. Faulty release of the fertility factor results in Sexduction and production of F (F prime) mal

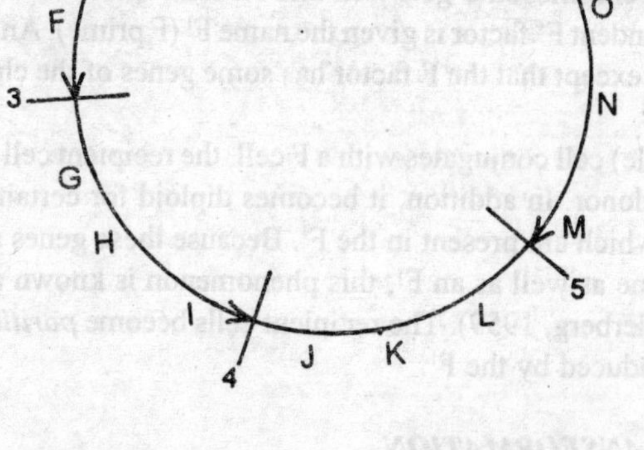

Fig. 9.8 Genetic Recombination

The linear order of genes in five hypothetical Hfr strains, resulting due to breakage of bacterial strain at different sites and subsequent orientation acquired by the resulting linear chromosome

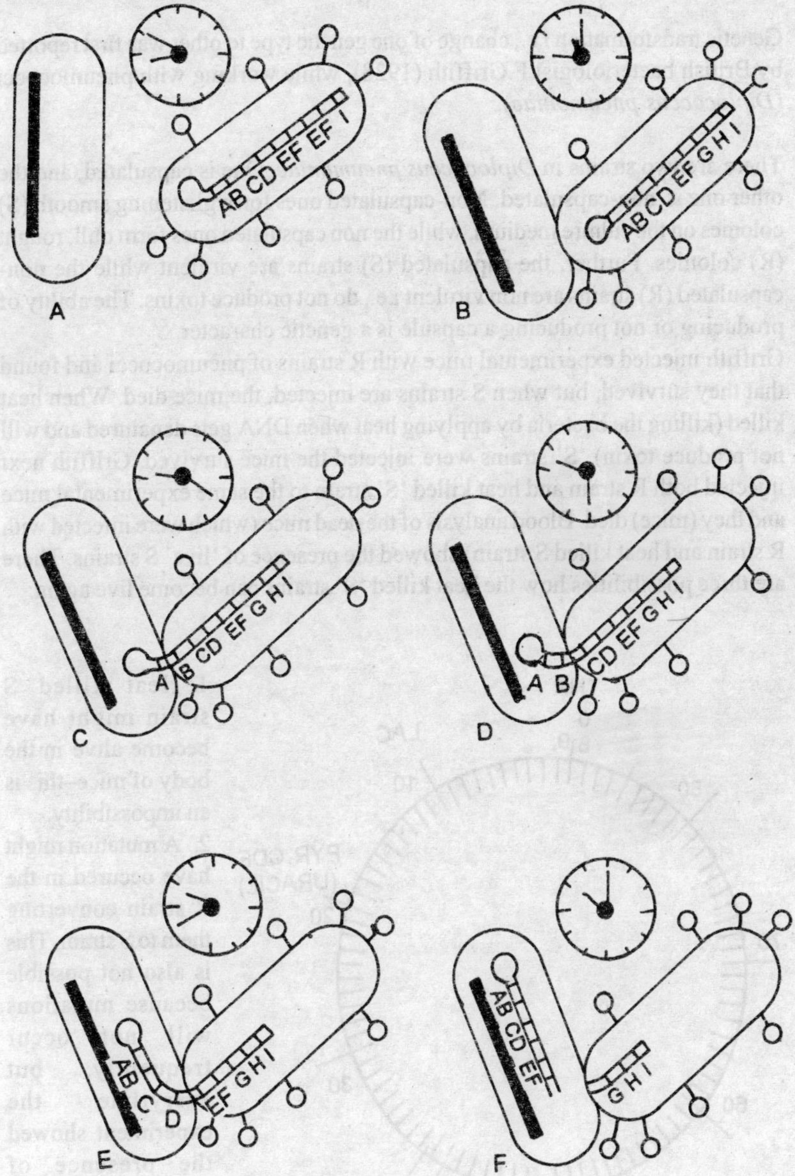

Fig. 9.9 Genetic Recombination

The transfer of genes in linear order from male bacterium (tagged with virus) to female; the length of transferred segment is dependent upon the time taken for transfer and the chromosome is, therefore, mapped in tems of time units

Genetic transformation i.e., change of one genetic type to other was first reported by British bacteriologist F.Griffith (1928), while working with pneumococci (*Diplococcus pneumoniae*).

There are two strains in *Diplococcus pneumoniae.* One is capsulated, and the other one is non-capsulated. Non-capsulated ones form glistening smooth (S) colonies on the culture medium, while the non capsulated ones form dull, rought (R) colonies. Further, the capsulated (S) strains are virulent while the non-capsulated (R) strains are non virulent i.e., do not produce toxins. The ability of producing or not producing a capsule is a genetic character.
Griffith injected experimental mice with R strains of pneumococci and found that they survived, but when S strains are injected, the mice died. When heat killed (killing the bacteria by applying heat when DNA gets denatured and will not produce toxin) 'S' strains were injected the mice survived. Griffith next injected both R strain and heat killed 'S' strain to the same experimental mice and they (mice) died. Blood analysis of the dead mice (which were infected with R strain and heat killed S strain) showed the presence of 'live' S strains. There are three possibilities how the heat killed 'S' strains can become live again.

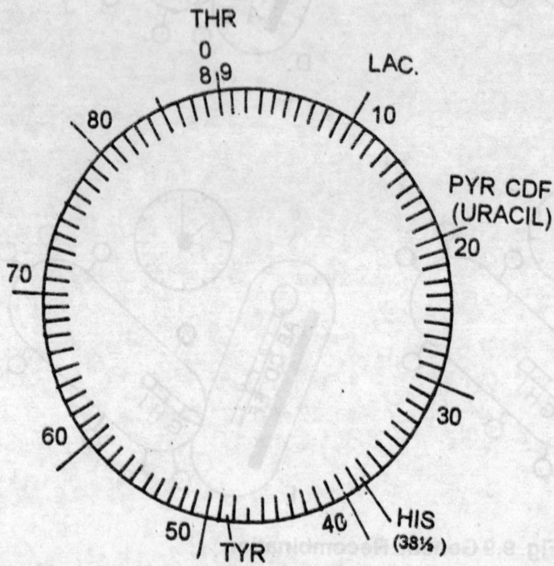

Fig. 9.10 Genetic Recombination
Circular linkage map of *Escherichia coli* chromosome the units of recombination shown as minute required for transfer

1. Heat killed S strain might have become alive in the body of mice–this is an impossibility.
2. A mutation might have occured in the R strain converting them to S strain. This is also not possible because mutations will not occur frequently, but everytime the experiment showed the presence of capsulated S strains in the blood of the dead mice.
3. Some factor in the heat killed strain has transferred the genetic

information for capsule synthesis to R strain making them produce the capsule. In other words, they become S strain. This is the only way a live S strain could be found in the blood of mice. Griffith (1928), thought that the transforming principle was a bacterial polysaccharide. But, Avery *et al* (1944), demonstrated that the transforming principle was DNA.

This experiment clearly demonstrates that a piece of DNA responsible for capsule synthesis recombined with the DNA of the R strain inducing it to produce a capsule. This can be called a natural method of genetic engineering.

Mechanism of transformation

This involves the following steps – competence, binding, penetration, synapsis and integration.

Competence

The recipient bacteria must be competent if they are to accept the foreign DNA and undergo transformation. Competence of recipient bacteria depends on several factors such as age, physiological condition etc. The duration and time of occurrence of competence varies with species also. During growth, competence is observed during the end of logarthmic phase and just at the beginning of stationary phase. Experiments of Euphrissi–Taylor and Freed (1964), have shown that in *Pneumococcus,* competence developed in 90–100 minutes old culture and lasted for just 7–15 minutes. In *Bacillus subtilis* competence lasts for several hours, but only a few cells in the culture become competent. Several external factors are known to induce competence in an otherwise non competent cells.

Binding

DNA of the transforming bacterium comes into contact with the cell surface of recipient cells as a result of random collision. Subsequently, the DNA binds to the surface at places called *receptor sites* which appear to be associated with mesosomes.

Penetration

After binding itself to the cell surface, the DNA penetrates the bacterial cell. It is believed that the double strand DNA (dsDNA) becomes single strand DNA (ssDNA) on entering into the recipient cell in *pneumococcus.* According to one of the theories when one end of dsDNA enters the cell, an endonuclease enzyme of the recipient digests one of the strands, while the other strand is pulled into the cell. Labelling experiments have shown, half the labelling is present in one strand of DNA, while the other half is in the degraded strand. The ssDNA of the donor migrates from the periphery to the chromosome of recipient. In *Bacillus subtilis*, this takes 15–60 minutes. Mesosomes seem to play a role in carrying the ssDNA on to the chromosome.

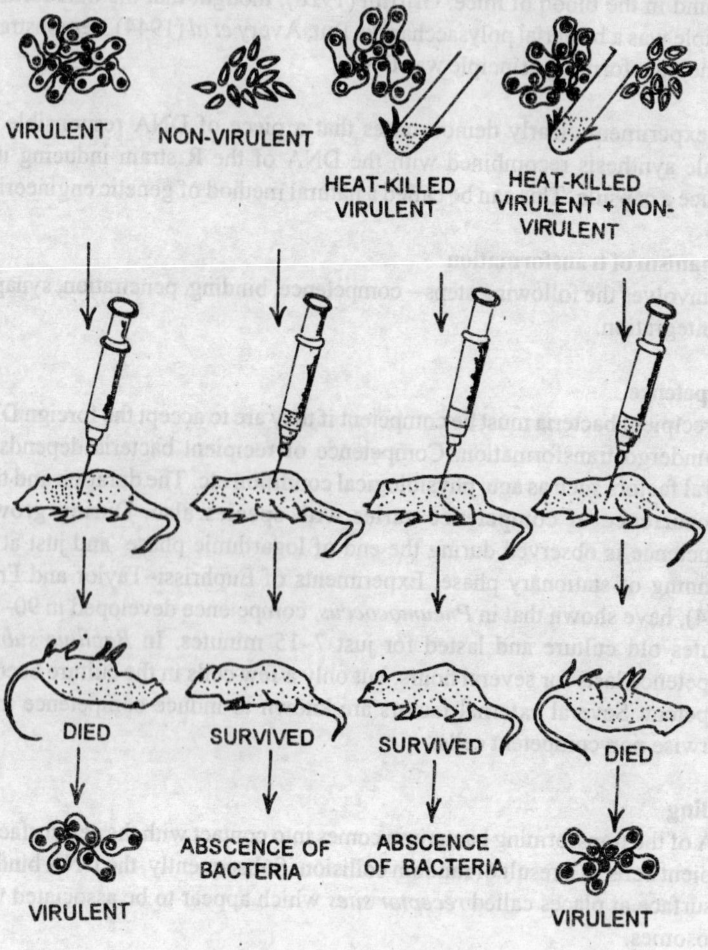

Fig. 9.11 Genetic Recombination
Schematic representation of Griffith's experiment to demonstrate transformation in *Diplococcus pneumoniae*

Synapsis

The dsDNA of the recipient unwinds and initiates pairing with the donor ssDNA. The donor ssDNA is identical, but not exactly homologous with the recipient DNA. This is necessary as otherwise there would be no variation.

Integration

The unpaired recipient DNA strand is cut in two places by an endonuclease enzyme. An exonuclease enzyme also degrades the unpaired portion of ssDNA. Integration of donor DNA is brought about when *repair synthesis* of DNA takes place to fill any gaps. These are joined by ligase. The transformed heteroduplex undergoes replication to produce copies when the recipient cell divides and expresses its phenotype ie., virulence in the case of *Pneumococci.*

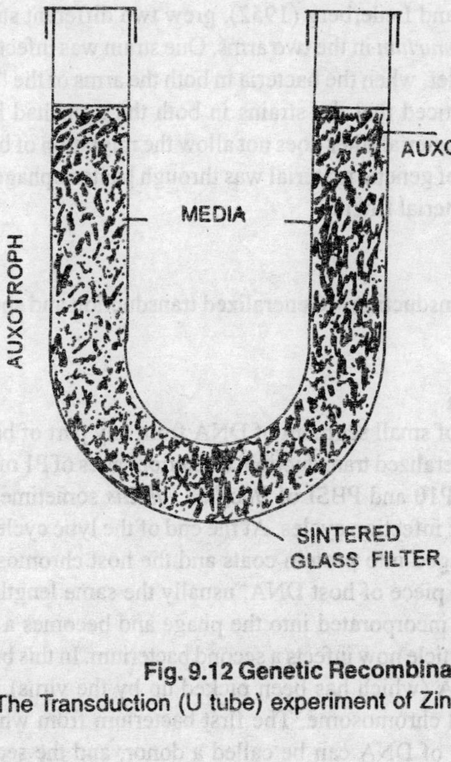

Fig. 9.12 Genetic Recombination
The Transduction (U tube) experiment of Zinder and Lederberg

TRANSDUCTION

Transfer of genetic material from one bacterium to another mediated by bacteriophages is known as transduction. Transduction was first discovered by Zinder and Lederberg (1952), in *Salmonella typhimurium,* and subsequently, by other workers in *Escherichia, Shigella, Staphylococcus* etc.

In transduction, when bacteriophages infect a bacterium they often incorporate a portion of the genetic material of bacterium into their genetic system. If such a bacteriophage infects another bacterium (of the same species), the bacterium genes (of the first infection) may get incorporated into these bacteria. These bacteria now exhibit some of the traits of the first strain. This shows that bacteriophages have picked up some genes from the first bacterium and transferred it on to the second bacterium.

In order to prove bacteriophage mediated genetic transfer. Zinder and Lederberg performed a 'U' tube experiment. The experimental set up consists of a 'U' tube where the two arms are separated at the base by means of a sintered glass filter (Bacterial filter), Zinder and Lederberg (1952), grew two different strains of bacteria *Salmonella typhimurium* in the two arms. One strain was infected with bacteriophages. A little later, when the bacteria in both the arms of the 'U' tube were analysed, it was noticed that the strains in both the arms had become recombinants. Since the bacterial filter does not allow the migration of bacteria, the only way for transfer of genetic material was through bacteriophages since they can move across bacterial filters.

Types of transduction

There are two types of transduction – Generalized transduction and special or restricted transduction.

Generalized transduction

Phage mediated transfer of small segment of DNA from any part of bacterial chromosome is called generalized transduction. Bacteriophages of P1 of *E.coli*, P22 of *Salmonella* and SP10 and PBSI of *Bacillus subtilis* sometimes make mistakes during their lytic infection cycles. At the end of the lytic cycle, when phage particles are packaged into protein coats and the host chromosome is getting degraded, a small piece of host DNA, usually the same length as the phage chromosome, gets incorporated into the phage and becomes a part of phage DNA. The virus particle now infects a second bacterium. In this bacterial cell, the fragment of DNA (which has been picked up by the virus), can be inserted into the bacterial chromosome. The first bacterium from which the phage picked out a piece of DNA can be called a donor, and the second in which the DNA fragment gets integrated can be called a recipient. The recipient will be a recombinant. This is called generalized transduction because random fragments of bacterial DNA get incorporated with the viral genome.

Specialized transduction

Also called restricted transduction, in this, the phage is able to transduce or transfer only certain segment of the bacterial genome. Specialized transduction takes place as follows.

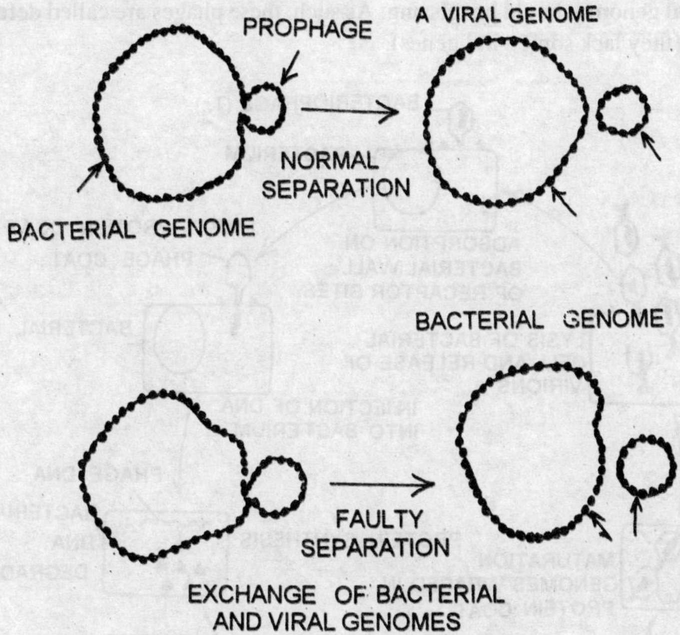

Fig. 9.13 Genetic Recombination
Showing normal and faulty separation of phage DNA from bacterial genome. A piece of bacterial DNA is incorporated into the phage DNA and vice-versa.

When prophages are induced to leave the bacterial chromosome, the viral genome carries with it only some bacterial genes covalently linked to the viral genes. This hybrid DNA packed in the phage coat is called *special transducing particle* because it can transfer from one bacterium to another only certain genes – genes that are adjacent to the prophage attachment site. Specialized transduction is induced by λ 80 prophage, which usually occupy a fixed site on the bacterial chromosome. The site is called prophage site. In *E.coli,* the prophage site is flanked on one side by bacterial genes gal K, T and ε genes of the galactose gene cluster, and on the other side by block of genes involved with the synthesis of biotin–bio A,B,F,C and D. Thus, λ prophage can transduce only *gal* or *bio* genes.

The total length of hybrid DNA in a transducing prophage is exactly equivalent in length to the normal viral genome (as otherwise it cannot be packed into the coat), which means to accomodate the bacterial genes, a corresponding length of viral genome should be missing. As such, these phages are called defective paste (they lack some viral genes).

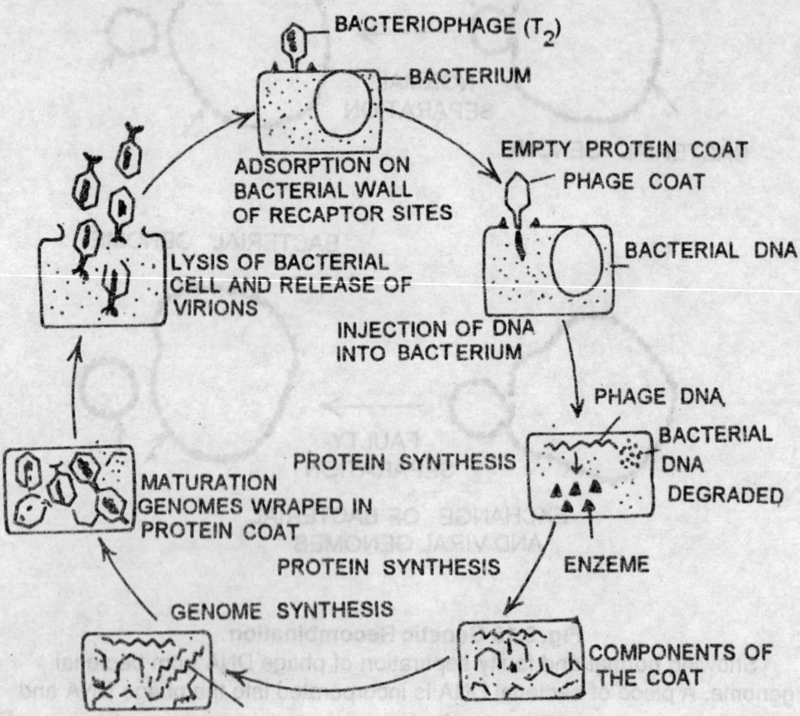

Fig. 9.14 Genetic Recombination
Lytic cycle of a bacteriophage (T_2).

Abortive transduction

In this, the foreign DNA (of the donor tagged to a viral genome) does not get incorporated into the genome of recipient bacterium. The DNA fragment remains free in the cytoplasm and cannot undergo replication. When the bacterial cell divides, only one cell in each generation will have the foreign DNA (exogenote), while all the others are non-transducted. The non -transduced cells might have the gene product (produced by the exogenote before cell division), but progressively it gets reduced as it is distributed among the cells. This phenomenon which brings about incomplete transfer of genes is called *abortive* or *incomplete transduction.*

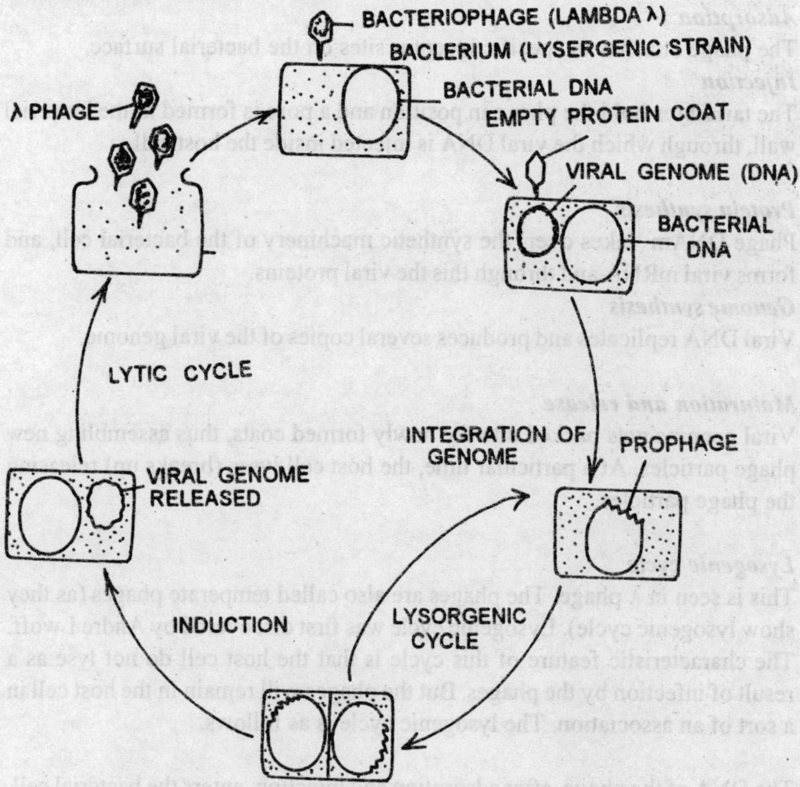

Fig. 9.15 Genetic Recombination
Lysogenic cycle of λ bacteriophage

Growth phases of transducing phages

Bacteriophages which a bring about transduction can be divided into two groups based on their growth cycles. These are **lytic cycle** and **lysogenic cycle.**

Lytic cycle

This is seen in phages called virulent phages typified by T series phages. Here, soon after infection, the host cell lyses releasing the newly formed viral particles. The lytic cycle seen in Tphages infecting *E.coli* bacterium has the following stages.

Adsorption

The phage attaches to specific receptor sites on the bacterial surface.

Injection

The tail fibres hold the phage in position and a pore is formed in the bacterial wall, through which the viral DNA is injected inside the host cell.

Protein synthesis

Phage DNAm 'takes over' the synthetic machinery of the bacterial cell, and forms viral mRNA and through this the viral proteins.

Genome synthesis

Viral DNA replicates and produces several copies of the viral genome.

Maturation and release

Viral genome gets packed into the newly formed coats, thus assembling new phage particles. At a particular time, the host cell lyses (breaks up) releasing the phage particles.

Lysogenic cycle

This is seen in λ phage. The phages are also called temperate phages (as they show lysogenic cycle). Lysogenic cycle was first discovered by Andre Lwoff. The characteristic feature of this cycle is that the host cell do not lyse as a result of infection by the phages. But the phages will remain in the host cell in a sort of an association. The lysogenic cycle is as follows.

The DNA of the phage, after adsorption and injection, enters the bacterial cell. But, instead of starting its degrading activity, the phage genome gets integrated with the host genome. This integrated condition of the viral genome is said to be **prophage.** When the bacterial cell divides, the integrated genome replicates in the normal way. The viral genome may be said to be in a hibernating state as it is not trying to express itself. Occasionally, the viral genome gets separated from the bacterial genome and is released into the cytoplasm where it starts multiplying and causes the lysis of cells. The dissociation of viral genome from bacterial genome (with, causes lysogeny to become lytic), is called *induction.*

10

GENE MUTATION

Evolution in the biological world is primarily dependent on two events to bring about hereditary variations. These are - **mutation and recombination**; while mutation provides the raw material for the changed potentiality of inheritance, recombination permits permutation combinations in the gene pool, thus stabilizing the mutational change in the environment.

What are mutations? To understand this we will briefly analyse how characters are trasmitted from parents to offspring. We know that the characters are trasmitted from parents to offspring through the genes. But the offspring will not have the same genes as those of the parents, rather they have copies of them. In other words during reproduction, parental genes produce copies of themselves to be passed on to the offspring. Gene copying or gene reproduction is a precise process and generally there are no mistakes. Occasionally however, mistakes may occur while copying, so that the newly formed gene gets an altered structure. This modified gene then goes on multiplying and produces its own copies just as the original gene does. This alteration in the changed gene structure is called **gene mutation.**

Broadly speaking, mutation includes all those phenotypic variations, which arise due to altered gene structure and are heritable. Hence the term encompasses, small and minute changes brought about by altered individual genes (gene mutation) and large changes brought about by the altered chromosome structure or number. These latter types of changes can also be called chromosome mutations, and they have been described under the chapter chromosomal aberrations. In this chapter we will study gene mutations. Gene mutations are also called point mutations as they occur at a specific gene locus altering its molecular structure. Sometimes chromosomal aberrations also can bring about a slight phenotypic change, in which case it is difficult to distinguish between these two.

History : The term mutation was first coined by Hugo devries in the early 1900 to designate sudden and drastic changes that occured in plants. Devries was conducting breeding experiments on the plant *Oenothera lamarckiana* (of the family Onagraceae) commonly called the *Evening primrose.* He noticed that some of the progeny possessed characters which were not to be found in either of the parental types. He further observed these changed types bred true. He therefore described these sudden changes as mutations and ascribed them to some changes occuring in the germplasm of parents.

Even Charles Darwin was aware of mutations, even though he did not use the term (mhtations). He was aware of sudden changes occuring in organisms both plants and animals and had used the term sports, to describe them.

The first recorded instance of mutation dates back to 1791, to the pen of Seth Wright, a New England farmer, where appeared a male lamb with short bowed legs. Wright reared this lamb and it bred true. A lamb with short legs was of great advantage as it could not jump even a low level fence. This short legged breed was called **Ancon breed.** This breed however became extinct after about eight years. But the same Ancon sheep appeared some fifty years later in the flock of a Norwegian farmer, representing probably the repetition of the same mutation. Since then a number of mutant types have been discovered. Hornless individuals in cattle, double toed cats, albino rats, mule footed swine, dwarf cupid sweet pea, double flowered, white flowered varities of many plants are all mutants.

The scientific study of mutations began with the work of Morgan on *Drosophila* in 1910. He observed that among a population of red eyed flies, some individuals with white eye appeared and they were predominant among the male individuals. He further found out that the white eyed gene is recessive to red eye and located on the X chromosome. Since the male individuals possessed only one X chromosome and the females had two X chromosomes, the recessive character appeared more frequently among males than among females.

After the original discovery of the white eyed mutant in *Drosophila* by Morgan and his co-workers, a detailed analysis of millions of these fruit flies for over 17 years was carried out by different geneticists and nearly 500 gene mutations have been discovered. Simultaneously other organisms like maize, rodents, pea, snap dragons, poultry, man etc. also have been investigated and scores of mutations have been discovered. Even in microorganisms also mutations have been detected.

Size or range of mutations : Size of mutation varies in different organisms and in different genes. These may be very minute and may not be detectable at all or as in some cases (Ancon breed), so striking that they can be easily made out. In some microorganisms like Bacteria, Fungi *(Neurospora),* Blue green Algae etc, mutations may not bring out any morphological effect but may bring about some changes in their nutritional requirements. In some cases mutations may be so large as to even result in lethality to the individual.

Fig. 10.1 Gene Mutation
Ancon sheep with ram and ewe

Frequency of mutations : The frequency is generally less unless there is a drastic change of the surroundings. According to Muller, in *Drosophila,* the mutation rate for any gene is less than one in one million. It has also been calculated that a particular gene in *Drosophila* may mutate only once in every 40,000 years.

The rate of mutation also varies between different individuals, organisms and sometimes even under different ages of the same organism. J.B.S. Haldane, a renowned geneticist has estimated that in a comparison between man and *Drosophila,* the former has less chances of mutation as most of them are harmful so there is a selection pressure in man against mutations. If the mutations in man were to take place at the same rate as in *Drosophila,* the whole genetic system of the human race would be impaired. As against the unaltered life expectancy of 40,000 years for a *Drosophila* gene, the human gene has a life expectacy of 25,000,00. In one of the case studies of Chondrodystrophic dwarfness (in these human beings, arms and legs are short, while other parts are normal), it was estimated that one in every 11,500 births haye this genetic defect (the data obtained from a Denmark hospital). This gives a rate of one mutation for every 23,000 genes at this locus. If we assume, the average age of the parents to be 30, who give birth to dwarf children, then one can estimate that the gene mutates once every 6,900,00 years. The following two tables give the frequency of mutations in man and Corn.

Table : Rates of mutation in some human genes

Sl. No.	Phenotypic effect	Genetic condition	Frequency in %	Occurrence oncein every gametes
1.	Colour cannot be distinguished	Colour blindness	0.0028	35700
2.	Blood clots very slowly	Hemophilia	0.0032	31250
3.	No pigmentation in hair, skin etc.	Albinism idiocy	0.0028	35700
4.	Mental faculties dull after birth	Amaurotic	0.0011	90909
5.	Limbs and arms dwarf	Chondrodystrophic dwarfism	0.0070	14300
6.	Face with red patches, tumours in Kidney, brains etc., at later stages	Epiloia	0.0012	83333
7.	Iris in the eyes absent	Aniridia	0.0005	2,00,000
8.	Abnormal WBC resulting in reduced immunity	anomaly	0.0080	12,500

Table : Rates of mutation in maize (After Stadler 1943)

Sl. No.	Gene	Phenotype	Frequency (1 in one million gametes)
1.	R	Plant colour	492
2.	Cl	Chlorophyll	106
3.	Dr	Red aleurone	11
4.	WX	Normal endosperm	0
5.	Su	Sugary endosperm	2
6.	Sh	Full endosperm (not shrunken)	1
7.	Ci	Aleurone colour inhibitor	2
8.	Y	Yellow endosperm	2

Mutational Unit : Earlier it was not known as to what constitutes a mutation in a gene. But recent researches have shown that the smallest amount of DNA within which mutation can take place is called a *muton.* At the molecular level mutation involves alteration in the base pair sequence.

Mutator genes and mutable genes : It is difficult to assess (among plants and animals) as to which gene frequently mutates and which gene is comparatively stable. But in some of the orgamisms like maize (Emerson), it has been observed that some genes change quite frequently, w-hile others are not so. Barbara McClintock (Nobel laureate, discovery of jumping genes) has shown in maize that some genes are not only stable, but influence others to mutate. The genes that influence others (to mutate) are called mutator genes and the genes that are so influenced are called mutable genes. These genes, mutator and mutable are more frequent in plants than in animals.

Classification of mutations : Broadly speaking, mutations are of two types - Chromosomal and gene. Chromosomal mutations actually do not change the gene structure, they bring about change in the phenotype either due to loss, addition, altered seqeuence or trasferred position of the genes. Gene mutation however changes the phenotype by basically altering the gene structure itself at the molecular level.

Gene mutations are basically of two categories - **Spontaneous** and **induced.** Spontaneous mutations apparently arise without the help of any identifiable external agency. They arise due to some internal changes. Induced mutations on the other hand are traceable to some external agency like radiation, temperature, chemicals, gas etc.

Mutations may also be further classified into various categories based on the effect, cell type inwhich it occurs, direction of mutation etc. The following is a brief classification of such mutations.

Gene mutations (both spontaneous and induced)

- Visible mutations
  - Germinal
    - Gametic
    - Zygotic
  - Somatic
  - Reverse
  - Spurious
- Biochemical mutations
- Lethal mutations

We shall briefly study a few examples of spontaneous and induced mutations and then study a few instances of the specific types of mutations above.

SPONTANEOUS MUTATIONS

As has already been pointed out, the agency which brings about spontaneous mutations is not known. They apparently occur due to natural causes. The first recorded instance of a spontaneous mutation is that of the Ancon breed sheep. A detailed study of the spontaneous mutation in *Drosophila melanogaster*, is that of T.H. Morgan, where he described the sudden appearance of white eyed flies in a population of red eyed individuals. Some of the spontaneous mutations noticed in plants and animals are as follows.

Spontaneous mutations in plants : 1) Double petunia, 2)Double rose, 3) Shirley poppy, 4) Single flower mutant in sunflower, 5) Multileaf formation in tobacco reported by Hayes, 6) Reduction in ray size in the ray florets in *Bidens pilosa* of Compositae(reported by the the author).

Spontaneous mutations in Animals : 1) Polydactyly in man, sheep, cats etc. 2) Multinippled condition in sheep, 3) Silky fowls, 4) Albinism in man, pig, pigeon, rodents etc. 5) Double cared cattle, 6) Mule footed swine, 7) White canaries, 8) Hornless cattle, 9) Hares' lip in man, 10) Widows peak in man, II) Lobster claws in man, 12) Cleft palate in man etc.

Spontaneous mutations are generally rare and may be of two types dominant and recessive. Usually recessive mutations are more frequent than dominant mutations.

Factors affecting spontaneous mutations : The rate of spontaneous mutations are known to change under the influence of various external and internal factors. One of the external factors which might bring about spontaneous mutations is a change in the temperature range (normally a higher temperature makes the genes susceptible for change). Disturbed metabolic rates may also bring about mutational changes. Age and sex may also have a bearing on the mutational rate. Increased age may render the genes prone for mutational change. *In Drosophila,* it has been observed that males are more susceptible to mutational alterations than the female.

One of the important factors bringing about spontaneous mutations may be, specific mutation inducing and mutation succeptible genes. As has already been pointed out, these are called mutator and mutable genes.

The mutator genes have been discovered by Me Clintock in maize. She described some genes which had the capacity to influence other genes to mutate. Rhodes has described the role of a mutator gene which influences the functioning of a colour gene. In maize the colour is governed by three recessive genes *a, c,* and *r,* while the dominant alleles of these govern the colour other than green. In addition to these there is a mutator gene)((dotted colour) which controls the variegated colour in the grain kernel. This gene influences the gene *a* to mutate so that red patches are produced. Usually the plants with the genotype *aa cc RR* will be green (all the three dominant gene pairs are required to produce colour other than green), but the presence of *Dt* makes it variegated.

Mutable genes other than the one mentioned above (a) have been reported in maize by Emerson. He investigated a variety of maize withwhite pericarp having several red spots of varying sizes; the homozygous recessive allele producing the white colour. According to Emerson, these red patches are produced by the unstable white gene undergoing spontaneous mutation. Each red spot in the endosperm indicates an area where a mutation has occurred. Jones further continued the studies of Emerson and found that there are six red spots, each having 10 - 45 coloured cells. These coloured cells were exactly identical with those produced by the pure red gene. Jones was conducting an experiment with a homozygous plant and he beleived that the six red spots indicated that the gene for white had mutated six times producing a red cell every time. These cells further multiplied and produced the red patches. The gene for white endosperm colour is thus a typical example of a mutable gene.

Several such mutable genes have been reported *in Drosophila,* sweeet pea, 4 o' clock plant, snapdragons etc. The genes for miniature wings, gene for purple eye colour in *Drosophila* are known to be mutable genes. In maize plant itself, Stadler has observed that the gene R for aleuron colour mutated some 492 times in about a million gametes that he studied.

INDUCED MUTATIONS

Various external agencies are known to bring about mutational changes in genes. All these are included under the general category of induced mutations. H.J. Muller (1927) may be regarded as pioneer in radiation genetics. In his studies on *Drosophila,* he showed that flies subjected to X radiation undergo mutations. Similarly Stadler (1928) demonstraded the mutational effects of X rays in barley and maize. Various kinds of agents causing mutational changes in organisms are known. These are of the following types, a) Radiation b) Temperature changes and c) Chemicals.

Radiation induced mutations : Various types of radiation (excluding visible light) are known to cause mutations in different intensity. Radiations are of two categories depending upon their capacity of penetration into a substance -*Ionizing* and *non ionizing.* Generally speaking, non ionizing radiations are not of a high potentiality to bring about mutations at a rate which is brought about by ionizing radiation. The Alpha rays, Beta rays and the gamma rays and X rays come under the category of ionizing radiations while low penetration rays lik U.V. rays belong to non ionizing radiation.

Mayor (1920) experimented on the effects of X radiation *in Drosophila* (and similar other radiations) to study the effects on germplasm. He demonstrated that as a result of radiation, there was non disjunction of the X chromosome resulting in a XXY individual, which was female phenotypically.

H.J. Muller also worked on the effects of X rays on *Drosophila* and published a classical paper 'The artificial transmutation of the gene'. He also devised a technique to detect induced mutations (see later in the chapter, for detailed explanation).

Stadler studied the effects of X rays on barley plant, while Gager and Blackslee induced mutants in the jimson weed *Datura stramonium.* It should be noted however that X rays are lethal and the dosage of the rays given to the individual has to be carefully monitored. On an average it has been found out that 80% of the X ray induced mutations are lethal.

Oliver, Tirnofeeff Ressovsky and many others have worked out the relationship between the frequency of mutation and the amount of X radiation absorbed by the organism. The intensity of the radiation is measured in terms of Roentgen units or *r*- units. The frequency of mutation seems to be proportional to the amount of *r*- units absorbed by the plant. The duration of exposure or the distance between the radiation source and the experimental organism, however do not seem to affect the frequency of mutations. The following table gives the rates of mutation in *Drosophila melanogaster* due to two different doses of X rays. In the table below t4 dose is twice as large as the t2 dose. (The data in the table below are based on Muller's experiment, quoted from Principles of Genetics, Sinnott Dunn and Dobzansky).

Experiment	Number of chromosomes observed	Number of mutations		
		visible	semilethal	lethal
Control	198	0	0	0
Xrays t_2	676	1	44	9
Xrays t_4	772	3	12	89

Some of the mutations on exposure to X radiation in *Drosophila* are as follows.

Gene wt to W (white), W^e (eosin), and W^a (apricot). Genes coral (W^{co}) buff (W^b), cherry (W^c) and apricot W^a) can mutate into white (W).

Chlorophyll deficiency has been induced in barley due to X radiation (Stadler). X rays can also induce mutations in the wasp *Habrobracon.* Among the other types of ionizing radiation Alpha, Beta, and Gamma rays are important. The First two however do not seem to have a profound mutational role as they cannot penetrate the body cells as deeply as gamma rays.

Mechanism of action of ionizing radiation : There are two theories which account for the mechanism of mutagenic effect of ionizing radiation. These are the **direct hit theory** and the **indirect hit theory**. According to the direct hit theory, the radiation directly strikes the genes bringing about a change in the chemical composition. The electrons that. strike the atoms of genes bring about a change in the chemical composotion, thus altering the gene. But in several experiments, it has been seen that , the same dose of radiation does not result in the same frequency of mutation indicating that direct hit theory cannot account for all radiation induced mutations.

According to the indirect hit theory, gene alteration is brought about by changing the molecular environment of the surroundings. Muller is a supporter of this theory. According to him when the high speed protons of X ray collide with the molecules of the cell, electrons are ejected out of the atoms rendering them positively charged. The electrons that come out move at high speeds and knock out other electrons that come in their way resulting in an environment which is ionized (positively charged). The electrons moving freely may attach themselves to other atoms rendering them negatively charged. Various chemical reactions and shifts take place to bring both the positive and negatively charged atoms to their normal state. During this process, alteration in the chemical composition takes place, thus mutating the gene. At the molecular level, base pair sequences are altered in the DNA molecule (for more details see mutation at the molecular level, in the same chapter).

Effect of non ionizing radiation : Ultra violet rays are non ionizing because they have a long wave length and therefore have less energy. They are generally non penetrative. They may bring about many changes in the traits, but are not as effective as X rays. Altenberg was able to accelerate the rates of mutations by using U.V. rays. Stadler also achieved pollen mutations induced by U.V. rays in maize; the same is the case in *Antirrhinum* as reported by Noethling and Stubbs. The U.V. rays are known to cause alterations in the bond characters of purines and pyrimidines. Between the two, the latter are more prone to mutational changes.

Chemical mutagens : A number of chemicals are known to have mutagenic properties. It was C.Auerbach who discovered during second world war that chemicals could be used to induce mutations. Thorn and Steinberg (1939) had also reported the chemical induction of mutations. They treated the Fungus *Aspergillus* with nitrous acid. Since then a long list of mutangenic chemicals have been identified. Some of the common chemicals used as mutagens are given in the following table.

Table : Some mutagenic chemicals

Class of chemicals	Name of individual compound	Common Abbreviation
1 Aziridine	1 Ethyleneimine	EI
2 Mustards	2 Nitrogen mustard	
	3 Sulphur mustard	
3 Nitrosamines	4 Dimethylnitrosamine	DMN
	5 Nitrosoguanidine	NG
	6 Nitrosomethyl urea	NMU
4 Epoxides	7 Ethylene oxide	Eo
	8 Diepoxybutane	DEB
5 Alkyl sulphonates	9 Diethyl sulphate	dES
	10 Methyl methane sulphonate	MMS
	11 Ethyl methanal sulphonate	EMS
6 Others	12 Nitrous acid	
	13 Colchicine	
	14 Maleic hydrazide	
	15 Hydrazine	
	16 Hydroxylamine	
	17 Caffeine	
	18 Hydrogen peroxide	
	19 Formaldehyde	
	20 Phenol	
	21 Urithane	

Hydrogen peroxide is known to bring about mutations in *Neurospora*. Urithane and phenol are capable of inducing mutations in higher plants and *Drosophila*. Formaldehyde may induce mutations during spermatogenesis in *Drosophila*.

Colchicine is an alkaloid extracted from the bulbs of *Colchicum autmnale* of the family Liliaceae. This is known to induce polyploidy by suppressing spindle formation. Colchicine is not known to bring about any point mutation. Mustard gas is another chemical widely employed as a mutagen. Its mutagenic effect has been tested with reference to *Drosophila, Neurospora*, bacteria, barley, mice etc.

Temperature induced, mutations : Child, Plough and Ires have studied the effect of temperature as a mutagenic agent in *Drosophila*. A range of temperature from 4°C to lethal range was used to find out the possible effect of temperature on the rate of mutation. A higher range of temperature is known to accelerate the rate of mutation. But it has to be borne in mind that a gradual increase in temperature might not be an effective mutagen. Any sudden increase or decrease in temperature might alter the gene copying mechanism thus altering the gene. The temperature effects largely remain as a somatic influence.

Vegetable oils as mutagenes : In one of their experiments concerning oil yielding plants, Swaminathan and Natarajan (1959) noticed that seeds with high oil content are generally resistant to mutagens. They soaked the cereal grains in oils of ground nut, mustard, and castor and wanted to know whether these oils provide any mutation resistance to the grains. Much to their surprise, the results were startling. Instead of the oils providing resistance to the cereals, they (oils) themselves had acted as mutagens as noticed by marked reduction in rates of germination, chromosome irregularities and many ear head variation mutations in wheat, one of the cereals used in the experiment. It was assessed that in wheat *(Triticum astivum)*, the rate of mutation, induced was much higher than that of X rays.

Detection of mutations : A variety of methods are available for detecting mutations, both in plants and animals. Some of these are described below.

CLB method for detecting mutations : This method perfected by Muller has been used to detect mutations in X chromosomes of *Drosophila*. Even though it is primarily meant to detect radiation induced mutations, it can be used in the detection of spontaneous mutations as well. Three characters on the X chromosome are chosen to study the mutation. These are -

C = Cross over suppressor. This represents a long inversion in the X chromosome, suppressing the cross over in the inverted portion. Cross over may still occur, but cross over gametes rarely survive due to deficiences and duplications.

L = Lethal gene. This is a recessive character. Hence homozygous females (l/l) and hemizygous (l/y) males do not survive.

B = Bar eye. This is a dominant gene. This is used as a phenotypic marker for the flies required to mate in the F_1 generation. The working mechanism of this is illustrated in the Fig. A CLB male is mated to a normal male which is subject to radiation. The idea is to find out whether there are any lethal mutations in the X chromosome of the male. Y chromosome is regarded as neutral as for as sex linked characters are concerned, hence the male will be hemizygous as for as genes on X chromosomes are concerned. Therefore any X chromosome of male which is irradiated can be followed in subsequent generations and lethal mutations if any can be detected.

In the X_1 generation (F_1 generation; called X_1 because the individuals are X radiated) the irradiated X chromosome of the male goes to the daughters, who have another X from the mother; hence in this generation any mutation induced can not be detected. The X_1 progeny will be as follows.

Females : 50% CLB; 50% Normal

Males : 50% CLB (die); 50% Normal.

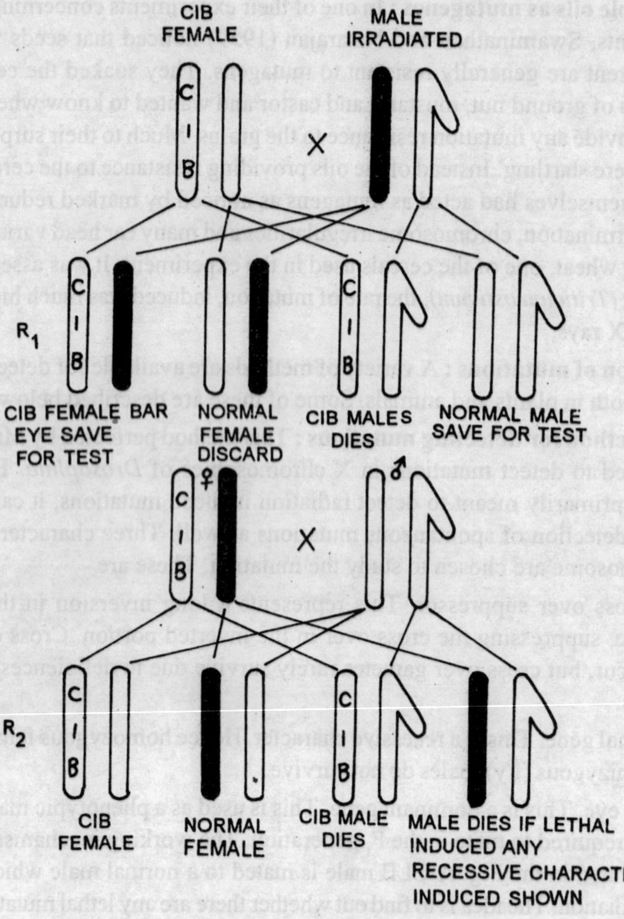

Fig. 10.2 Gene Mutation

Muller's CIB method to detect induced mutations drosophila melanogaster

A cross is now made between the normal male (with X also normal because it is obtained from female) and the CLB female. This CLB female has one of its X chromosome from the male parent which is irradiated and the other is CLB. In this (X_2 or F_2) generation all the males get one of the X chromosome, they die (50%) because there is already a lethal gene. If they get the irradiated X chromosome, and if a new lethal mutation is induced then the other 50% also will die. Hence if there is a lethal mutation on the X chromosome in the F_2, no male progeny will survive and all will be females. This is a clear cut evidence for the presence of a lethal mutation on the irradiated X chromosome. In the X_2 generation the progeny will be as follows.

Females : 50% CLB; 50% Normal (with an irradiated X chromosome)

Males : 50% CLB (die); 50% carry irradiated X chromosome; die if there is a lethal mutation.

The Muller 5 method : .Like the CLB method this is also devised by Muller. The experiment involves crossing a homozygous Muller 5 female *Drosophila* with an irradiated male. The Muller 5 female *Drosophila* has two marker genes - apricot eye (w^a/w^a) and Bar eye (B/B) in a long inversion so that there is no cross over. A homozygous apricot (eyed) Bar eyed female is crossed with an irradiated male. As shown in the figure, the F_1 generation female offspring which gets the irradiated X from the male parent is crossed with a wild male. If a lethal mutation has occurred in the X chromosome, no wild types of male would survive, because they would get the irradiated X.

The absence of wild males in the F_2 is an indication of lethal mutation. Muller was awared Noble prize in Medicine in 1946 for this and many other discoveries that he had made in Genetics.

Detection ofAutosomal mutations (Balanced lethals)

Autosomal mutations can be detected by making use of balanced lethal genes in organisms. A balanced lethal stock is one which carries a lethal gene in one of the homologues and its wild allele on the other homologue. Homozygous individuals will die, so what survives is always heterozygous. Balanced lethals can be studied by using a specific strain of *Drosophila melanogaster* called the **curly lobe plum.**

In this, a stock of flies carry on one of the autosomes two dominant genes *C* (curly wing) and *L* (lobed eye) and a recessive lethal gene. An individual with two *CYL* chromosomes will not survive because the recessive lethals will be in double dose and its homologue carried the *Pm* (plum coloured eye) gene which is also dominant.

When two individuals with the genotype *{(CYL/Pm)* X *(CyllpM)}* are crossed, the heterozygous individuals only would survive, as homozygous for *(CYL/CYL)* would carry lethal genes in double dose, because *CYL* chromosome carries a lethal gene. There is a cross over suppressor to prevent the transfer of the gene *CY to* the chromsome carrying *Pm* gene.

A female *CYL/Pm* is crossed to another normal male which is irradiated.

Four types of progeny are produced in the F_1. These are -

(a) Curly lobe (with an irradiated autosome A)

(b) Curly lobe (with another irradiated autosome B)
(c) Plum (with an irradiated autosome A)
(d) Plum (with another irradiated autosome B)

One of these flies is selected (because each one has one irradiated chromsome) and back crossed to curly lobe plum.

The progenies will be of the following types.

(a) homozygous curly lobe (dies)
(b) Curly lobe plum.
(c) Curly lobe with an irradiated chromsome.
(d) Plum with an irradiated chromsome.

Among the above progeny, (c) flies are crossed among themselves to obtain -

(a) homozygous curly lobe (dies)
(b) Heterzygous curly lobe (with an irradiated chromosome)
(c) same as b.
(d) A fly with both irradiated chromosomes.

It is in the (d) category that any mutation induced would appear. If the induced mutation is lethal, it also dies and only the curly lobed individuals survive. If the mutation induced is not lethal, but a recessive trait, that would be expressed, since it is homozygous. The same test can be done by selecting the plum with an irradiated chromosome and crossing it with a curly lobe plum with an irradiated chromosome.

The significant aspect of this cross is to bring together the same irradiated chromosome in the homozygous condition to detect all the recessive mutations.

Detection of induced mutations in plants

L.J. Stadler of the University of Missouri and Ralph Singleton of the University of Virginia have worked on the methods to detect mutations in maize and barley.

Stadler (1942) has made a detailed study of spontaneous mutations in maize. The following are the important steps.

(a) A genetic stock dominant for a number of genes is grown as the female parent. The tassels are removed before pollen discharge (to avoid self pollination).
(b) A multiple recessive plant in every fifth row is used as the pollen providing parent for (a) plants.
(c) Seeds were observed in large numbers for endosperm characters. Most of them show dominant characters. If any recessive character (of the male parent) appeared it is due to mutation in the female gametes.

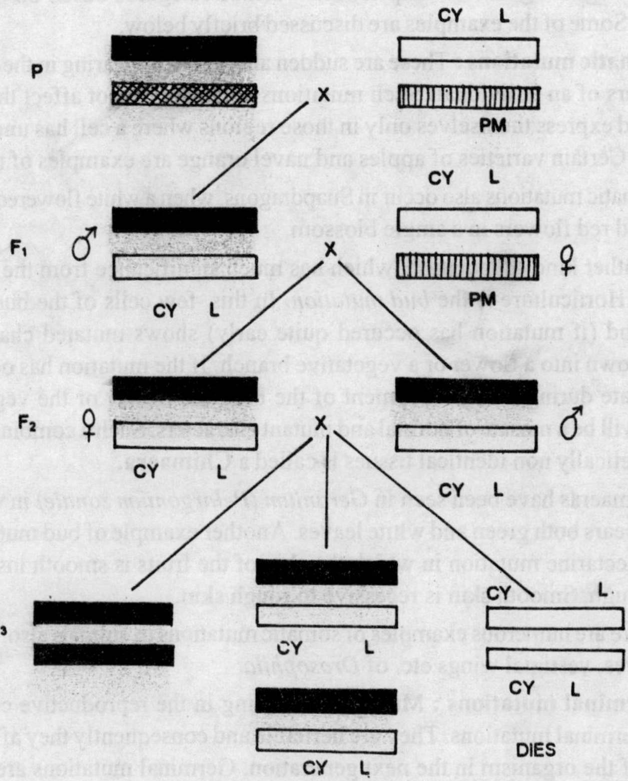

Fig. 10.3 Gene Mutation

Detection of lethal mutations in a balanced lethal system

Singleton (1962) used a techinque similar to that of Stadler do detect both spontaneous and induced mutations in the pollen of maize. His method involves collecting pollen from plants dominant for many genes and cultivated in a field irradiated with cobalt 60 and transferring them (pollen) on to the silks of plants with all recessive traits and cultivated in a normal field. The seeds of the progeny would normally show dominant characters. If any mutation has been induced in a dominant gene, the recessive trait would appear in the progeny. In this way Singleton studied four genes *Su -su,* on chromosome 4, *Pr - pr* on chromosome 5, *Y-y* on chromsome 6 and *Sh -sh* on chromosome 9. The phenotypes of these genes are *Su* - normal endosperm, *su* - sugary endosperm; .Pr- Purple aleurone,pr-red aleurone; *Y*- yellow endosperm, *y* - white endosperm and *Sh* - full endosperm, *sh* - shrunken endosperm..

Apart from spontaneous and induced, mutations may also be classifed (see the beginning of this chapter) into various categores based on different criteria. Some of the examples are discussed briefly below.

Somatic mutations : These are sudden alterations appearing in the somatic characters of an individual. Such mutations normally do not affect the entire plant and express themselves only in those regions where a cell has undergone change. Certain varieties of apples and navel orange are examples of this.

Somatic mutations also occur in Snapdragons, when a white flowered variety produced red flowers in a single blossom.

Another kind of mutation, which has much significance from the poin of view of Horticulture is the *bud mutation.* In this, few cells of the bud or the entire bud (if mutation has occured quite early) shows mutated characters, when grown into a flower or a vegetative branch. If the mutation has occurred a little late during the development of the bud, the flower or the vegetative branch will be a mosaic of normal and mutant characters. Such a combination of two genetically non identical tissues is called a **Chimaera.**

Chimaeras have been seen in *Geranium (Pelargonium zonale)* in which a branch bears both green and white leaves. Another example of bud mutation is that of nectarine mutation in which the skin of the fruits is smooth instead of being rough. Smooth skin is recessive to rough skin.

There are numerous examples of somatic mutations in animals also; for e.g. white eyes, vestigial wings etc. of *Drosophila.*

Germinal mutations : Mutations occuring in the reproductive cells are called germinal mutations. They are heritable and consequently they affect the whole of the organism in the next generation. Germinal mutations are called gametic when they occur in gametes and zygotic when they occur in the zygote. The gametic mutation will express itself readily when either the mutation is dominant or if recessive two mutant gametes should unite. Zygote mutations on the other hand express themselves readily. The ancon breed sheep is a typical example of germinal mutation.

Biochemical mutations : Mutations bringing about alterations in metabolic pathways, nutritional requirements etc. are called biochemical mutations. These are mostly noticed in microorganisms. Biochemical mutations in *Escherichia coli* can be induced by repeated sub culturing. In *E.coli,* the normal strains are called *Paratrophs* while the mutants are called *Auxotrophs*. These mutations are for some of the nutritional requirements like Biotene, Methionine, Leucine etc. The paratroph is called genetically B+, M+ and L^+ because it has the capacity to synthesize all these. On mutation the auxotroph loses the capacity to synthesize these and consequently it is known as $B^- M^- L^-$

Several metabolic disorders due to biochemical mutations have been reported in man also. They are commonly called **Inborn errors in metabolism'**. The first such biochemical disorder to be reported was Alkaptoneuria (Garrod 1902). This relates to a metabolic disorder concerning the amino acid phenyl alanine. **Alkaptaneuric** individuals have a mutated gene which does not produce the .enzyme homogentisate oxidase, resulting in the accumulation of homogentisic acid in the cell, which is excreted along with the urine. Homogenticic acid is a strong reducing agent and gets oxidised in the presence of air into a black pigment, resulting in the black colouration of the urine.

Another biochemical mutation in man is phenyle ketoneuria. It is caused by a recessive mutant gene in which due to failure of conversion of phenylalanine to tyrosine, the former accumulates in the blood and excreted through the urine. This leads to mental retardation. In addition to the above, Tyrosinosis, Albinism and Cretinism are the other metabolic disorders due to biochemical mutations.

Reverse mutations : The back mutation of a gene to come back to its wild state is known as reverse mutation. In *Drosophila,* the wings are normally long and due to a mutation, the wing becomes miniature. In some instances, all of a sudden, a few long winged flies appear in a population of miniature winged flies. This is due to the reverse mutation of the miniature winged gene back to the wild state. Another reverse mutation in *Drosophila* is the mutant, fork bristles back to normal bristles. There are many such reverse mutations in bacteria and fungi.

Lethal mutations : Mutations of gene, from normal to lethal leads to the death of the individual. Normally lethal genes are recessive and nothing happens to the organism in the heterozygous condition. But occassionally there may be dominant lethals in which case individuals do not survive even in the heterozygous condition. The latinum colour in Fox, yellow coat colour (see chapter on Interaction of factors for a detailed discussion) in mice are such examples.

Many lethal mutations are also known in plants *likeAntirrhinum majus* (Snapdragon). The golden plant stock has variegated leaves. On self pollination this variety produces Golden and green in the ratio 2:1 (not 3:1), obviously the golden is heterozygous and green breeds true. The other one (golden) is missing because pure golden can not exist due to homozygosity of lethal genes. Lethal genes are also known in paddy and maize plants.

MUTATIONS AT THE MOLECULAR LEVEL

The basic hereditary material namely DNA is a highly stable biological molecule. The hereditary information is present in the form of sequence of the four nucleotides Adenine, Thymine, Guanine and Cytosine. During the inheritance of characters from one individual to the other, the sequence is duplicated and transmitted in a precise and accurate manner.

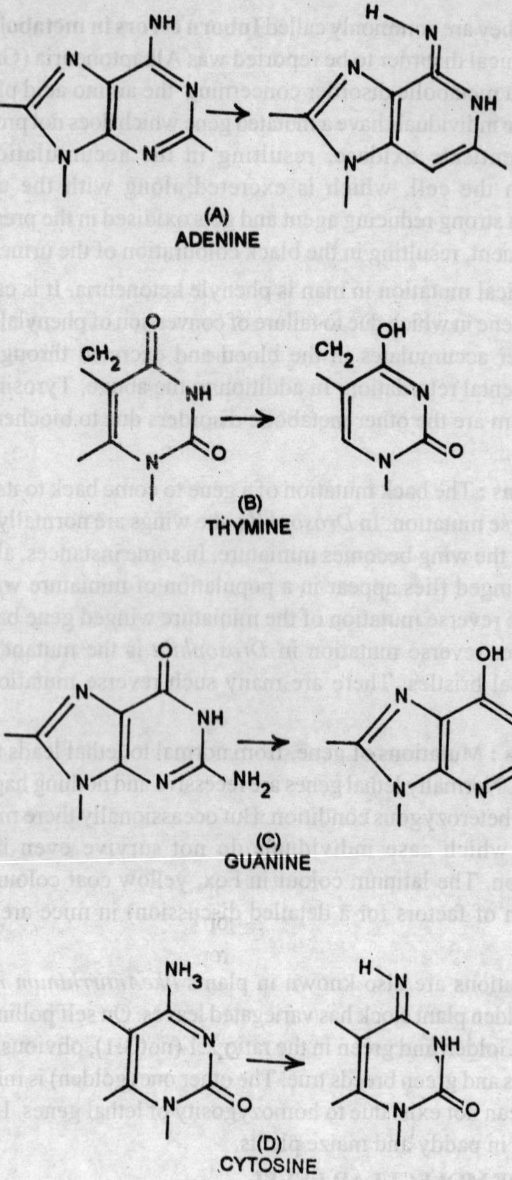

Fig. 10.4 Gene Mutation
Uncommon forms of DNA bases

A slightest mistake at this stage will alter the base pair sequence and results in an entirely different type of genetic code resulting in the formation of an altered protein which may have far reaching effects on the organism. Such mistakes or faults occuring in the DNA molecule during duplication are ultimately the basis of mutation at the molecular level. The following two basic types of sub nucleotide changes are noticed in the DNA molecule - Frame shift mutation and substitution mutation.

Frame shift mutations : Addition or deletions of nitrogen bases either from DNA or mRNA are called frame shifts, because, from the site of the alteration, the reading frame of the message alters. In some cases frame shifts are neutralized ie. deletion may be neutralized by the addition at the same site. Frame shift mutations are of two categories. These are deletion mutations and insertion mutations. In deletion a single nucleotide in a triplet in deleted. These are found frequently in T_4 bacterophages. In insertion mutations an extra nucleotide is added to a gene. These insertions can be induced by chemicals such as acirdine, Proflavin etc.

Substitution mutations : This generally occurs during the replication of DNA, when one nitrogenous base is replaced by another changing the triplet codon. The altered codon now codes for a different amino acid resulting in an altered protein. There are two kinds of substitutions - Transitions and Transversions.

Transitions : The replacement of one purine by another purine or one pyrimidine by another pyrimidine is called trasition. In other words A is replaced by G and T by C or *vice versa*.

Transitions may be brought about by tautomerization, ionization, base analogs or by deamination. In tautomerization, a normal base, pairs with its tautomer partner [A tautomer is an alternate state of existence in a molecule. For e.g. In cytosine and adenine tautomers, the normal amino group NH_2 is converted into the NH (imino) group and the keto group (C = 0) of thymine and Guanine is converted into the OH group in their tautomers. Normally this does not happen]. Such pairing is called *forbidden base pair.* Watson and Crick (1953) opined that the existence of a tautomer to DNA provides a mechanism for mutation.

Ionization is another mechanism by which transition can be brought about. Ionization of a base at the time of DNA replication involves the loss of H from a nitrogen base. Such ionized bases pair with normal bases.

Another method by which transition may be brought about is by the replacement of normal bases by their analogs. These compounds called *base analogs* are derivatives of the natural bases. For eg. cytosine may be replaced by *5 hydroxyl methyl cytosine* or by *5 glucosyl hydroxy methyl cytosine.*

There are naturally occuring base analogs of cytosine found in T even phages of *E.coli. Uracil* has a naturally occuring analog - 5 *hydroxy methly uracil* found in some Viruses. Besides the naturally occuring analogs, artificial analogs like 5 *bromouracil, 5 iodouracil, 5 bromocytosine, 5 methyl cytosine* may be used as analogs of thymine (first two) and cytosine (last two) which may be used to induce mutations.

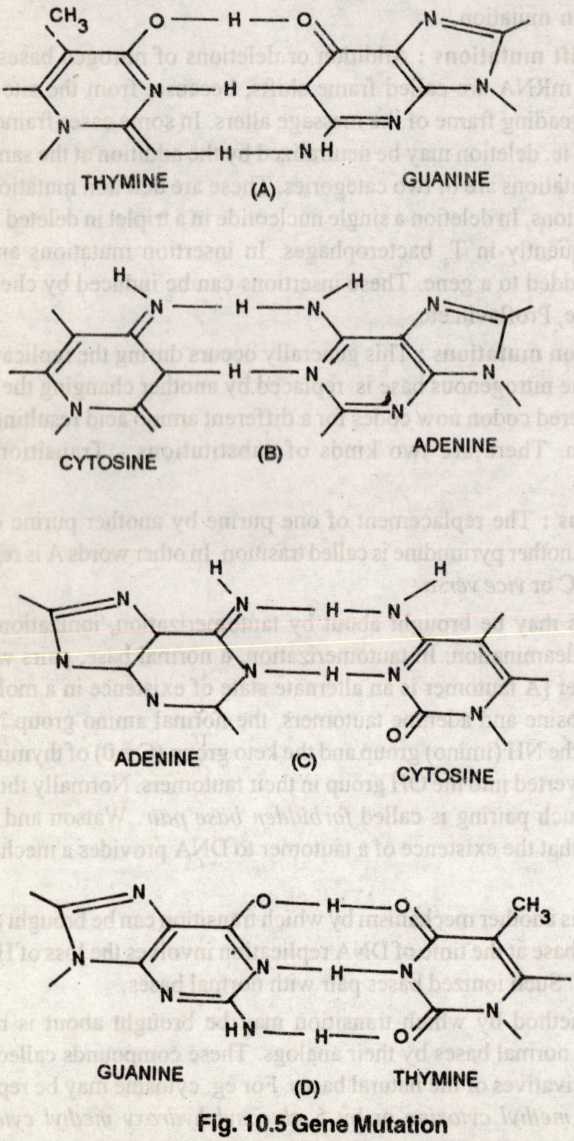

Fig. 10.5 Gene Mutation
Unusual pairing of tautomeic bases in DNA

Deamination is another method of induction of transition, by replacing the NH_2 group of bases with OH group. This may be induced by chemials such as nitrous acid, hydroxylamine, diethyl sulphate etc. Deamination of cytosine leads to the formation of uracil, and that of adenine forms hypoxanthine. During replication U pairs with A and hypoxanthine pairs with C. As a result A = T is substituted by G = C and G=C byA=T.

Trasversions : Unlike m transition, a purine is replaced by a pyrimidine and *vice versa* ie. CG replaced by GC and AT by TA. This change can be induced by alkylating agents like ethylmethane sulphate (EMS) and Methylmethane sulphate (MMS). The alkylating agents alkylate the nitrogen in the 7th position in a purine and lead to its separation from the DNA strand. This gap can be filled by any of the four bases at the time of replication. Trasversions were first postulated by E.Freese in 1959.

PRACTICAL APPLICATION OF MUTATIONS

Generally mutations are lethal and recessive and possibly of no help except as a scientific curiosity in understanding the basic funcitoning of the hereditary material. But there are specific instances where both spontaneous and induced mutations are of selective value from the point of view of Evolution and also from the point of view of human beings. Ancon breed sheep is one such example of a spontaneous mutation. Mutations are also of value in plant breeding, when mutant varieties can be stabilized in the gene pool by cross breeding. Gustaffson has estimated that about one in thousand mutations may be used in breeding experiments. Several mutations in wheat, like branched ears, resistance to lodging, high lysine content, amber seed colour, awned spikelets have been useful inbreeding. The amber seed colour which is originally seen in the mexican variety has been induced in wheat of Indian varieties.

Many useful mutations in rice are also known. Some of these are *Reirnei* variety mutant, for reduced duration of the crop. In barley also mutants like *erectoids* and *eceriferum* have been induced for increased yield. The *Todds Mitchan* variety of mint plant *and-Aruna* variety of castor serve as good examples of the role of mutations in crop improvement. The above mentioned variety of mint has disease resistance as well as high oil content. In the castor variety duration of the crop was reduced by 130 days (originally 270 days). In a survey conducted by FAO in 1974 it has been estimated that 98 varieties of crops have been improved by induced mutations.

11

CELL DIVISION

The unique and fundamental property of protoplasm is its power to grow and multiply by cell division. Thus new cells arise only by the division of preexisting cells in plants and animals. In unicellular organisms like bacteria, some algae, and fungi and protozoans cell division is the fundamental means of asexual reproduction. In multicellular organisms division of the single celled zygote contributes to the growth of the body. The gametes of the sexually reproducing organisms develop by means of cell division. Three types of cell division occur in living organisms. These are

1. Amitosis or direct cell division
2. Mitosis or indirect cell division
3. Meiosis or reduction division

Amitosis : This is found mainly is prokaryotic organisms like cyanobacteria, bacteria, yeasts etc., (among plants) Protozoa among animals are also known to divide by amitosis. Mostly found in unicellular organisms, amitosis is a means of reproduction. During amitosis, the nucleus elongates and a constriction appears in the centre resulting in a dumb bell shaped structure. The constriction deepens and the nucleus gets divided into two bits. There is no spindle formation and no formation of chromosomes. The nuclear division is followed by cytoplasmic division and the cell gets divided into two daughter cells. Thus amitosis is only a quantitative division.

MITOSIS

Occuring in somatic cells of multicellular organisms mitosis is the chief means of multiplication of cells contributing to the increase in size of the individual. It is the chief means of embryonic development into the adult. Thereafter the capacity for division is restricted to the cells at the tip of the stem, root tips and the cells of the cambium in flowering plants. In animals it is limited to the replacement of worn out cells like blood cells, skin, intestinal epithelium, and the germinal epithelium of the gonads. Mitotic division also occurs during the repairing of wounds in plants and animals.

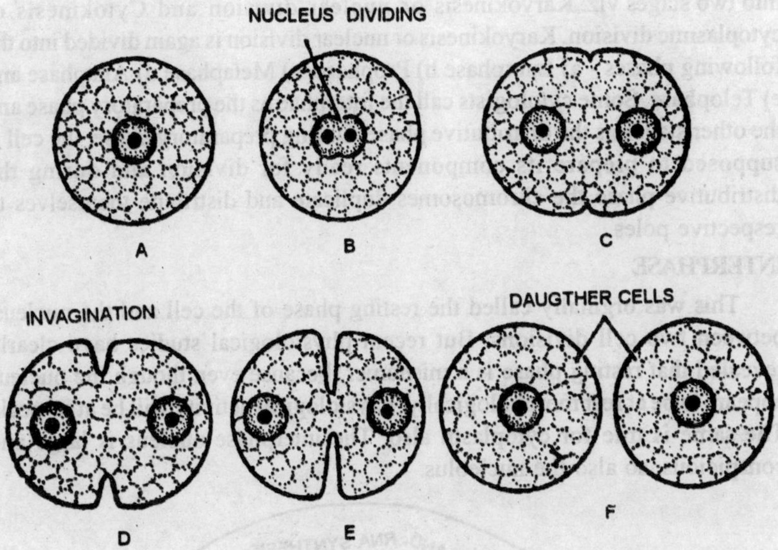

Fig. 11.1 Cell Division
Stages in Amitosis

Definition : The division of the somatic cells is generally called mitosis (Gr *mitos* = thread). It is a complex process involving series of changes leading to the formation of two equal daughter cells from a single parental cell. The term mitosis was first proposed by W.FIemming (1882). It is also called somatic cell division.

Cause of cell division : Why do cells divide? It is very difficult to obtain an answer to this question, although Kinetin a hormone has been isolated which acclerates cell division even in extremely small quantities. The possible reasons are - a) To maintain a relative volume of nucleoplasm and cytoplasm, b) To maintain the nuclear surface to nuclear volume.

Initiation of cell division : Chromosomes may release certain substances which initiates the nucleaı changes and cause cytoplasmic division. Cell division may also be resulted by a periodically released hormone called cell division hormone. According to Heilbrunn and Wilson (1948), changing viscosity of the cell may initiate division, the viscosity being lowest at the beginning and gradually increasing and once again becoming low. While it is true that viscosity changes occur during division whether these changes are cause or effect is still not known.

Stages of Mitosis : For the convenience of study, mitosis maybe divided into two stages viz. Karyokinesis or nuclear division and Cytokinesis or cytoplasmic division. Karyokinesis or nuclear division is again divided into the following phases - a) Interphase b) Prophase, c) Metaphase d) Anaphase and e) Telophase. Some cytologists call the Interphae as the preparatory phase and the other stages as the distributive phase. During preparatory phase, the cell is supposed to prepare its components ready for division and during the distributive phase the chromosomes duplicate and distribute themselves to respective poles.

INTERPHASE

This was orginally called the resting phase of the cell and the nucleus, between two cell divisions. But recent physiological studies have clearly revealed that resting phase is a misnomer, because even though the nucleus appears to be at rest morphologically, physiological activity will be at its peak. The same is true for cytoplasm also. The interphase nucleus is large and conspicious, so also the nucleolus.

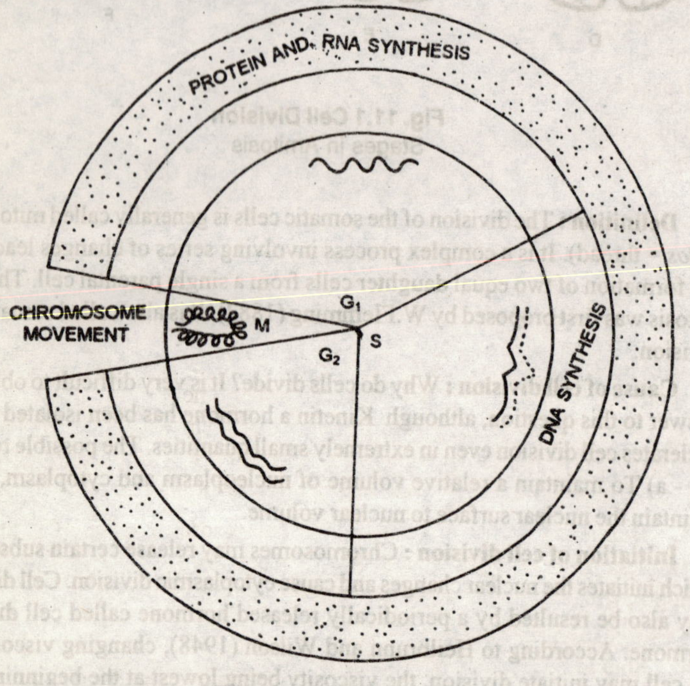

Fig.11.2 Cell Division
Stages in the inter phase (between mitotic circle)

Three important stages are recognised in the interphase. These are G_1, phase, S phase and the G_2 phase.

The G_1 phase also known as the post mitotic gap phase occurs at the end of every cell division (at the beginning of subsequent division). During this period the daughter cells grow in size. Proteins and RNA are synthesised during this period, but there is no synthesis of DNA.

The S phase, also called the synthetic phase is the most active stage from the point of view of DNA synthesis. It is in this phase that new DNA is formed by replication. Cell division will not proceed if S phase is inhibited.

In the G_2 phase also called the premitotic gap phase, RNA and protein synthesis continues. The volume of the nucleus increases with the increased synthesis of various types of RNA. DNA synthesis however stops in this phase.

In the division cycle G_1 phase is the longest with G_2 and S phase lasting equal amount of time.

Some cytologists use the term M phase or Mitotic phase for the entire distributive phase involving prophase, metaphase, anaphase and telophase and the following cytokinesis.

PROPHASE (Gr. *pro* = before, *phase* = appearance).

This is the first stage of nuclear division. This is the longest stage in the mitotic phase lasting from one to several hours. In the case of onion root tip cells, it lasts for 71 minutes, while in grasshopper neuroblast-102 minutes and in the cells of pea endosperm about 40 minutes. The following are the important features and events taking place in prophase.

1. The cell becomes spheroid (in animals) with increase in its viscosity and refractivity.

2. **Differentiation of chromosomes :** Chromonemata start condensing and as aresult a long coiled net work called spireme is formed. Continued coiling of the spireme produces thick threads called chromosomes. By the end of prophase the chromosomes contract up to about 1/25 of their length. Specific number of chromosomes are formed (specific to each species).

3. **Duplication of chromosomes :** Each chromosome splits longitudinally into two identical chromatids connected by a centromere. Two types of coils develop in the chromatids namely somatic coils and minor coils. While the former are larger, the latter are smaller.

4. **Shortening of the chromosomes :** As the prophase progresses, the somatic coils increase in size but decrease in number. As a result of this,the chromosomes become thick and short.

5. Coiling of the chromatids : The two chromatids are coiled round each other and cannot be easily separated. The coiling is called plectonemic coiling unlike the paranemic coiling of meiosis.

6. Condensation of chromatids : The nuclear matrix begins to accumulate on the chromatids and the chromosomes assume a hazy outline due to irregular accumulation of nuclear matrix

7. Production of achromatic figure : The centrioles which have undergone duplication during interphase and situated above the nucleus start moving in opposite directions. Spindle fibres appear in between the two centrioles. At the same time, radiating fibres called astral rays arise from each centriole forming the aster. The spindle fibres between the two centrioles constitute the amphiaster. The amphiaster is achromatic and does not take stain. Taylor has shown that in Newts, the centrioles move at the rate of one micron per second. When the centriole is present and particpates in cell division it is said to be centric. In acentric mitosis centrioles are absent. Cell division in which aster is formed from the centroile is called astral mitosis (animals and lower plants), when aster is absent it is said to be anastral mitosis (higher plants)

8. Dissolution ofnucleolus : As the coiling continues, the pars amorpha region of the nucleolus decreases in size and finally disappers. But the nucleolonema expands and gets associated with the chromosomes. There is also increase in the RNA content of the chromosomes

9. Dissolution of nuclear envelope : Towards the end of prophase, the nuclear membrane breaks down and gets dispersed into cytoplasm to be absorbed by it. The chromosomes, now free of the nuclear envelope get collected inside the spindle fibres.

PROMETAPHASE

Some cytologists (Threadgold 1968) regard this to be the last stage of prophase, when nuclear envelope disintegrates.According to White (1963), prometaphase is the stage when the spindle is formed and the chromosomes try to reach the equatorial region of the spindle fibres.

METAPHASE

In the cell cycle the metaphase lasts only for a short duration. The following are the main events during metaphase

1. In early metaphase the highly shortened and condensed chromosomes show characteristic radial arrangement on the equatorial plane of the spindle. This pattern of orientation is known as the metaphasic plate or the equatorial plate. It is formed at right angles to the spindle axis.

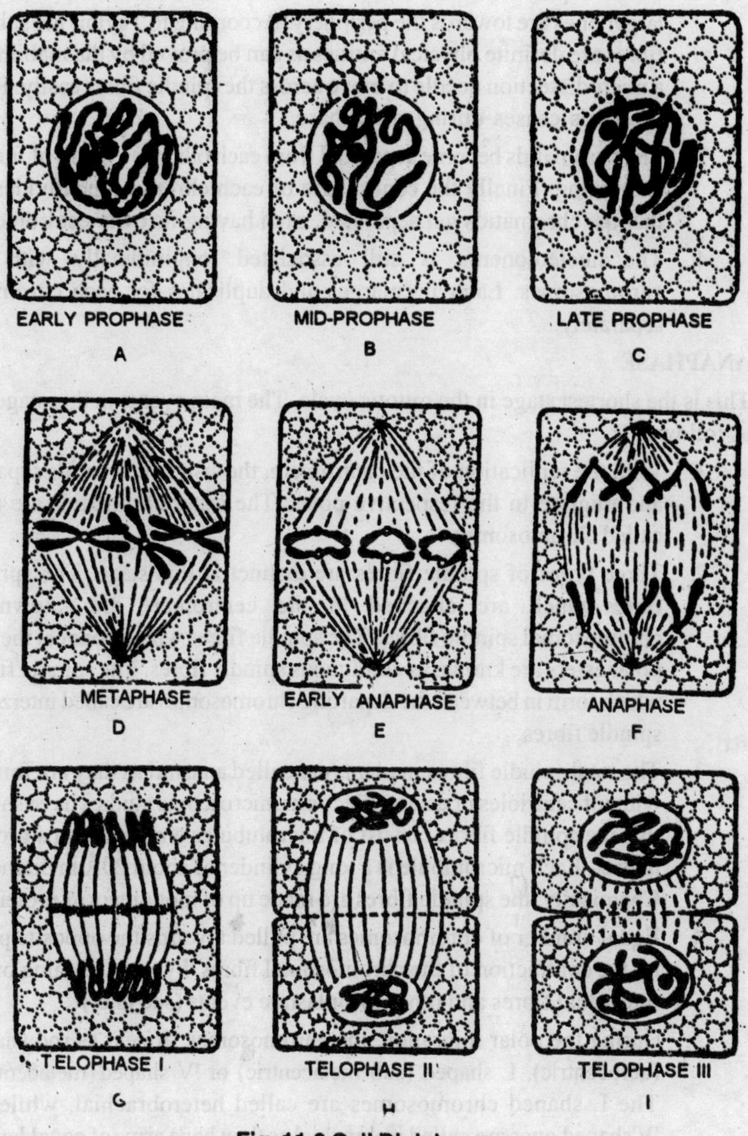

Fig. 11.3 Cell Division
Stages of mitosis in a plant cell

2. The chromosomes lie at the equator with the centroiles pointing towards the poles attached to the spindle fibres, and the arms lying freely in the cytoplasm. In plant cells, chromosomes are irregularly

scattered on the equatorial plate. Smaller chromosomes lie at the centre, while larger ones are towards the periphery. According to Taylor, the spindle fibres are definite physical entities as can be proved by the inability of a microdissection needle to move across the spindle fibres easily. RNA synthesis ceases during metaphase.

3. The chromatids become separated from each other as a result of further shortening. Finally the centromere of each chromosome divides, so that the chromatids get separated, each having its own centromere
4. The nucleolonema is still associated longitudinally with the chromosomes. Later it enlarges and duplicates for each chromatid separately.

ANAPHASE

This is the shortest stage in the mitotic cycle. The main events in this stage are as follows-

1. With the replication of the centromere, the sister chromatids separate and migrate to the respective poles. The sister chromatids are now called chromosomes
2. Three types of spindle fibres are distinct at this stage. The spindle fibres which are attached to the centromere are known as chromosomal spindle fibres. The spindle fibres which connect the two polar asters are known as continuous spindle fibres. The spindle fibres which form in between the departing chromosomes are called interzonal spindle fibres.
3. The total spindle fibres in plants are called anastral as they are formed without centrioles or asters. Electrom microscopic studies have shown that the spindle fibres consist of microtubules embedded in a protein matrix. Each microtubule is a long cylinder of about 20Å in diametre. Chemically, the spindle fibres are made up of monomers of proteins.
4. Equal number of chromosomes are pulled towards the opposite poles by the contraction of the chromosomal fibres, by the stretching of the interzonal fibres and other cytoplasmic events (see later)
5. During the polar migration, the chromosomes, appear as rod shaped (acrocentric), L shaped (sub metacentric) or 'V shaped (metacentric). The L shaped chromosomes are called heterobrachial, while the 'V'shaped ones are called isobrachial as they have arms of equal length.
6. In this manner, the genetic material is equally distributed between the two parts of the cell, which eventually become two daughter cells.
7. Since the centromere is essential for movement,those chromosomes which are without a centromere fail to move to the opposite poles. Such chromosomes are called laggards.

8. The velocity of movement of chromosomes during anaphase ranges from 0.2 μ per minute to about 5 μ per minute. The rate of movement is independent of the size of the chromosomes.
9. Generally all the chromosomes move simultaneously except for the sex chromosomes which lag behind. Some chromosomes move very early before the group has started migrating. Such movement is said to be precocious movement.
10. End of the anaphase is marked by the appearence of clusters of chromosomes at each pole of the cell.

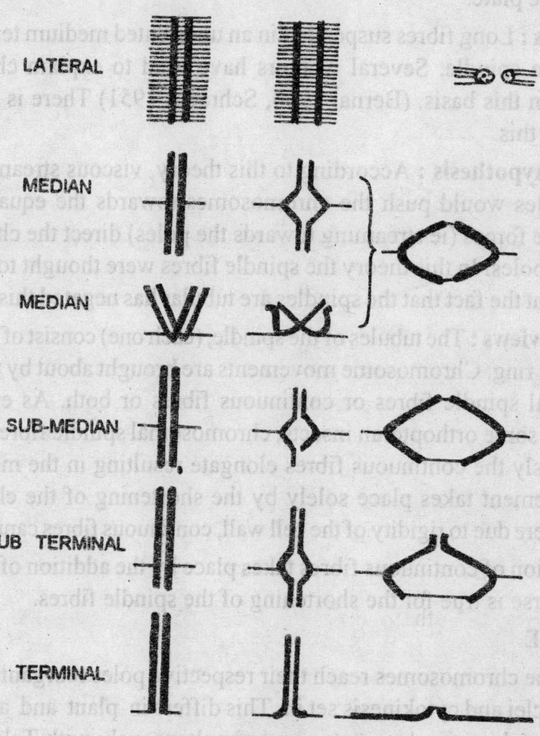

Fig. 11.4 Cell Division
Movement of Chromosomes during anaphase according to the mode of attachment

Chromosome movement during anaphase : Several theories have been proposed to account for the anaphasic movement of the chromosomes. Some of these are-

Electrostatic force : According to Lillie (1905), the chromosome movement is due to electrical charge. The cytoplasm is electronegative, while the nucleus (not chromosomes) is electropositive. Due to this difference in charge, the chromosomes move. This theory however does not explain how chromosomes having same charge as cytoplasm can move.

Hydrodynamic theory : According to this theory proposed by the physicist Bjerknes (1919), two spheres pulsating in a liquid medium set up hydrodynamic forces resembling electric or magnetic forces. In the cell, the spindle poles represent the pulsating forces and they attract the neutral spheres ie., chromosomes. this theory has not been acepted as it will not explain the formation of metaphasic plate.

Tactoids : Long fibres suspended in an unoriented medium tend to line up parallely in a spindle. Several workers have tried to explain chromosomal movement on this basis. (Bernal 1940, Schrader 1951) There is however no evidence for this.

Streaming hypothesis : According to this theory, viscous streams ernnating from the poles would push the chromosomes towards the equatorial plate, while reverse forces (ie streaming towards the poles) direct the chromosomes towards the poles. In this theory the spindle fibres were thought to be lamellar structures, but the fact that the spindles are tubular has negated this hypothesis.

Present views : The tubules of the spindle, (each one) consist of 13 filaments arranged in a ring. Chromosome movements are brought about by the action of chromosomal spindle fibres or continuous fibres or both. As evidence has indicated, in some orthopteran insects, chromosomal spindle fibres shrink and simultaneously the continuous fibres elongate resulting in the movement. In plants, movement takes place solely by the shortening of the chromosomal spindles, where due to rigidity of the cell wall, continuous fibres cannot elongate.

Elongation of continuous fibres takes place by the addition of protein unit and the reverse is true for the shortening of the spindle fibres.

TELOPHASE

When the chromosomes reach their respective poles reorganisation of the daughter nuclei and cytokinesis set in. This differs in plant and animal cells. In plants, cytokinesis does not occur simultaneously with Telophase as in animal cells. The following are the nuclear events in telophase.

1. The chromosomes begin to uncoil into their strands and become indistinct.
2. Nucleolous begins to appear as the nucleolonemal strands become thicker and accumulate.
3. The *gel* state of the spindle reverts to the sol state and the spindles disappear.

4. In animal cells, the asters lose their fibres and reconverted into centrioles. The centrioles take up their natural position outside the nucleus.
5. The nuclear membrane reappears forming a regular nucleus.

CYTOKINESIS

In plants, cell plate is formed between the two groups of chromosomes, by the accumulation of phragmoplasts at the equatorial region. The cell plate is formed at the centre, they grow on either side and join the cell wall. The cell plate formed first will have only middle lamella.

In animal cells, cytokinesis takes place by furrowing, a cleavage furrow appears at the early telophase which becomes deeper as the spindle starts disappearing. The furrows begin from either side and both join at the centre separating the two daughter cells. Various views have been put forward to explain the mechanism of furrowing. Some of these are -

1. **Contractile ring theory :** (Swann and Mitchison 1958) The ectoplasm contracts like the non motile part of amoeba resulting in the formation of a furrow.

2. **Expanding surface theory :** According to this view, some secretions from the nucleus allow the cytoplasm to enlarge at the poles resulting in their contraction at the equational region and a furrow results. But the phenomenon of nuclear secretion has been questioned by Naehtwey (1965).

3. **Spindle elongation theory :** (Dan and Dan 1947, 1958) According to this theory, spindle and astral Fibres are responsible for furrowing. The elongation of the spindle results in contraction at the centre.

4. **Astral relaxation theory :** (Wolpert 1963) According to this, the surface of the cell has uniform tension, but the tension is reduced at the poles when the asters reach the poles. However the tension (being retained) at the equator results in the contraction.

5. **Vescicle formation theory :** (Threadgold 1968) The cell elongates during anaphase causing increased electron density. With the plasma membrane also increasing its electron density, a dielectronicfibrllary plate is formed in the equator. As the furrow progresses, small membrane lined vesciles appear on either side of the plate and coalesce with one another. These seem to arise from the ER. Finally the fusing vesciles form the two daughter cells.

Thus the two daughter cells are formed from the parental cell. During division, the cell organelles like mitochondria, plastids, golgi complex, lyososomes and the cytoplasm are distributed equally to the two daughter cells. Of these organelles, the plastids and mitochondria reproduce themselves, while the behaviour of the other organelles is not known. The two daughter cells are genetically identical with each other. Soon they enter into interphase.

Mitotic apparatus : The mitotic apparatus consists of spindle, centrioles and the aster. The spindle Fibres are of three types (for details see anaphase) The centrioles lie just outside the nuclear membrane.

Frequency and Duration of mitosis : Varies according to the type of organism, tissue, nutrition, temperature and other physical and chemical forces. Bacteria divide every 20 minutes. Paramecium divides every six hours. The cells of higher plants and animals divide once or twice a day. In man about 1-2% of worn out somatic cells are replaced daily. How long does it take a cell to go through the entire process of mitotic division? This can be determined by examining a dividing cell (this is possible in cells grown under tissue culture). In human fibrocytes a duration of 18 hours is required from one interphase to another. But the actual division process from prophase to telophase requires only about 45 minutes. This means that a cell will be in prepration for cell division for over 17 hours.

In the neuroblast cells of the grass hopper embryo, the time required (in minutes) for each stage of division (at 38°C) is as follows. Prophase 102, Metaphase 13, Anaphase 9, Telophase 57, Interphase 27. In this cell, surprisingly the interphase is short compared to human cells. The duration however is dependent on temperature. Lower temperature increases the duration, while higher temperature shortens the duration, subject however to a limit.

In *Tradescantia*, the cells of the staminal hair reveal strikingly the influence of temperature on cell division. At 10°C a complete division takes about 135 minutes. Root tip cells generally take about 22 hours to complete one division at ordinary temperatures, but they can divide at 0°c also. In warm blooded animals however cell division ceases below 24°C and above 46°C.

Significance of mitosis:

1. Mitosis is an extremely regular process. It maintains the qualitative and quantitative distribution of hereditary materials to daughter cells.
2. It maintains constancy of the diploid number of chromosomes of species
3. It maintains the constancy and balance in the DNA and RNA content in the cell
4. It maintains the nuclear and cytoplasmic balance in the cell. In all the cells, the cell cannot expand indefinetly; it has to maintain a balance between the cytoplasm and nucleus called the nucleoplasmic index.
5. Mitosis balances the surface volume ratio of the cell. When the cell increases in size, the surface area in relation to increased volume becomes less. By mitotic division, cells become smaller in size and the surface/volume ratio is restored.

6. Mitosis results in the growth and development of the organism. An idea of growth brought about by mitosis can be obtained from the fact that adult human body consists of 60,000,000,000,000 cells all derived from a single cell.

7. Repair and replacement of dead cells is ensured by mitosis. In human body, dead cells of upper epidermis, gut, RBC are constantly being replaced. According to an estimate 500,000,000,000 cells are replaced daily.

8. It plays an important role in the asexual reproduction and in the formation of gametes in sexually reproducing organisms.

Role of Mitosis in cancer : The cell division in organisms is a highly controlled process and takes place at a specific rate. Usually,after every division there is a gap of time so that the cells physiologically mature, develop polarity and individuality. In other words the cells are allowed to differentiate. This differentiation is the single most important character that is responsible for the integrated and orderly growth of organs and organisms. Strictly speaking, all cells derived from a parental cell by mitosis should be alike but they will not be so due to differentiation. After two cell divisions it is not just four cells that are formed, but four cells with individual characters. If there were to be no phenomenon like differentiation, all organisms produced by zygote should have been nothing but lumps of cells.

If for some reason the mechanism of differentiation operating between cell divisions fails, cells divide at a rapid rate there being no opportunity for them to differentiate. The result will be uncontrolled, unintegrated growth thus producing a cluster of undifferentiated cells. This is called cancerous or tumorous growth. {The rate of cell division is generally uniform for a particular cell type, and cancerous cells violate this}. Also called neoplastic growth, cancerous cells bring disorder and chaos in the structure and physiology of the organ where they are produced. They consume more nutrition, more energy than is the normal case, thus depleting other parts,of the vital ingredients of life.

Mitosis is of great significance in cancer because, the latter is the result of uncontrolled division. What makes the regular rate of mitosis to alter, resulting in the hazardous growth is as yet fully not known. There are many mitotic abnormalities in cancerous cells. The chromosomes differ in their size and number. Chromatid separation is delayed resulting in multinucleate cells. There is a high quantity of DNA in the cancerous cells. The cancerous cells show loss of contact inhibition ie,, ability to grow even in contact with others indefinitely, while normal cells stop growing after they establish a contact with other cells. Much of the cytological details of cancerous cells are understood by working on a human cell culture maintained since a long time called *Hela cells*; these are cells oblained from Henrita Lacks who died of cancer.

Hypotheses on cancer : What makes mitotic division to go haywire? or why the cells become cancerous? The answer to these questions are badly needed in order to control cancer. There are many hypotheses. The following are some of the important ones.

Somatic mutation hypothesis According to this,a mutation occuring in the somatic cells (non germ ells) will alter the control mechamism or inhibit the production of repressor proteins, which in turn block the functioning of the gene.

Viral gene hypothesis : Viruses may be responsible in altering the mitotic mechanism thus producing cancerous cells. This was first discovered by Rous (1911), but people did not beleive it at that time. While there is enough evidence of Viruses causing cancer in animals, hitherto there is no concluding evidence regarding Virus induced cancer in human beings. There is however one evidence of Virus cancer relationship in Burkitt's lymphoma, a tumour occuring in certain regions of Africa.

Carcinogens : Certain compounds like coal tar, nicotine etc are known to produce cancer.

Defective immunity hypothesis : The body has its own defence mechanism or immune system which destroys cancerous cells even when they are formed. This is called immunological surveillance. If for some reason this fails, cancer cells survive and produce tumours.

In spite of the above theories, it has to be conceded that none of these satisfactorily explains what causes the cells to divide. All these merely attempt to explain the various influences (external agencies) that induce rapid cell division. But the basic question that has remained unanswered is which machinery in the cell (either at morphological or at molecular level) gets altered by mutation or immuno difficiency or whatever cause resulting in unregulated division. Possibly the ultimate answer to cancer lies in understanding the basics of interphase. It is here that the cell prepares itself for division. Even in interphase perhaps it is the S phase (as the important stage of DNA synthesis) which commits a cell for division needs to be studied in more detail to understand the rapid sequence of divisions.

MEIOSIS

In higher plants and animals, sexual reproduction takes place by the union of specialized sex cells called sperms and ova. The fusion of these sex cells that prevents the doubling of chromosomes in every generation. The answer lies in a special type of nuclear division called meiosis. It precedes the formation of gametes, so that each gamete acquires only half the number of chromosomes of the somatic cell. Thus when two gametes unite, the resulting zygote has the normal number of chromosomes. For example human body cells have 46 chromosomes in 23 pairs. But the human ovum and sperm have only 23 chromosomes, as aresult of meiosis. The reduced number of chromosomes in gametes is termed haploid, while the full complement of chromosomes in somatic cells is called diploid.

Definition of meiosis : Meiosis is a special type of division occuring in the diploid germ cells during which the chromosomes divide once, while the nucleus and cytoplasm divide twice resulting in the formation of four haploid cells. Hence meiosis is also *reduction division* as the number of chromosomes is reduced in comparison with the parental cell. These terms were coined by J.B-Farmer and J.E.Moore in 1905.

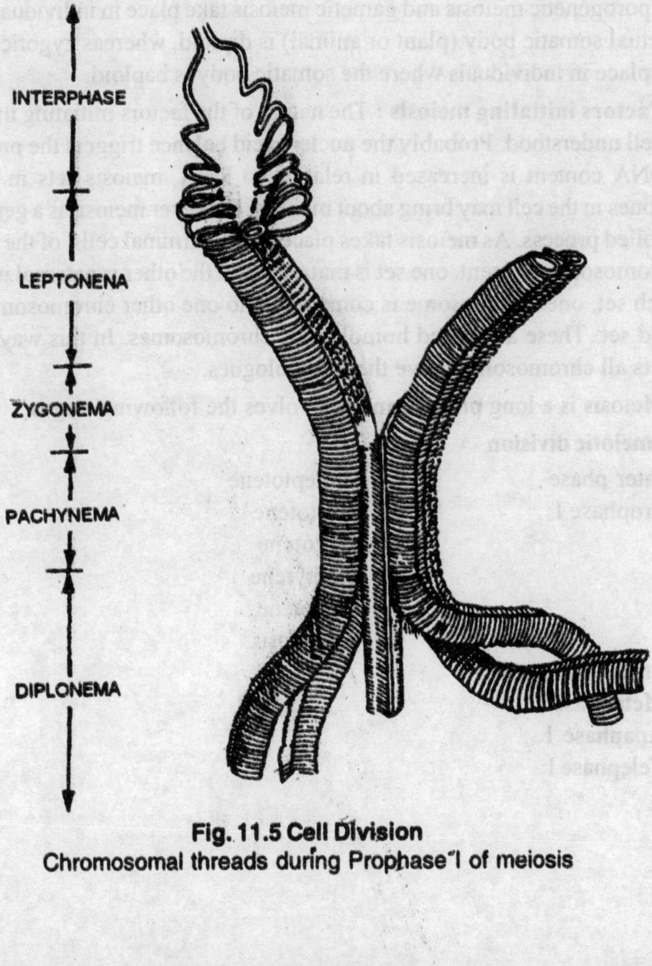

Fig. 11.5 Cell Division
Chromosomal threads during Prophase I of meiosis

The events of meiosis are uniform both in plants and animals. But depending on the time of occurrence, at different stages of life cycle the following types of meiosis may be identified.

1. Sporogenetic meiosis : In certain plants meiosis takes place in the sporophyte (2x) at the time of formation of spores. These spores develop into gametophyte.

2. Gametic meiosis : In most animals and some plants, reduction division takes place during the formation of gametes, so that haploid gametes are produced.

3. Zygotic meiosis : This is seen in lower plants where two haploid gametes produced by mitosis fuse to form a diploid zygote. The zygote undergoes meiosis to form haploid cells which develop into the haploid plant body.

Sporogenetic meiosis and gametic meiosis take place in individuals where the actual somatic body (plant or animal) is diploid, whereas zygotic meiosis takes place in individuals where the somatic body is haploid.

Factors initiating meiosis : The nature of the factors initiating meiosis is not well understood. Probably the nucleic acid balance triggers the process. If the DNA content is increased in relation to RNA, meiosis sets in. Certain hormones in the cell may bring about meiosis. However meiosis is a genetically controlled process. As meiosis takes place in the germinal cells, of the two sets of chromosomes present, one set is maternal and the other is paternal in origin. In each set, one chromosome is comparable to one other chromosome in the second set. These are called homologous chromosomes. In this way in both the sets all chromosomes have their homologues.

Meiosis is a long process and it involves the following stages.

First meiotic division

Inter phase — 1 Preleptotene

Prophase I — 2 Leptotene
3 Zygotene
4 Pachytene
5 Diplotene
6 Diakinesis

Prometaphase I
Metaphase I
Anaphase I
Telephase I

INTERKINESIS

Second meiotic division

Prophase II
Metaphase II
Anaphase II
Telophase II

Meiosis I is also called heterotypic division because here a diploid nucleus gives rise to two haploid nuclei. Meiosis II is called homeotypic division because here, the chromosome number is maintained. Meiosis II is actually nothing but mitosis.

INTERPHASE : The interphase consists of the same Gl phase, S phase and G_1 phase same as in mitosis. The essential events are as follows-

1. There is replication of RNA and synthesis of proteins during G_1 phase.
2. DNA is synthesized only in S phase and not during G_1 and G_2 phases.
3. DNA replication takes place by a semiconservative process. As a first step, an incision enzyme initiates a nick in one of the strands. Both the strands of DNA participate in the replication process under the influence of DNA polymerase. Together with DNA, histones are also synthesised. DNA is synthesised in short segments (Okazaki segments) which later join together to form large molecules.

PROPHASE I : This is the longest phase in meiosis. It has the following stages-proleptotene, leptotene, zygotene, pachytene, Diplotene and Diakinesis.

Proleptotene : Replication of DNA having already taken place in interphase, the chromosomes appear double stranded. The other events in proleptotene are-

1. Nucleus enlarges in size
2. Chromosomes are thin and hazy except perhaps for sex chromosomes

Leptotene (Leptonema) : This is also called the slender thread stage. Chromosomes appear distinct and the double strand (chromatid) nature can be seen clearly. The following are the other structural details of leptotene chromosomes.

1. Chromosomes are distinct,long and less coiled.
2. Chromosomes appear as threads with regularly placed beads or granules called chromomeres.
3. Electron microscopic observations have revealed that leptotene chromosomes have an axial filament with chromatin fibres attached to it in a series of lateral loops.
4. In certain areas the lateral loops cluster together to form chromomeres. The axial filament has two protein fibrils. These fibrils later get transformed into a part of the *synaptonemal complex*.

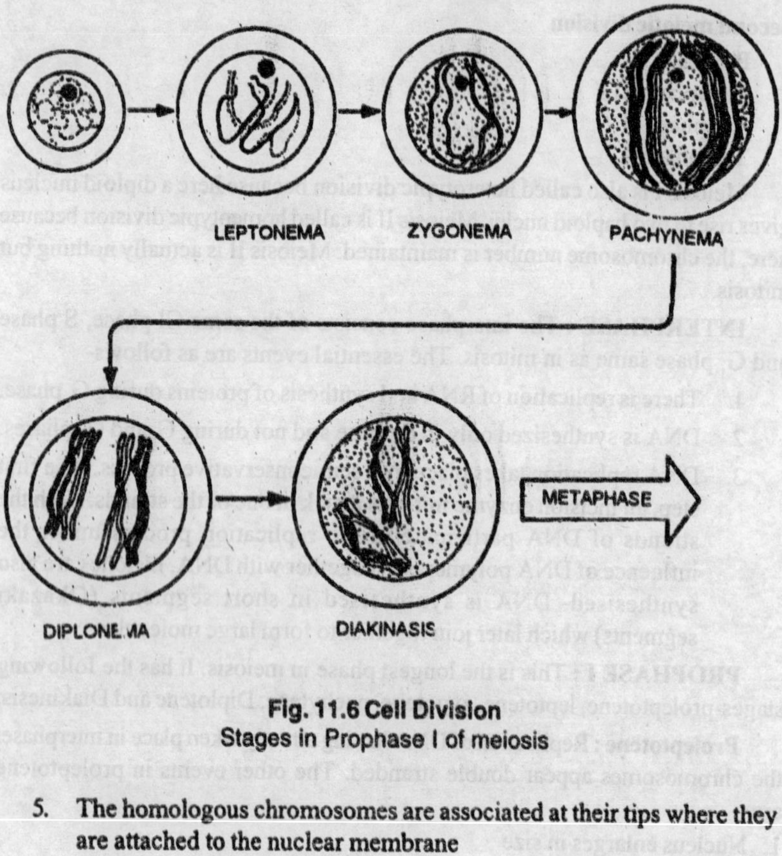

Fig. 11.6 Cell Division
Stages in Prophase I of meiosis

5. The homologous chromosomes are associated at their tips where they are attached to the nuclear membrane
6. Leptotene chromosomes may be uniformly distributed or localized into *bouquets*.
7. In some plants, chromosomes may form a tangled mass of fibres called *synizetic knot*.
8. Cytoplasm of a leptotene cell has many polyribosomes but vesicles are less in number. Nucleolus increases its size to the maximum due to increases in RNA synthesis.
9. The centriole duplicates and each daughter centriole migrates to the opposite poles of the cell. On reaching the pole the daughter centriole duplicates again.

Zygotene (zygonema) : This is a very important stage where pairing of the homologous chromosomes takes place. The following are the important events with reference to zygotene chromosomes.

1. The chromosomes become shorter and thicker.
2. Lengthwise pairing of the homologous chromosomes begins as they are attracted towards each other.
3. Of the two homologues, one is maternal in origin and the other is paternal. These are called bivalents.

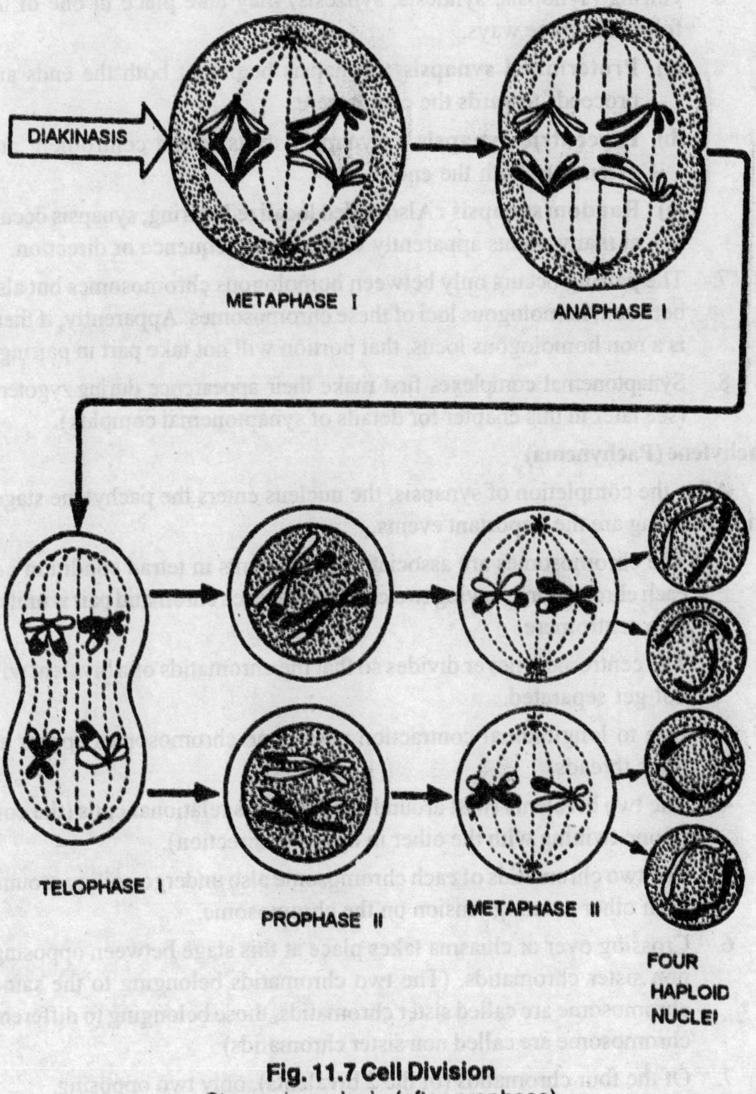

Fig. 11.7 Cell Division
Stages in meiosis (after prophase)

4. Pairing also called synopsis is precise and takes place chromomere - for chromomere. As a result,the chromosomes (homologues) align themselves parallel to each other throughout their length.
5. The bivalent has two homologous chromosomes; each homologous chromosome has two chromatids i.e. totally there will be four chromatids (tetrad).
6. Pairing (synopsis, syndesis, synzesis) may take place in one of the following three ways.
   (a) **Proterminal synapsis :** Synapsis begins at both the ends and proceeds towards the centromere.
   (b) **Procentric synapsis :** Synapsis starts at the centromere and extends to both the ends.
   (c) **Random synapsis :** Also called localized pairing, synapsis occurs at many points apparently without any sequence or direction.
7. The pairing occurs only between homologous chromosomes but also between, homologous loci of these chromosomes. Apparently, if there is a non homologous locus, that portion will not take part in pairing.
8. Synaptonemal complexes first make their appearence during zygotene (see later in this chapter for details of synaptonemal complex).

Pachytene (Pachynema)

After the completion of synapsis, the nucleus enters the pachytene stage. The following are the important events.

1. The chromosomes are associated as bivalents in tetrad condition i.e. each chromosome having two chromatids; each chromatid pair is united by a centromere.
2. The centromere never divides so that the chromatids of a bivalent will not get separated.
3. Due to longitudinal contraction pachytene chromosomes appear as thick threads.
4. The two bivalents twist around each other as relational coils (the coil of one twining with the other in opposite direction).
5. The two chromatids of each chromosome also undergo coiling around each other creating tension on the chromosome.
6. Crossing over or chiasma takes place at this stage between opposing non sister chromatids. (The two chromatids belonging to the same chromosome are called sister chromatids, those belonging to different chromosome are called non sister chromatids)
7. Of the four chromatids (of the 2 bivalents), only two opposing

chromatids take part in crossing over, while the other two (which are away from each other) do not participate in crossing over.

8. Crossing over takes place because of breakage and reunion at one or two points between opposing non sister chromatids. It takes place as follows. First one or two transverse breaks occur at the same level in the two non sister chromatids opposing each other. (The breaks will never occur in both the sister chromatids) The break is followed by reunion, but their union will not join the same segments of a sister chromatid. The chromatid which breaks at one point will unite with the segment of the opposing non sister chromatid. Break and union occurs at 1, 2 or 3 points. As a result after the cross over, each chromatid has 1, 2 or 3 pieces of the other non sister chromatid. If break occurs at one point it is called single point cross over, at two points is called double point cross over etc. Due to crossing over, the chromosomes produce cross shaped configurations called chiasmata. The number of points at which cross over can take place depends on the length of the bivalents. Thus as a result of cross over in the bivalents, there will be two unbroken or original chromatids (which are away from each other) and two broken and reunited chromatids opposing each other which have exchanged their segments.
9. Exchange of genes takes place between homologues due to exchange of segments.
10. Synaptonemal complex is very well organised at this stage. It has a central element and two lateral elements, (see later for detailed explanation)
11. Nucleolus remains unaltered; it may increase in size. It is found to be associated with the nucleolar organizing region of certain chromosomes.

Diplotene (Diplonema)

The diplotene nucleus is characterised by repulsion of the homologues, while there is coupling in pachytene. The following are the important features of diplotene stage.

1. Homologues start separating (repulsion), but are held together at one or more points called chiasmata.
2. Chiasmata are very clear in diplotene than in pachytene stage because of increased condensation of chromosomes.
3. Chiasmata are found in meiosis of almost all eukaryotic organisms. However in certain Dipteran insects including *Drosophila*, there is no chiasma in males. Meiosis here is termed non cross over meiosis or

achiasmatic meiosis. Achiasmatic meiosis has also been reported in oocytes of Copepoda and Lepidoptera.

4. Chiasma may be formed at the ends (terminal chiasma) or in tshe middle (interstitial chiasma) of the chromosomes.

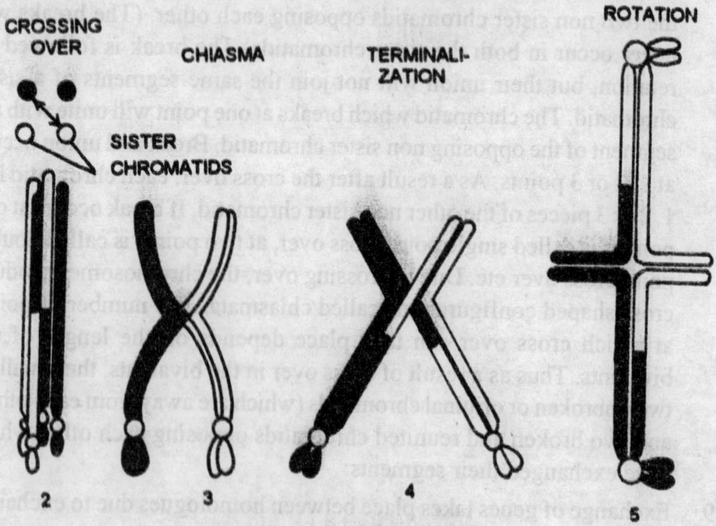

Fig. 11.8 Cell Division

5. The most important event occuring in diplotene is the teminalization of the chiasma. Terminalization or uncoupling of chiasma begins due to repulsion in homologous chromosomes. The chromatids begin to separate at the centromere and gradually progress towards the ends. As a result the terminal chiasma slip off or uncouple and progressively the chiasma towards the centromere move towards the end. When the terminalization is almost complete, the two homologues remain in contact through the terminal chiasma.

   The degree of terminalization may be expressed in the form of terminalization coefficient (T).

$$T = \frac{\text{Number of terminal Chiasma}}{\text{Total number of bivalents}}$$

   Chiasma frequency (Fq) is the average number of chiasma in bivalents of a nucleus. This may be calculated as follows

$$Fq = \frac{\text{Total number of chiasma}}{\text{Total number of bialents}}$$

During terminalization there is rotation of chromosomes; when there is a single chiasma, the chromosomes rotate through 180° and form a cross; when there are two chiasmata a loop is formed; when there are more than two, a number of loops are formed and rotation is brought about by the forces of repulsion.

6. Synaptonemal complexes usually disappear, but occasionally may persist towards the end of bivalents particularly where they are attached to the nuclear membrane.
7. There is increase in the endoplasmic vesicles.

Diakinesis

This is the Final stage of prophase 1. The structural details of this stage are -

1. Bivalent chromosomes are very thick and highly contracted.
2. Bivalents are distributed more towards the periphery of the nucleus.
3. Terminal chiasma may be retained maintaining contact between homologous chromosomes.
4. The nucleolous detaches from the chromosomes and finally disappears.

PROMETAPHASE I

1. The nuclear membrane disappears. The centrioles are joined by spindles which develop from tubules.
2. The chromosomes reach their maximum contraction.

METAPHASE I

At the end of diakinesis, the amphiaster (spindle) becomes clearly visible and it becomes more pronounced now. The bivalents arrange themselves on the equatorial plate. The centromeres of the bivalents lie towards the poles, wilile the arms lie towards the equator. The centromere is attached to the spindle fibres. These are called chromosomal fibres.

ANAPHASE I

It is at this stage, the homologous chromosomes move to the opposite poles. The following are the main features of anaphase I.

1. Each homologous chromosome of each bivalent moves towards the opposite pole. The movement is random unlike in mitosis.
2. The homologues do not separate quickly. While short chromosomes separate easily, longer ones take a little more time (due to interstitial chiasmata).
3. The homologous chromosomes migrating to the opposite poles are different from the ones that undergo pairing during prophase in the sense in each chromosome one chromatid is unaltered, while the other is a mixture of paternal and maternal segments.

4. The centromere does not divide (It divides in mitosis).
5. The chromosomes assume a V shape as they are pulled towards the opposite poles.

TELOPHASE I

The following are the important events during telophase I.

1. The chromosomes reach the respective poles. The spindles persist for sometime.
2. Chromosomes undergo despiralization, becoming indistinct.
3. Nucleolus reappears, so also the nuclear membrane. In some instances nucleolus may not appear at this stage.
4. Cell membrane is formed in animal cells generally by furrowing, and in plant cells, cell wall is laid down in the form of plate. In some of the plant cells cytokinesis does not take place at this stage. It takes place when both meiotic divisions are complete. In pollen mother cells of flowering plants cell wall is laid down after meiosis I and II is called successive and if laid down only after meiosis II is called simultaneous. After Telophase I, two cells or two nuclei are formed.

INTERKINESIS

The two daughter nuclei or daughter cells undergo a resting period. This intervening phase is called interphase or Interkinesis. This phase is of a very short duration. In some instances interkinesis is absent, the incompletely formed nuclei of telophase I may enter prophase II. This is seen in plants like *Trillium*.

MEIOSIS II

The second meiotic division is essentially a mitosis in which both the chromosomes and cell divide.

But the main difference is DNA does not duplicate unlike in the interphase of mitosis. And also (as has been pointed out already) the interphase or interkinesis is short.

PROPHASE II

This is essentially similar to the mitotic prophase. The salient features of this stage are -

1. The centrioles divide into two and two pairs of centrioles are formed. They migrate to the opposite poles.
2. The microtubules get themselves arranged at right angles to the previous plane (during meiosis I).
3. The arms of the chromosomes remain separate and there is no relational coiling.
4. Nucleolus as well as nuclear envelope disappear with the formation of the achromatic figure.

METAPHASE II

1. It is of a very short duration like in mitosis.
2. Chromosomes (dyads) with their (two) chromatids arrange themselves on the equatorial plate.
3. The centromeres lie on the equator while the arms are directed towards the poles.
4. The centromere divides, thus separating the two chromatids (monads) which eventually become separate chromosomes. At this stage they can be called daughter chromosomes.
5. The centromere is attached to the chromosomal spindle fibres in the achromatic figure.

ANAPHASE II

1. The daughter chromosomes start their migration towards the opposite poles. The movement is brought about due to the contraction of chromosomal spindle fibres and stretching of continuous spindle fibres.
2. The chromosomes are not very thick and short like in anaphase, but are like the anaphase daughter chromosomes of mitosis.

TELOPHASE II

1. The chromosomes having reached the opposite poles uncoil and become less distinct.
2. Nuclear membrane and nucleolus reappear resulting in the formation of a daughter nucleus. The nucleus gets into the interphase state.
3. The nuclei are haploid; they have half the number of chromosomes in comparison with the parental cell.

CYTOKINESIS

Wall formation begins either as a furrow or as a plate and two daughter cells are formed. Thus totally at the end of meiosis four daughter cells are formed. In some instances cell wall (in some plants) would not have been laid after meiosis I. As a result at the end of Telophase II four daughter nuclei are formed.Cell walls are laid down now resulting in the formation of four cells.

SYNAPTONEMAL COMPLEX

This is a special structure visible during meiotic prophase - specially during zygotene and pachytene. Observed first with the help of electron microscope in the spermatocytes of cray fish- **Procambams darkii** (Moses 1956) and in Pigeon, Cat, Man etc (Faweett 1956), the synaptonemal complex consists of ribbon like structures associated with the chromosome core.

As these were observed during synapsis, Mosel (1958) named them synaptonemal complex (SC) subsequent to this, SCs have been found in a wide variety of organisms like Algae, Fungi, Gymnosperms, Angiosperms Protozoans, insects, molluscs, amphibians, aves, mammals etc. Hence SCs can be said to be of universal occurrence.

Structure : These are three partite ribbon like structures situated between the homologous chromosomes. Each SC is made up of a dense central axis from which arise dense lateral filamentous structures. Each lateral element on either side of the axis is attached to the inner side of the homologous chromosomes. The space between the central axis and lateral elements are traversed by a series of transverse units or LC fibres. Each lateral unit has a width ranging from 300-500 Å and the space between the lateral elements is between 900- 1200 Å. The width of the central element is betwee 150-500 Å. The SC is attached to the nuclear membrane through the lateral elements at the ends.

Comings and Okada (1971) have reported the emergence of a serie of chromatin loops from the lateral elements, which make up the chromosome material. The central axis (central element) is formed of the fusion of a series of small loops.

Each lateral element consists of two protein fibres one for each non sister chromatid. The lateral elements and LC fibres are also made up of proteins.

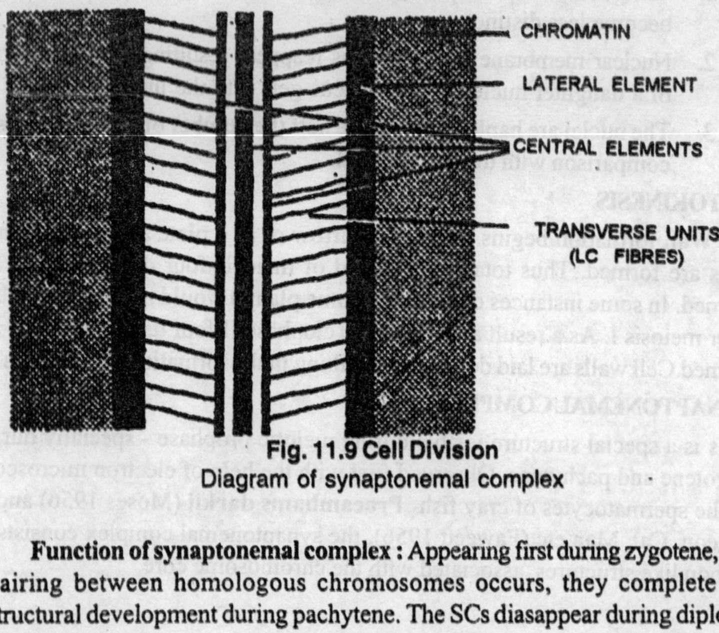

Fig. 11.9 Cell Division
Diagram of synaptonemal complex

Function of synaptonemal complex : Appearing first during zygotene, when pairing between homologous chromosomes occurs, they complete their structural development during pachytene. The SCs diasappear during diplotene, even though remnants may be seen during diakinesis. The above facts clearly attribute an important role to SCs in crossing over and exchange of segments between homologous chromosomes. According to Meyer (1964), the absence of SC in achiasmatic meiosis (as in males of dipteran insects) is proof enough of SC's role in chiasma formation. But in some instances SCs have been found in achiasmatic meiosis also.

According to Comings and Okada (1971), SCs help in chiasma not only at the morphological level but even at the molecular level also.

The SCs disintegrat gradually into cytoplasm as they disassociate from the chromosomes.

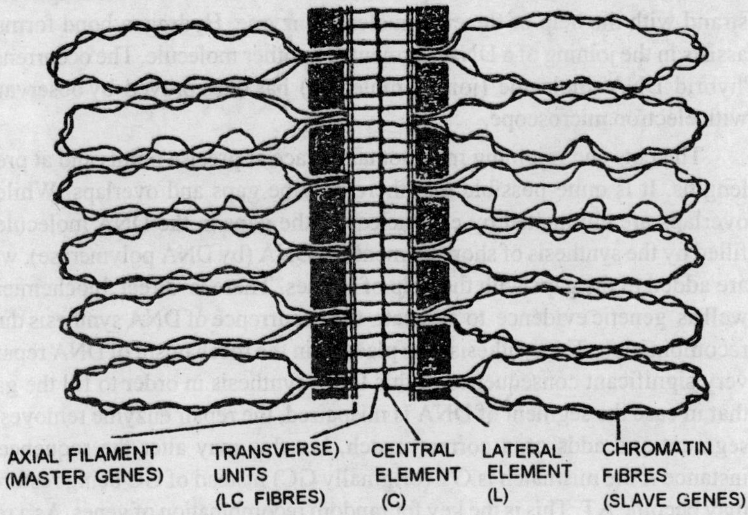

Fig. 11.10 Cell Division
Synaptonemal complex (Composite diagram showing details)

Significance of Meiosis

1. Helps to maintain constant chromosome numbers in organisms.
2. Crossing over which is the central focus in meiosis brings about recombination of genetic material leading to variation amongst individuals.
3. Meiosis in fact provides with the basic material (variation) for Evolution.
4. The random recombinations among homologous chromosomes give rise to a variety of mix up of parental characters and new characters.

Molecular mechanism of crossing over : several models have been proposed to account for recombination of DNA molecules of the homologous chromosomes. The following is the general theme of the crossing over at the molecular level.

It is generally believed that the specific interactions between two DNA molecules of homologous chromosomes leading to recombination are mediated by single stranded elements. Recombination begins probably by having an endonuclease enzyme make a 'nick' at one of the strands of the adjoining duplexes. As a result, a section of the single stranded DNA becomes displaced from that helix' and interacts with the complimentary strand of the other duplex. In order to facilitate this interaction, a second nick occurs in the neighbouring strand with the help of the endonuclease enzyme. Hydrogen bond formation assists in the joining of a DNA segment to another molecule. The occurrence of 'hybrid DNA' molecule (joined molecule) has been proved by observations with electron microscope.

The cuts and rejoining may not take place at precise points and at precise lengths. It is quite possible that there may be gaps and overlaps. While the overlaps are eliminated by exonucleases, the gaps in the DNA molecule are filled by the synthesis of short segments of DNA (by DNA polymerase), which are added to the gaps with the help of ligases. There is direct biochemical as well as genetic evidence to indicate the occurrence of DNA synthesis during recombination. The synthesis take place as in the mechanism of DNA repair. A very significant consequence of this DNA synthesis in order to fill the gap is that in case the segment of DNA is mispaired, the repair enzyme removes this segment and adds on a correct match, but this may alter the sequence for instance if the mismatch is GT (originally GC) instead of GC being restored it may become AT. This is the key for random recombination of genes. As a result of DNA synthesis during recombination one allele may get converted into the other. This may be called gene conversion and has been reported in *Neurospora.*

COMPARISON BETWEEN MITOSIS AND MEIOSIS

	MITOSIS	MEIOSIS
1.	Occurs in somatic cells	Occurs in germ cells
2.	There is one nuclear division and one cytokinesis	There is one nuclear division but two cytokinesis.
3.	Chromosome number is maintained.	Chromosome number is halved.
4.	Prophase is short and there are no substages many	Prophase I is very long and has substages

	MITOSIS	MEIOSIS
5.	The chromosomes replicate and the centromere also divides with the result, chromatids become daughter Chromosomes and they migrate to opposite poles.	The chromosomes do not replicate during meiosis I. The çentromere does not divide; as a result among existing chromosomes half migrate to one pole and the other half to the other pole.
6.	No pairing and there is no association between homologous chromosomes	Chiasma takes place between homologous chromosomes resulting in exchange of genetic material
7.	Chromosomes duplicate during prophase.	Duplication of Chromosomes takes place during prophase II.
8.	Chromosomes occur as diads in prophase.	They occur as tetrads in metaphase - I.
9.	Centromere is directed towards the equator in the metaphase plate with the arms towards the poles.	Centromere is directed towards the poles with the arms towards the equator
10.	In anaphase the chromosomes move as monads.	Chromosomes move as dyads.
11.	During anaphase chromosomes are short and thin	During Anaphase - I Chromosomes are long and thick.
12.	Telephase always occurs.	Telephase - I is omitted sometimes.
13.	Equational division.	Reduction division.
14.	Brings about growth.	Brings about reproduction.

12

LINKAGE AND CROSSING OVER

Mendel, while enunciating ,his law of independent assortment, ascertained that two or more elements, that constitute a hybrid, undergo independent seggregation without being influenced by the other. Mendel's principles of independent assortment apply not only to genes but to chromosomes as well, so that paternal and maternal hromosomes are seggregated to separate gametes during meiosis. It is for this reason that genes carried on different chromosomes undergo complete seggregation. Fortunately for Mendel, all the seven charcters that he chose for study were on different chromosomes (there are seven pairs of chromosomes in pea plants).

There are hundreds of characters in organisms and if there is a gene for every character as Mendel visualised (which is a fact also) there must be hundreds and indeed thousands of genes, and if genes are situated on the chromosomes there must be hundreds or thousands of chromosomes. But this is not so; while there could be thousands of genes, the number of chromosomes in most of the individuals hardly exceeds hundred, in fact it is much less. In *Drosophila*, where hundreds of characters have been studied there are only four pairs of chromosomes. In human beings with thousands of characters, there are only 23 pairs of chromosomes. If genes are situated on the chromosomes, it means then several genes responsible for so many characters must be situated on a single chromosome. When this is so, genes or characters on a single chromosome cannot segregate. They cannot be assorted independently instead they tend to be inherited together.

It was Sutton (1903) who first pointed out the anomalous situation in beleiving total independent assortment for all the characters, when the number of chromosomes are far less than the number of genes. For eg: if we identify two genes located on a chromosome and perform a dihybrid cross, the chances are the two genes will tend to go together, instead of being assorted independently. The two traits controlled by these will appear together in an individual, instead of seggregating to two individuals. In a dihybrid cross of this type, Mendel's 9:3:3:1 ratio will be modified. This phenomenon of genes (present on a single chromosome) showing the tendency to be inherited together is called linkage. However until Morgan's work in 1910, Genetics did not have a satisfactory explanation for the tendency of the genes to assort together in the offspring. We shall consider some of the hypotheses concerning the above mentioned behaviour of genes.

Sutton's view : As has been pointed out already, Sutton (1903) opined that several genes are present on the chromosome and they tend to be inherited together. However he did not visualise the concept of linkage. Instead he said that there are groups of genes, and these groups in an individual correspond to the number of chromosomes. Sutton did not experimentally confirm his views.

Coupling and repulsion hypothesis : This hypothesis was put forward by Bateson and Punnett (1906), who observed in a variety of sweet pea the 'violation' of the law of independent assortment. The experiment involved flower colour (purple/red) and pollen shape (long/round). When two varieties of sweet pea were crossed-one purple and long and the other red and round, the F_1 plant was purple and long. On selfing, the F_2 progeny should have exhibited the 9:3:3:1 ratio. But this did not happen so. Instead the purple long and red round plants appeared in large numbers ie., more frequently than what was expected in a dihybrid cross assuming independent assortment. These are the parental combinations ie., characters present together in one individual. The new combinations ie., purple round and red long did appear in the F_2 but in a lesser frequency. Here parental combinations are more than new combinations or recombinations. The following example illustrates this phenomenon.

Purple long X Red round
PPLL ppll
↓ ↓
PL → Pp Ll ← pl (Purple long) F_1 hybrid
selfing

F_2 Progeny	Purple long	Purple round	Red long	Red round
Expected	1199	400	400	133
Observed	1528	106	117	381

Data from K.B. Ahluwalia, Genetics, 1985

The experiment was repeated by Bateson and Punnett in a different way. This time they selected a new combination of characters in the parents. The parents were Purple round and Red long. the F_1 showed the dominant purple long characters. In the F_2. generation again parentalcharacters (Purple round and Red long) were more than the recombined characters (Purple long and Red round). This experiment also showed the tendency of the parental genes to remain together instead of seggregating.

Test Cross : Bateson and Punnet also performed a double cross. They crossed the F_1 dihybrid - Purple long (PpLI) with the double recessive parent (red round - (rrll). In a Mendelian inheritence, the result would be 1:1:1:1 of Purple long, purple round, Red round and red long. But Bateson and Punnett obtained a ratio of 7:1:1:7 of Purple long, purple round, red long and red round respectively.

Bateson and Punnett explained this deviation as the phenomenon of coupling of the characters coming from a single parent to remain together. Similarly the tendency of the characters belonging to two different parents to remain seperate is called **repulsion.** The coupling and repulsion hypothesis of Bateson Punnett did not visualise the actual reason for the coupled genes to remain together.

Bateson pointed out an explanation to coupling and repulsion hypothesis as due to "differential multiplication of cells containing different combination of characters of genes". The combination of genes seggregate before gametes are formed. After seggregation and gamete formation one set of gametes multiply at a faster pace than the other. This takes part in fertilization in large numbers than the other kind of gamete, hence the modification in the ratio of a typical dihybrid cross.

Morgan's concept of Linkage : The explanation of Bateson for coupling and repulsion was not very satisfactory. It was T.H. Morgan (1910), who was working on *Drosophila* together with his associates A.H. Sturtevant, H.J. Muller and C.B. Bridges, offered a plausible reason for the phenomenon of coupling and repulsion. According to Morgan, coupling and repulsion are but two facets of the same phenomenon. He further stated that pairs of genes from homozygous parents tend to enter the same gamete, while the same genes from heterozygous parents tend to enter different gametes and remain apart. The tendency of the genes to retain their original combination is due to their location on the same chromosome. Another significant aspect of linkage underscored by Morgan was that the strength of linkage (tendency to go together) depends on the distance between the linked genes, on the chromosome. The closer the genes, the stronger will be the linkage, and the farther the genes, the weaker will be the linkage. Tendency to remain together will be more if the linkage is stronger and the tendency to seggregate more will be there if the linkage is weak. Morgan and Castle formulated the chromosome theory of linkage.

The main features of this theory are-

1. Genes showing linkage are on the same chromosome.
2. Genes are arranged on the chromosome in a linear fashion. Linkage of genes is also linear.
3. The distance between the genes determines the strength of the linkage. Strong linkage indicates proximity of the genes, while weak linkage indicates that they are farther apart.
4. Linked genes will remian together during inheritance.

Cytological studies have given the physical proof of linkage. Linkage of genes as observed by the inheritance of traits has helped cytologists to map the chromosomes (for details see later in the chapter).

Arrangement of the linked genes : In a heterozygous individual with two pairs of linked genes, the linkage can be of any one of the following types.

(a) The dominant alleles of both characters are on one chromosome and similarly, the recessive alleles are on another chromosome. For example if the gene paris are Aa and Bb, A and B are on one chromosome and a and b are on another chromosome (AB/ab). This arrangement is called *cis* arrangement and the heterozygous individual is said to be a cis heterozygote.

(b) This arrangement is *vice-versa* of the *cis* type. Here, of the two pairs of genes one chromosome has dominant allele of one and the recessive allele of the other (Ab); like wise the other chromosome has recessive allele of the fyst pair and the dominant allele of the second pair (aB). This arrangement (Ab/aB) is called the *trans* arrangement in which a chromosome has one dominant and one recessive allele. The heterozygotes are said to be *trans heterozygotes*. According to Singleton (1967), *cis* arrangement exhibits the coupling phase, while the *trans* arrangement denotes the repulsive phase.

We shall now study a few examples of linkage in plants and animals.

Linkage in Maize : The phenomenon of linkage can be easily demonstrated in maize, as recombination also can be seen and traits in the seeds can be easily observed. Each ear having hundreds of seeds makes the observation easy. Hutchison has demonstrated linkage in two varieties of maize. One had coloured and full seeds, while the other had colourless shrunken seeds. The gene for colour is *C* and *f* is dominant over the colourless condition *c*. Similarly the full seed *F* is dominant over shrunken condition. The genotypes of the parents which represent the pure line is *CCFF* and *ccff*. In a cross between the two, as expected the F_1 is coloured and full. The genotype is *CcFf*. The genotype however is written as *CF/cf* to indicate C & *F* and *c* & *f* are linked. The gene combinations are written in the manner in which they enter the zygote. If the genotype is- written as *CcFf*, it is indicative of independent assortment.

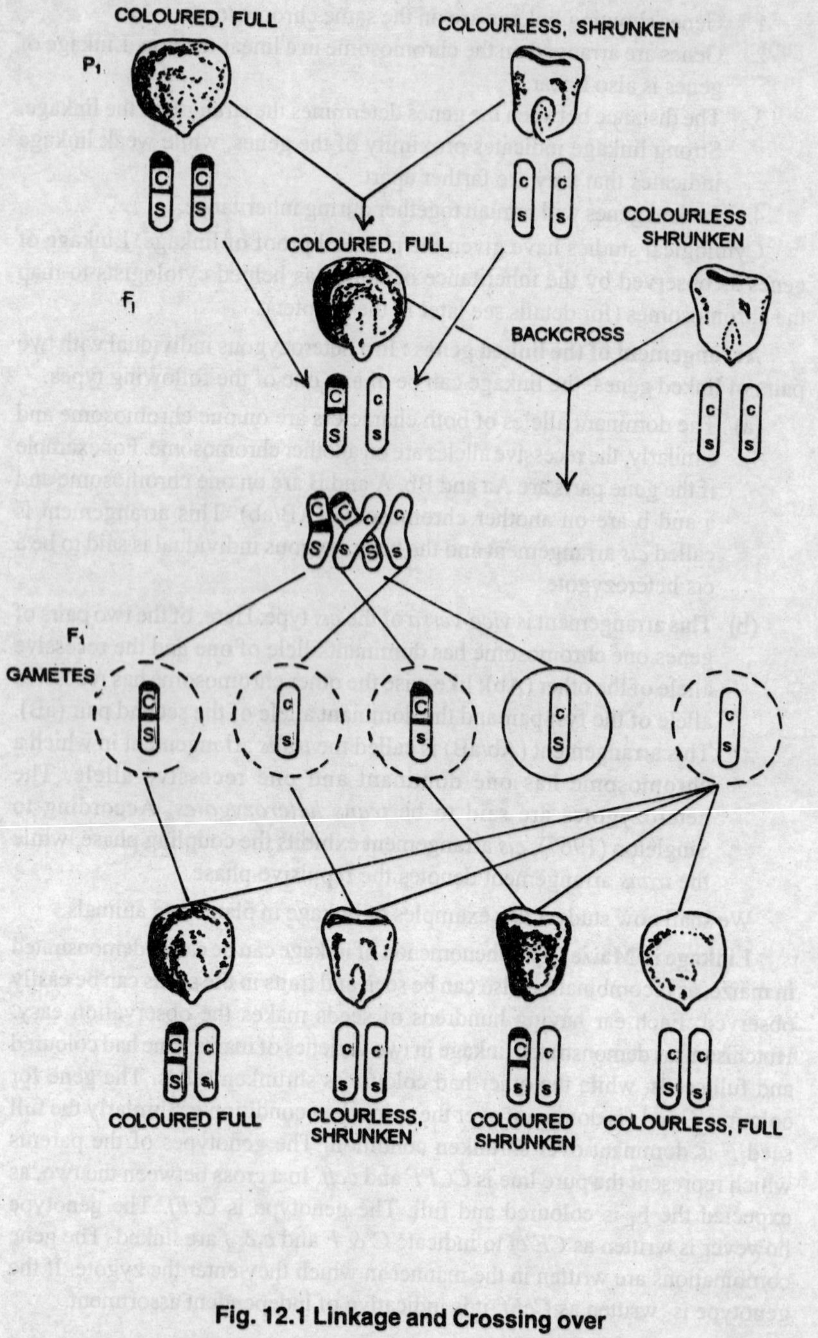

Fig. 12.1 Linkage and Crossing over

If the F_1 follows the Mendelian pattern of seggregation, four kinds of gametes should be -produced in equal proportions. These are *CF, Cf, cF*, and *cf*. In order to find out, whether these gametes are formed in equal numbers a test cross in done and the F_1 dihybrid is crossed with the double recessive parent *(CF/cf* x *cf/cf)*. If the F_1 ihybrid forms the 4 types of gametes in equal proportions it should yield the four classes of progenies. (Coloured full, coloured shrunken, colourless full and colourless shrunken) in the ratio of 1:1:1:1. But the actual number of progenies obtained are as given in the following table.

TABLE

Phenotype	Genotype	Number of individuals
Coloured full	*CF/cf*	4032
Coloured shrunken	*Cf/cf*	149
Colourless full	*cF/cf*	152
Colourless shrunken	*cf/cf*	4035
Total number of individuals of all categories =		8368

The parental combinations namely coloured full and colourless shrunken are in large numbers among the progeny than the recombinations viz., coloured shrunken and colourless full. If the assortment were to be independent all the four categories should have been produced in equal proportions. Instead, parental combinations abound. Out of a total of 8368 individuals about 96.4% (4032 + 4035) are parental combinations and only about 3.6% (149 + 152) are recombinations. This shows clearly genes *C* and *F* have not assorted independently, so also *c* and *f*. In other words, the linkage of 96.4%, while recombination i.e., break of linkage is seen in only 3.4% of the progeny.

In another cross in maize plant dealing with the same characters, different parental combinations were selected. The two parents selected for crossing were coloured shrunken and colourless full. Even in this cross, as the table below shows parental combinations were more than the recombinations. But here the parental combinations are different from those of the first experiment.

TABLE

Phenotype	Genotype	Number of individuals
Coloured full	CF/cf	639
Coloured shrunken	Cf/cf	21379
Colourless full	cF/cf	21906
Colourless shrunken	cf/cf	672

Total number of progenies	=	44595
Parental combinations		97.06%
Recombinations		2.94%

(Tables adopted from Principles of Genetics, Sinnott Dunn and Dobzansky, 1958)

This experiment shows clearly, that whatever parental characters are there, they tend to be inherited together and not assorted independently. This example in maize, however is not an example of complete linkage, because at least in some instances recombinations are seen indicating the break in linkage. While it is true linkage is there in 96.4% and 97.06% cases in the above two experiments, it is also true that the genes have assorted independently in at least 3.6% and 2.94% cases. Thus the type of linkage seen in maize is called **Incomplete linkage.** In a complete linkage, parental combinations are retained in 100%, there are just no recombinations.

Linkage in Drosophila : Genes for body colour and wing shape are known to be linked in *Drosophila.* Morgan and his associate Bridges (1910) experimented with grey body and long winged fly. They were able to obtain by mutation another stock of fly which had black body and vestigial wings. These mutations were located on the same chromosome. The normal allele can be indicated by + and the recessive allele for body shape (*b*) and vestigial wing (*v*). When a pure line grey bodied long winged fly was crossed with another pure line black and vestige (*++/++ X bv/bv),* the offspring has grey body and long wing (+ +/*bv*). If Mendelian independent assortment takes place, the F_1 individual should produce gametes of all types in equal proportions. The four gametes are + +, +*v*, +*b*, and *bv*. When a F_1 female was crossed with a recessive male *(bv/bv),* four kinds of offspring are produced not in equal proportion but in the ratio given below. In the genotype of the offspring, the male contributes to each only the recessive genes *bv*, rest coming from the female *parent*.

1.	Grey wing	+/*bv*	41%
2.	Black vestigial	*bv*/*bv*	41%
3.	Grey vestigial	+*v*/*bv*	9%
4.	Black long	*b*+/*bv*	9%

In the above, 1 and 2 accounting for about 82% of the progeny are parental combinations, while 3 and 4 accounting for about 18% are recombinations. The strength of linkage is 82% and of cross over is 18%. In another experiment the two characters body colour and wing shape were taken in different combinations. A black long fly (b+/bv) was crossed with another grey vestigial (+*v*/*bv*). Among the four classes of progeny obtained, once again the parental combinations were more. In this experiment, the parental combinations are - black long and grey vestigial.

1. Grey vestigial $+v/bv$ 41%
2. Black long $b+/bv$ 41%
3. Grey long $++/bv$ 9%
4. Black vestigial bv/bv 9%

Again here parental combinations are 82% and recombinations are 18%. Morgan pointed out that the two genes *B* and + and *b* and *v* are linked to each other (*B* to + and *b* to *v*). This is because they ae situated on the same chromosome. In the case of *Drosophila* males, complete linkage between genes is observed as there is no cross over. Consequently the characters always appear together. When a grey long male is mated with black long double recessive female, the progeny has grey long and black vestigial.

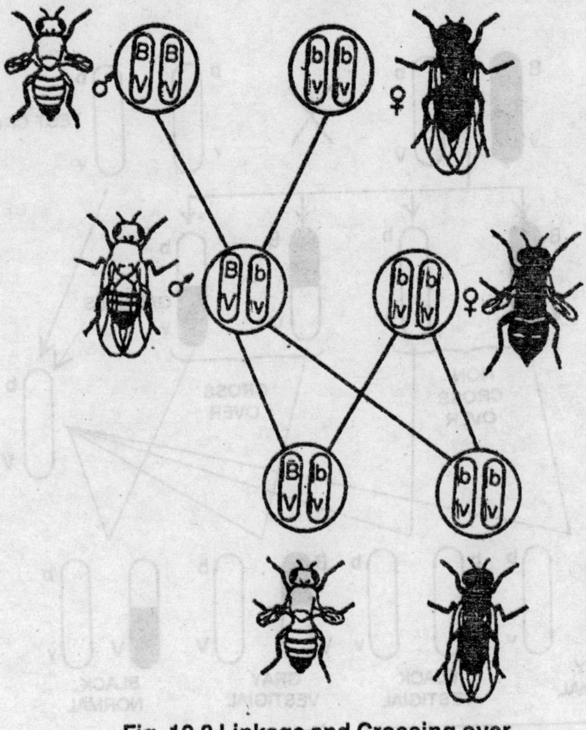

Fig. 12.2 Linkage and Crossing over
Cross in Drosophila showing complete linkage

Linkage in mammals : In several mammals linkage groups have been identified and they have been identified with their phenotypic traits. In rabbits out of 22 pairs of chromosomes, about six linkage groups have been identified. In guinea pig one linkage group has been found in human beings, but generally

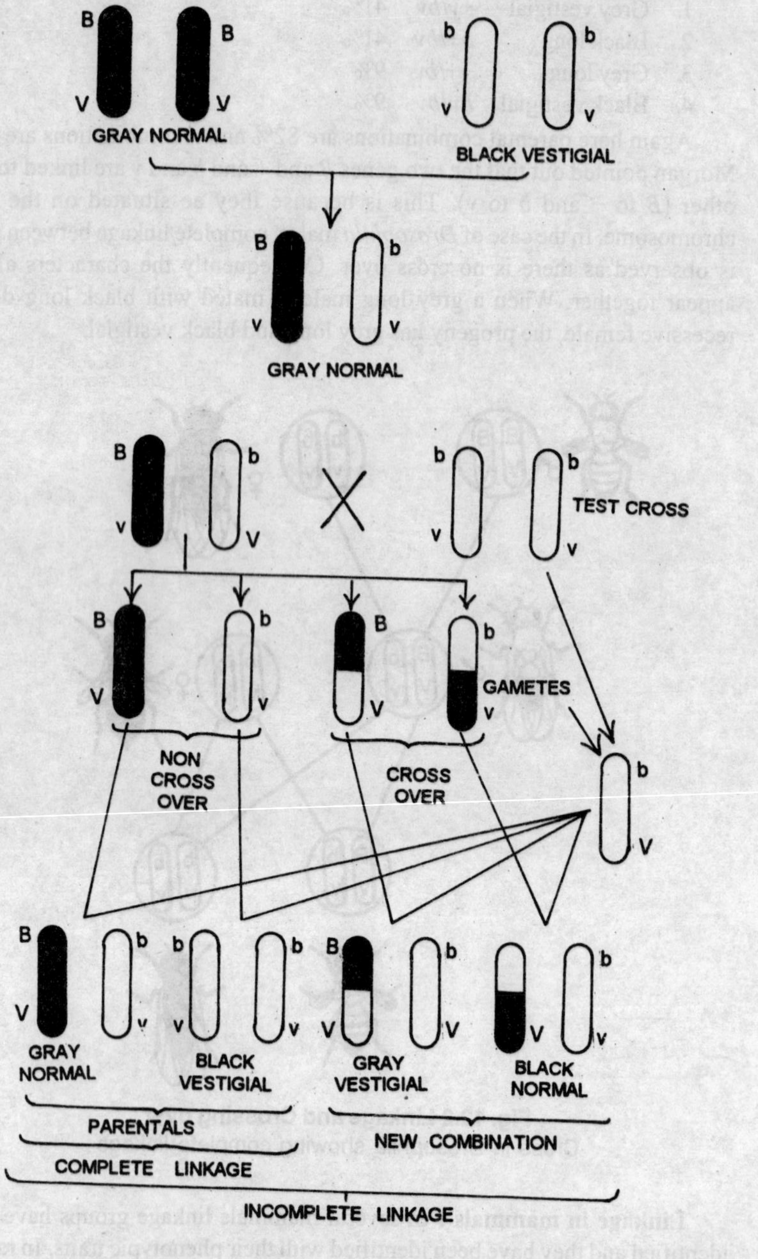

Fig. 12.3 Linkage and Crossing over
Linkage in Drosophila showing coupling

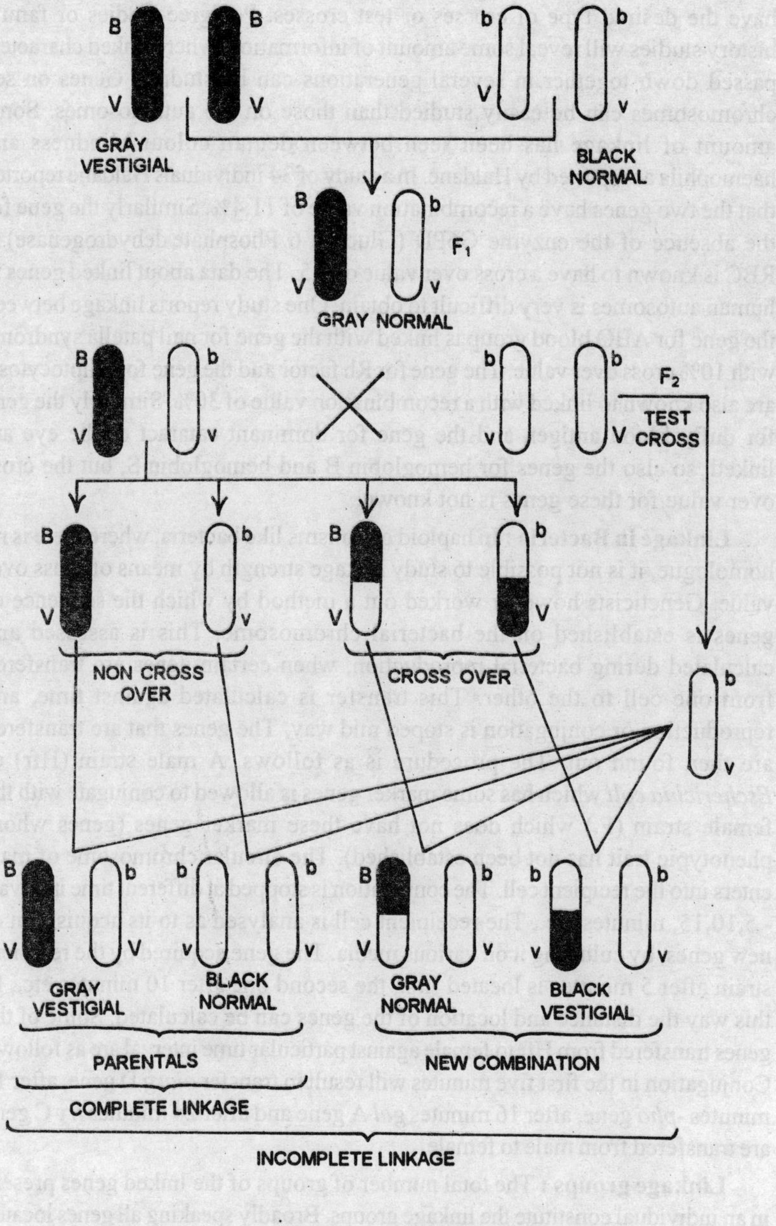

Fig. 12.4 Linkage and Crossing over
Linkage in Drosophila showing repulsion

it is difficult to study linkage or any genetic phenomenon in man as one cannot have the desired type of crosses or test crosses. Pedigree studies or family history studies will reveal some amount of information, where linked characters passed down together in several generations can be studied. Genes on sex chromosomes can be easily studied than those on the automosomes. Some amount of linkage has been seen between deutan colour blindness and haemophila as reported by Haldane. In a study of 34 individuals Haldane reported that the two genes have a recombination value of 11.4%. Similarly the gene for the absence of the enzyme G6PD (Glucose 6 Phosphate dehydrogenase) in RBC is known to have a cross over value of 6.5. The data about linked genes in human autosomes is very difficult to obtain. One study reports linkage between the gene for ABO blood group is linked with the gene for nail patella syndrome with 10% cross over value. The gene for Rh factor and the gene for ciliptocytosis are also known to linked with a recombination value of 30%. Similarly the gene for dufty blood antigen and the gene for dominant cataract of the eye are linked; so also the genes for hemoglobin B and hemoglobin S, but the cross over value for these genes is not known.

Linkage in Bacteria : In haploid organisms like bacteria, where there is no homologue, it is not possible to study linkage strength by means of cross over value. Geneticists however worked out a method by which the sequence of genes is established on the bacterial chromosome. This is assessed and calculated during bacterial reproduction, when certain genes are transfered from one cell to the other. This transfer is calculated against time, and reproduction or conjugation is stoped mid way. The genes that are transfered are then found out. The procedure is as follows. A male strain (Hfr) of *Escherichia coli* which has some marker genes is allowed to conjugate with the female strain (F.) which does not have these marker genes (genes whose phenotypic trait has not been established). The circular chromosome of male enters into the recipient cell. The conjugation is stopped at different time intervals - 5,10,15, minutes etc., The receipient cell is analysed as to its acquisition of new genes, by culturing it on various media. The gene acquired by the recipient strain after 5 minutes is located first, the second one after 10 minutes etc., In this way the distance and location of the genes can be calculated. Some of the genes transfered from Hfr to female against particular time interval are as follows. Conjugation in the first five minutes will result in transfer *oiarg* D gene, after 10 minutes *-pho* gene, after 16 minutes *gal* A gene and after 24 minutes *try* C gene are transfered from male to female.

Linkage groups : The total number of groups of the linked genes present in an individual constitute the linkage groups. Broadly speaking all genes located on a chromosome are linked to some or other degree. The strength of linkage varies. The closer they are, the chances are less of their seperation. The total linkage groups in an individual equals the number of chromosomes. If there are 7 chromosomes, there are 7 linkage groups. *Dorsophila* with four chromosomes has four linkage groups, Maize with 10 Chrosomes has 10 linkage groups etc. Sometimes in exceptional cases as in the *Gaudens* and *Valens* varities of *Oenothera* all chromosomes may form one single large linkage group. This however is a special situation brought about by translocation.

Classification of linkage : There are two types, complete and incomplete. In complete linkage, there is virtually no recombination. The same characters appear in unison for several generations. This type of linkage however is rare in nature and occurs due to absence of cross over. Male individuals of *Drosophila* and certain other insects show complete linkage. In incomplete linkage which is quite prevalent in biological world, the linked genes get seperated due to crossing over during meiosis. The extent of seperation of the linked genes depends upon the distance between the two genes.

In addition to this, interchromosomal linkage or false linkage as Longley (1945) calls it is brought about due to cytological irregularities like translocation.

Measurement of linkage : The easiest method to obtain linkage in terms of percentage value is to cross two individuals differing in two pairs of genes. The F_1 is back crossed to the double recesssive parent to obtian the parental and recombinations. The percentage of parental combinations in the total progeny gives the strength or percentage linkage between the two genes responsible for the characters.

A higher percentage of parental combinations indicates a strong linkage between the two genes and a higher percentage of recombination is indicative of weak linkage. The strength of linkage between any two pairs of genes is inversely proportional to the distance between them, lesser the distance stronger is the linkage and longer the distance, weaker is linkage.

Factors affecting the strength of linkage : There is no doubt that the distance between the two genes is the single most important factor governing the strength of the linkage. Besides this, there are several other physiological and environmental factors influencing the strength of linkage - these are

1. **Age of the organism :** Generally, increase in age decreases the chance of recombination thus strengthening the bonds of linkage. In *Drosophila melanogaster* the amount of recombination becomes smaller with increasing age.
2. **Temperature :** Initially increase in temperature increases the cross over value, hence it reduces the linkage.
3. **Radiation :** Organisms subject to ionising radiation show reduced linkage strength among the linked genes.

4. **Chemicals :** Chemicals and food consisting of various types of chemicals are also known to affect the strength of linkage.

Importance of linkage : Linked genes particularly of the desired trait play a great role in hybridization programme. As linkage helps to hold the parental characters together, selection of an individual for breeding program is based on the combination of characters which the breeder wants to remain together.

Another significance of linkage among genes in terms of evolution is that they tend to retain the identity and individuality of species by the clustering of characters.

Crossing over : It is an established fact that the genes are located on the chromosome and they (genes) tend to follow the inheritance of the chromosome on which they are situated. This means that the morphological traits whose genes are situated on a chromosome must always be inherited together. In other words in generation after generation, parental characters must repeat and the genetic composition of the chromosome must remain unaltered. But contarary to this expectation, it has been observed in almost all cases, linked genes do not always stay together even though they are on the same chromosome. This is particularly true in case of genes which are situated farther apart on a chromosome. The disruption of the linked genes has been traced to a phenomenon that generally occurs during meiotic cell division, where homologous chromosomes exchange the segments in a reciprocal fashion. This reciprocal exchange of parts of chromosomes between homologous chromosomes is called Crossing over (Morghan) and is primarily responsible for genetic variation between the individuals of the same species. It is as a result of crossing over that in the progeny, recombinations occur together with the parental combination of characters.

Crossing over may be defined as *'reciprocal exchange of segments between homologous chromosomes (generally occuring during meiosis, bringing the recombination of the linked genes particularly of those which are not very closely situated.*

There are two types of cross over depending on the type of cell where it occurs. The first type called **germinal or meiotic** cross over is of universal occurrence and is primarily responsible for variation among the individuals of a species. This type of cross over occurs, when the reproductive cells undergo meiosis during gametogenesis.

The second type of cross over which is of rare occurrence is the somatic or mitotic cross over taking place in the somatic cells during mitotic cell division. This is of no significance from the point of view of inheritance. Somatic crossing over has been observed *Drosophila* (Stern), *Aspergillus nidulans* (Pontecorvo) etc. After a somatic cross over, chromosomes will have unequal chromatids and in about 50% of the cases identical alleles go to the same pole. In the rest non identical alleles reach the same pole and if the genes involved are recessive, the cell (or its products) can be morphologically recognised.

Crossing over (meiotic) is of universal occurrence in all sexually reproducing organisms. The instances of incomplete linkage in *Drosophila* and maize also serve as illustrations of meiotic cross over.

Recombination in maize : As has already been pointed out under linkage, in a cross between coloured full and colourless shrunken a coloured full (heterozygoos) individual is produced. This F_1 hybrid when crossed with the double recessive parent the following progenies are produced.

1. Coloured full 48.2%
2. Colourless shrunken 48.2%
3. Coloured shrunken 1.8%
4. Colourless full 1.8%

Of the above 1 and 2 represent parental combinations while 3 and 4 represent recombination. The two recombinations have arisen as a result of exchange of gene loci for colour and shape between the homologous chromosomes. In this example 96% is linkage and 3.6 is cross over value. Similarly cross over value can be demonstrated in *Drosophila* also, when a grey vestigial fly is crossed with a Black long fly. (For details see linkage).

Meiotic cross over : In the meiotic cross over there are four steps. These are - a) Synapsis, b) Duplication of chromosomes c) Crossing over and d) Terminalization of chiasma.

(a) **Synapsis :** The homologous chromosomes present in the germinal cells come closer and pair longitudinally. This process called pairing or synapsis takes place during the zygotene of prophase 1. This is very important from the point of view of exchange of segments. The pairing is very specific and allelic loci in each homologue only undergo pairing. The paired chromosomes are called bivalents. Synapsis, which begins during zygotene continues up to pachytene, when chromosomes maintain physical contact at one or more points called chiasmata. Mutual attraction brought about by allelic gene loci is believed to be responsible for synapsis. The force of attraction is called the synaptic force. In order to account for the synaptic force Darlington has proposed a theory. The theory is known as the precocity theory.

Precocity theory : This theory proposed by Darlington C.D. (1937) assumes, that the chromosomes, to have an independent behaviour during division, must possess two chromatids. In other words they must replicate as it happens during mitosis. However, during early prophase the chromosomes do not replicate and they have only one chromatid which represents an unbalanced or unsaturated state. As a result, the chromosomes seek their partners and pair. During pachytene, when the chromosomes start exhibiting the double chromatids, their pairing need is satisfied and the chromosomes move apart. Darlington opines that synapsis is due to the precocious movement of the chromosomes.

Recent studies at the molecular level of the synaptic phenomenon have indeed supported the precocious movement theory of Darlington. The studies of Hotta, I to and Stern (1966) have indicated that, some amount of DNA (about 0.3%) is actually synthesised during zygotene and pachytene in the pollen mother cells of *Lilium*. The bulk of the DNA however is synthesized during premeiotic interphase. This shortage of DNA (0.3%) according lo Stern in responsible for pairing. The studies of Bogadnov *et al* (1968) is *Glyclus domesticalis* have also revealed a similar synthesis of DNA during meiotic division.

Synaptonemal complex : The appearence of a complex anastamaose of filamentous structures during the process of pairing has been reported in a number of organisms. This structure called synaptonemal complex is thought to function as a promoter of synapsis (For details see cell division).

(b) **Duplication of chromosomes :** The bivalents undergo duplication during pachytene. Each chromosome now consists of two chromatids (but the centromere will not divide); thus there will be four chromatids (tetrad).

(c) **Crossing over :** This takes place at the tetrad or the four strand stage. At the time of cross over the two opposing non sister chromatids (chromatids belonging to two different chromosomes), have a break at identical points. This is brought about due to the action of the enzyme endonuclease. The two chromatids exchange an identical length of the genome. After exchange the segments fuse with the chromatids due to the action of the enzyme called ligase (Stem and Hotta, 1969). There is some synthesis of DNA during this stage, as it is necessary to repair the broken points of the chromosome. The crossing over is physically demonstrable in the form of **chiasma.**

(d) **Terminalization :** After the exchange of segments, the two chromosomes start moving away from each other as the synaptic force lapses. The separation begins from the centromere and moves towards the ends of the chromosomes. It is this uncoupling of chiasma that is called terminalization. During diakinesis the homologous chromosomes get separated except at their ends.

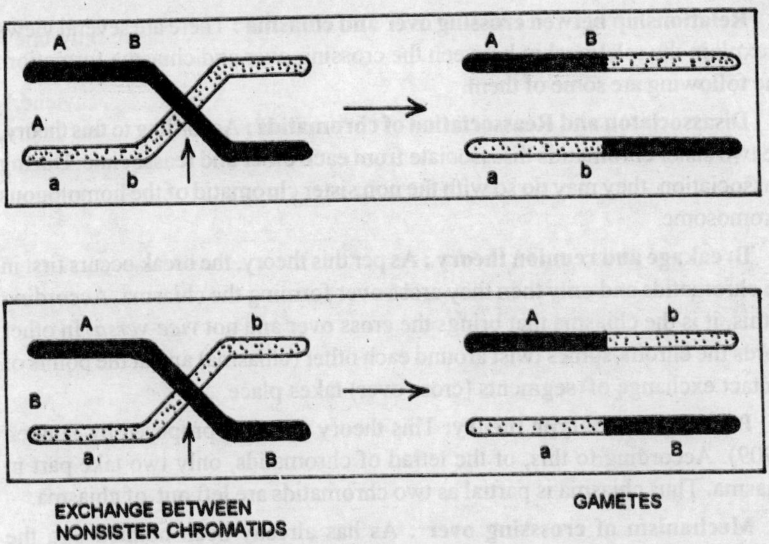

Fig. 12.5 Linkage and Crossing over
Mechanism of crossing over

Salient features of crossing over

1. **Crossing over always takes place at the tetrad stage of bivalents.**
2. **Out of four chromatids only the two opposing chromatids participate in cross over.**
3. **Cross over always takes place between non sister chromatids; never between the two chromatids of the same chromosome.**
4. **The number of chiasmata (ie. points of contact between chromosomes) depends upon the length of the chromosome - longer chromosomes will have more number of chiasmata.**
5. **Linked genes get seperated particularly those which are not very close.**

Chiasma : The process of chiasma development was first studied and explained by the Belgian cytologist Jaussens (1909) who showed the formation of a cross shaped configuration during chiasma. Subsequently a number of cytologists have studied chiasma formation, of whom the name of C.D. Darlington (1937) merits a mention. The exchange of segments is a highly sophisticatd and precise proces, where the two segments exchanged are of identical length, so that the participating chromatids neither lose nor gain genes. But even today, the precise mechanism which causes the breakage in the chromosome is still a matter of speculation. The position of chiasma on a chromosome is never fixed. The number of chiasma per chromosome varies from one to eight.

Relationship betwen crossing over and chiasma : There are several views to explain the relationship between the crossing over and chiasma formation. The following are some of them.

Disassociaton and Reassociation of chromatids : According to this theory, the two sister chromatids disassociate from each other and reassociate. During reassociation, they may do so with the non sister chromatid of the homologous chromosome.

Breakage and reunion theory : As per this theory, the break occurs first in the chromatids and only then they cross over forming the chiasma. According to this, it is the chiasma that brings the cross over and not *vice-versa.* In other words the chromosomes twist around each other (chiasma) and at the points of contact exchange of segments (cross over) takes place.

Partial chiasmatype theory: This theory was first proposed by Jaussen (1909). According to this, of the tetrad of chromatids, only two take part in chiasma. Thus chiasma is partial as two chromatids are left out of chiasma.

Mechanism of crosssing over : As has already been pointed out, the precise mechanism, which brings about breakage and reunion at specific points on the chromosome is not fully understood. Several theories have been put forward to account for the mechanism of crossing over. The following are some of them.

1. **Classical theory :** This view proposed by Karl Sax in 1932 states that, chiasma is formed due to alternate opening out between sister and non sister chromatids. At the points of overlap breakage and reunion occur resulting in cross over.
2. **Duplication theory :** When the chromosomes duplicate, longitudinal splitting occurs and the two non sister chromatids, which are close to each other may over lap, leading to break, exchange and reunion.
3. **Breakage and exchange theory :** Stern and Hotta (1969) have described the involvement of two enzymes viz. endonuclease and ligase in the break and reunion respectively of the chromatids.
4. **Breakage theory :** This theory proposed by Muller argues that there is a mechanical break in non sister chromatids first, leading to the exchange and then reunion. No cross over would have taken place before breakage.
5. **Copy choice theory :** This theory was first proposed by Belling (1932). Belling opines that the mechanism of crossing over involves two steps. First step is the duplication of chromatids resulting in the formation of new genes. Second step relates to the new connections between genes and chromosomes. When DNA synthesis takes place during

pachytene, the portion of the new helix synthesized may copy a non sister helix rather than a sister helix thus bringing about recombination. This is called the copy choice. The main objection to this theory is that it asssumes DNA duplication after synapsis, while it is known that, duplication would have taken place much earlier to synapsis. According to the copy choice theory, the recombination of the chromatids is entirely due to the DNA synthesised anew. It beleives that when the chromosomes duplicate, the new chromatid will use the chromatid of one parent as a template and then switches on to the chromatid of another parent; as a result the newly formed chromatid is a recombined product.

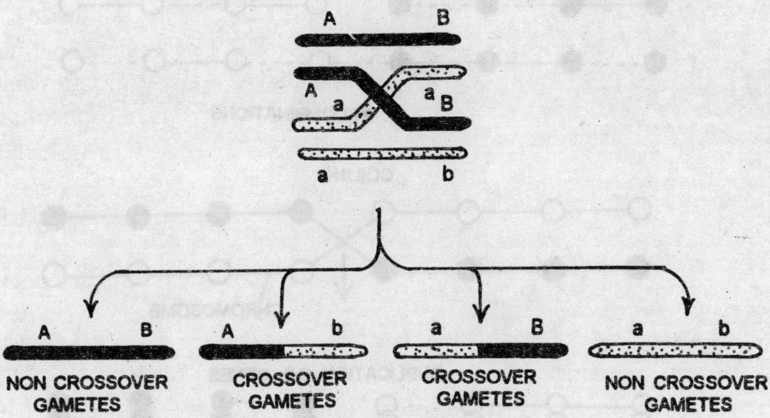

Fig. 12.6 Linkage and Crossing over
Crossing over between non sister chromatids

There is one basic difference between the breakage reunion theory and copy choice theory. The latter assumes the recombined chromatid to be a newly synthesised one whil- he former beleives that recombination is due to the break and exchange of existing genetic material. The copy choice mechanism however cannot be supported in view of the semi conservative nature ofDNA duplication. Meselson (1964) has shown in lambda bacteriophages, recombinant chromosomes have resulted from the mutual exchange of segments between unreplicaled DNA.

6. **Strain theory of Darlington :** According to Darlington, the specific relational coiling in the chromatids is responsible for breakage and reunion. The homologous chromosomes are rationally coiled around each other in zygotene; with the sister chromatids of each chromosome also being coiled, but in opposite direction to that of the coiling of the chromosomes. When chromosomes pair and seperate, the unravelling of the chromosomes and chromatids apparently takes place in opposite directions, thus exerting a strain on the latter resulting in breakage.

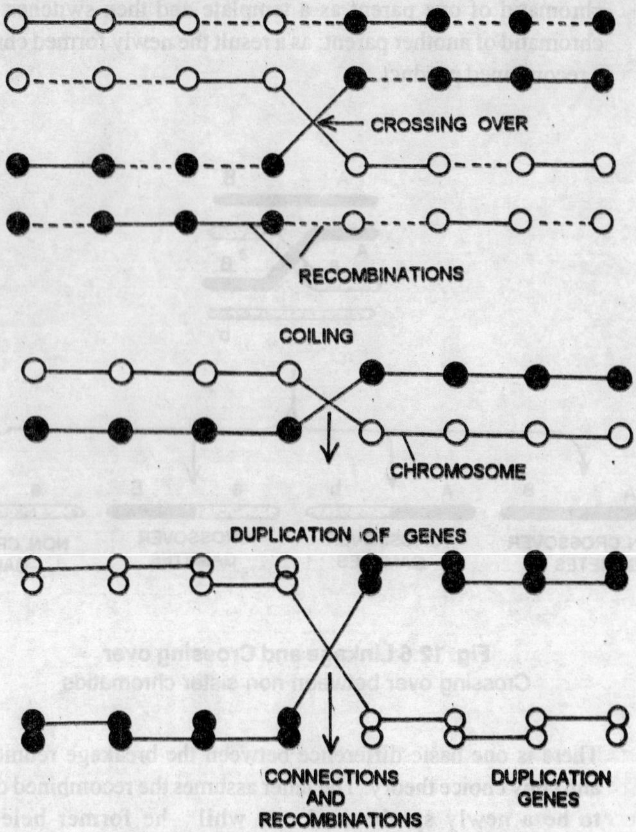

Fig. 12.7 Linkage and Crossing over
Copy choice theory of Belling

7. **Contact theory :** According to this theory, the homologous chromosomes touch each other first and as a result cross over. Breakage and reunion take place as a consequence of contact.

Molecular model for crossing over : Whitehouse and Hastings (1965) have proposed a molecular model of the crossing over phenomenon. According to them, DNA replication is practically completed prior to crossing over. When the chromosomes synapse, along their length, breakage is possible at several sites. Usually breakage occurs in one of the chains of the double helix induced by the endonuclease enzyme (Good enough and Levine 1974). The free end of the DNA unwinds from its original helix. The gap is now filled by the newly formed DNA which however uses the DNA of the non sister chromatid as a template. As a result, the newly formed DNA will be a hybrid. This attaches firmly to the original DNA. The piece of DNA of the original helix is removed and eliminated. The amount of DNA synthesised is equvivalent to the amount eliminated, so that there is no net gain.

Crossing over at the four strand stage : Analysis of the meiotic products, particularly in haploid individuals such as *Neurospora* have clearly revealed that exchange of segments has taken place during the four strand stage affecting only two chromatids. If some genes or traits have been located on these chromosomes, they can be tested by allowing the ascospores to germinate and observing their marker characters. In the ascus, out of eight ascospores adjacent ones are always identical. Such an analysis of the traits of meiotic products is called tetrad analysis

Cytological demonstration of crossing over : Crossing over between homologous chromosomes can be physically and visually demonstrated (in addition to genetic) when the chromosomes have special physical parameters. Stern (1931) has demonstrated this very clearly in *Drosophila* where the two members of the homologous pair can be identified as a result of translocation. A translocated chromosome has a modified configuration and can be identified visually. In one of the strains of *Drosophila,* a portion of the Y chromosome had been attached to one of the X chromosomes resulting in an L shaped chromosome, while the normal X chromosome is rod shaped. In another strain, the X chromosome had broken into two parts. One part had been translocated to one of the small fourh chromosome. These two strains were crossed and females were produced. In these females, one of the X chromosomes was Lshaped and the other was very small.Stern identified two genes, carnation eye (recessive mutant) and bar locus in the short X chromosome while the normal alleles were present on the L shaped X chromosome. Such a fly has red and bar eye. When such a fly was crossed with a male having carnation and bar gene located on its X chromosome, four classes of offspring were produced. Among the females of the offspring, there were two cross and two non cross over categories.

Cross over category - Carnation and normal shape red with bar shape.

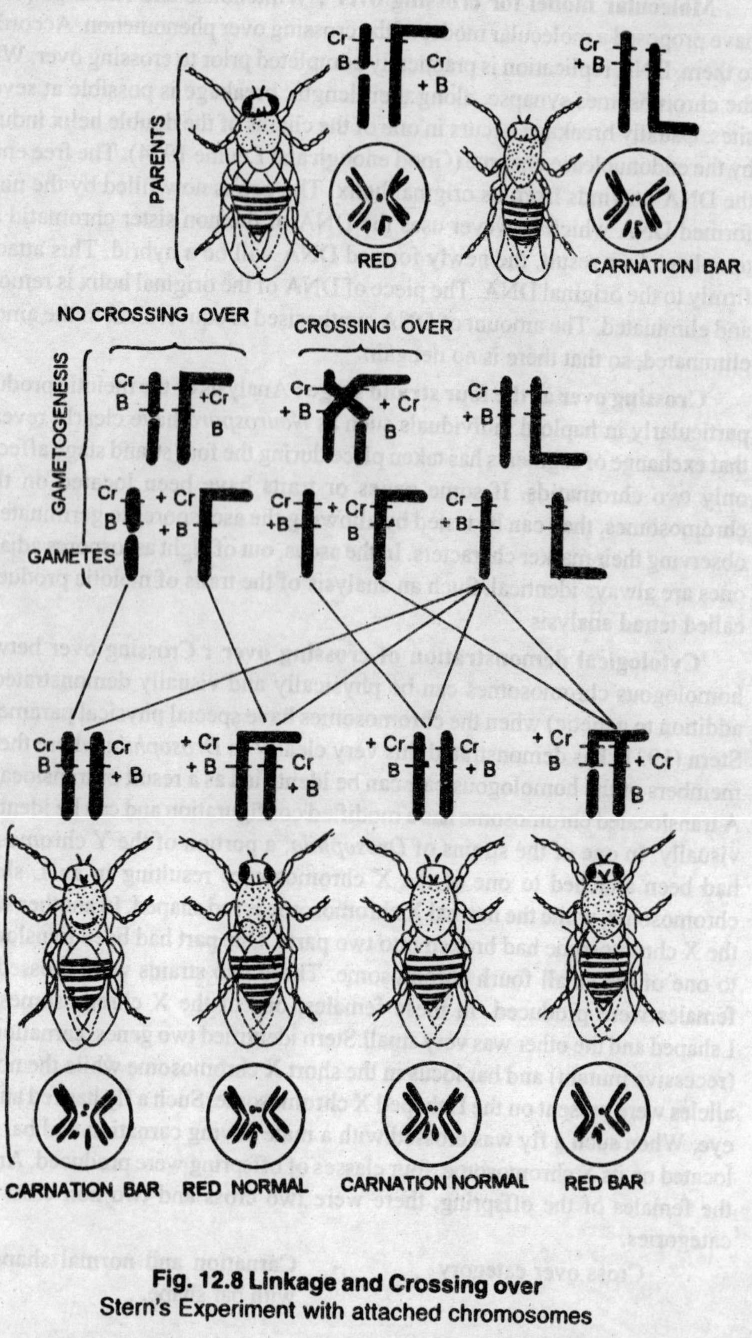

Fig. 12.8 Linkage and Crossing over
Stern's Experiment with attached chromosomes

Non cross over category - Carnation and normal shape carnation and bar.

Cytological analysis of the X chromosomes in the above flies clearly shows that, in the bar and non carnation eye, the X chromosome is evidently a result of a cross over, where the fragment of Y is now attached to the small X chromosome. In carnation and red eyes a normal X chromosome is present. This is possible only when the upper portion of the X chromosome has been exchanged during meiosis. This is a clear example where genetic recombination has been demonstrated cytologically.

Types of cross over : Crossing over is divided into three types depending on the number of points of contact. These are single point cross over, double point cross over and multiple cross over.

In the single cross over there is only one chiasma. In other words there is only one point of contact between the homologous chromosomes. In double cross over, contact between homologous chromosomes is established at two points. Recombination will be more in double cross over than in a single cross over as more quantity of genome is exchanged between homologous chromosomes. Depending on the number of strands involved, double cross over has been classified into the following categories.

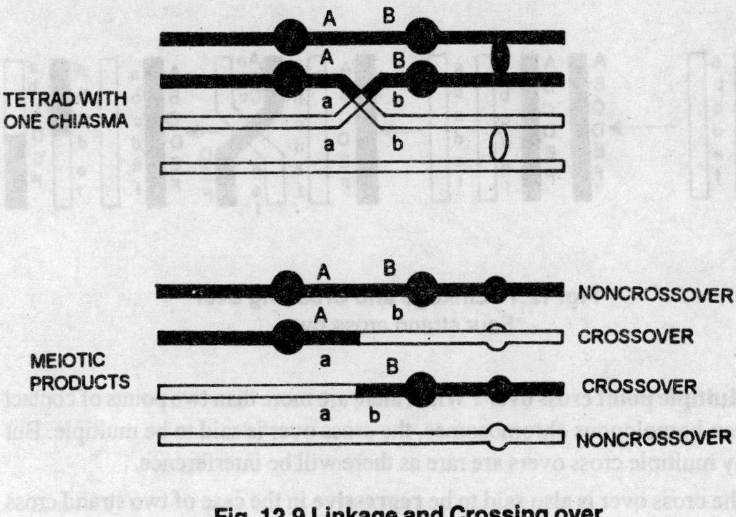

Fig. 12.9 Linkage and Crossing over
Single crossover

(a) **Two strand cross over :** In this type of double cross over, the same (two) non sister chromatids are involved in the cross over at both the points. As a result, two chromatids are cross over chromatids and the other two non cross over chromatids.

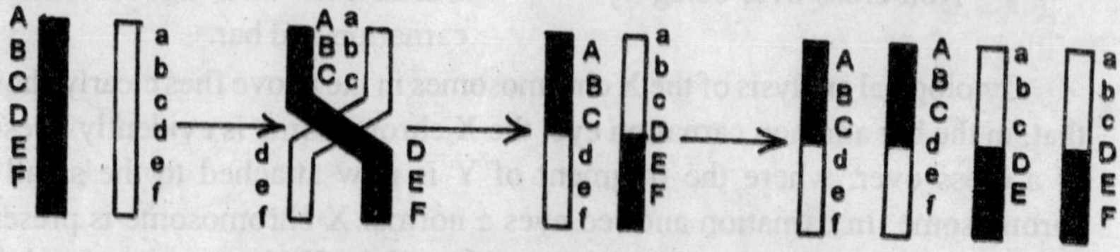

Fig. 12.10 Linkage and Crossing over
Two strand cross over

(b) **Three strand cross over :** In this case, one of the points of contact is between the two chromatids while the second contact is between one of the chromatids of first contact and another non sister chromatid. In this way three chromalids are involved in cross over, while only one will be a non cross over chromatid.

(c) **Four strand *cross over* :** In this type of cross over, all the four chromatids are involved in the cross over. First chiasma involves two chromatids, while the second chiasma involves the other two chromatids of the tetrad with the result there is no non cross over chromatid. Such chiasmata are called **complementary chiasmata.**

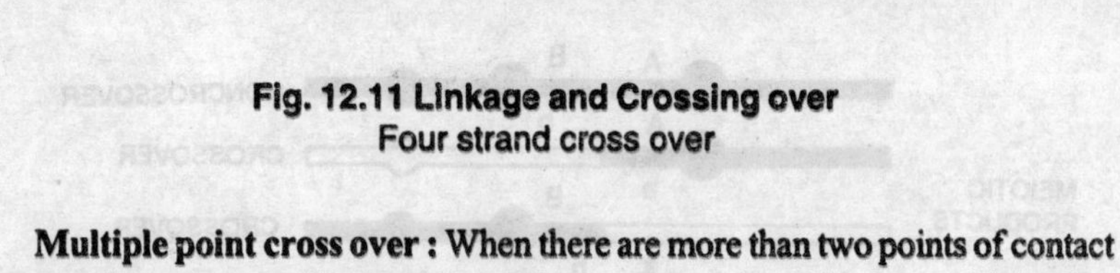

Fig. 12.11 Linkage and Crossing over
Four strand cross over

Multiple point cross over : When there are more than two points of contact between homologous chromosomes, the cross over is said to be multiple. But usually multiple cross overs are rare as there will be interference.

The cross over is also said to be **regressive** in the case of two strand cross over, **digressive** if four strands are involved and **progressive** when three strands are involved.

Certain special types of cross over have been reported in some organisms. According to Sturtevant (1952) who has reported them in *Drosophila,* an unequal cross over produces two chromatids which are unequal in the sense genes, the other suffers from deficiencies. Another type of cross over is illegitimate cross over which takes place in haploidor polyploid individuals, between the homologous segments of otherwise non homologous chromosomes.

Mitotic crossing over : This is of rare occurence and generally regarded as an abnormality - This has been reported in the somatic cells of certain strains of *Drosophila* (Stern 1936). Stern observed actual pairing and cross over in some somatic cells during the course of development. Such a cell may multiply and produce a patch of a tissue. Such a patch of cross over tissue may form a mosaic among the non cross over tissues in the body.

Factors affecting crossing over. : The frequency of cross between genes is basically dependent upon the distance between the two, but a number of other genetic, environmental and physiological factors are known to affect the frequency of cross over. The following are some of them.

1. **Sex of the organism :** In certain individuals, cross over may be reduced, suppressed or even totally eliminated altogether. The following table gives a comparison between the male and female individuals of a species with regard to the frequency of cross over.

	Organisms	Male	Female
1.	Drosophila	Absent	Present
2.	Garden pea (anther & ovary)	Present	Present
3.	Silk worm	Present	Absent
4.	Pigeon	Higher than female	Present
5.	Mouse	Present	Higher than male

The table given above clearly illustrates the role of sex in influencing the frequency of cross over. Among the instances cited above we-will discuss in some detail the situation in *Drosophila* male as it has been extensively studied. When a cross is made between gray bodied vestigial winged fly with a black bodied long winged one, the F_1 hybrid will be gray and long. Here gray body colour *G* is dominant over black body colour *b* and normal wing *V* dominant over vestigial wing *v*. When the F_1 males of the above phenotype are crosseed with double recessive females (black vestige), only two kinds of progeny are produced. Gray vestigial and black long. The recombinants namely gray long and black vestigial do not appear at all. However if an F_1 female is crossed with a double recessive (black vestige) male all the four classes of progeny (black vestige, black long, gray vestige, and gray long) appear with 83% parental combinations and 17% recombinations. The percentage of recombination is quite high indicating that the genes for wing and body colour are situated quite far apart. Hence the non occurrence of recombinants in a cross between F_1 male and double recessive female is not due to the closeness of the genes but due to the absence of cross over in males during spermatogenesis.

2. **Mutation :** The studies of Growen (1933) have indicated that in some strains of *Drosophila*,mutation has resulted in complete suppression of cross over.
3. **Age of the organism :** Bridges (1917) in his studies on *Drosophila* has shown that with the increasing age in the flies, the frequency of cross over particularly of those genes close to centromere would be substantially reduced.
4. **Temperature :** The rate of cross over will be fairly normal at a temperature range between 17 to 29 (Plough 1917). Any alteration in this range,either decrease or increase will correspondingly decrease or increase the frequency of cross over. This is however subject to certain limit.
5. **Decreased Hydration :** Lowering of water content in the surroundings is known to reduce the frequency of chiasma. It may even result in complete suppression of cross over in extreme dehydrated conditions (Sybenga 1972)
6. **Distance between the genes :** This is perhaps the most important factor influencing the frequency of cross over between two gene loci. As has been repeatedly pointed out, proximity between genes reduces the chances of cross over while distance increases the rate of cross over.
7. **Ionizing radiation :** Organisms subject to the effect of ionizing radiations show increased cross over frequency as seen in the case of *Drosophila*. One of the reports (Sybenga 1972) indicate that even in *Drosophila* males, crossing over may be induced by radiation. Work on *Lilium* and *Chlamydomonas* (Lawrence 1965) have indicated radiation effects on chiasma frequency.
8. **Chemicals :** Certain chemicals which induce chromosomal aberrations are also known to induce or suppress crossing over. Ethyl methane sulphonate is known to induce somatic cross over. Colchicine also prevents cross over by preventing pairing between chromosomes. High dose of selenium will also reduce crossing over frequency.
9. **Chromosomal aberrations :** Inversion in chromosomes alters the gene sequence and thus reduces pairing and crossing over. This is particularly true in inversion heterozygotes, when any cross over in the inverted portion would result in duplications and deficiencies in the meiotic products.

In addition to the above, some of the cytoplasmic factors are also known to have an influence on the frequency of cross over.

Significance of crossing over : The following are some of the significant features of cross over which make it a unique phenomenon in providing variation within a parameter among the individuals of a species.

1. Intraspecific variation among individuals is due to crossing over. Crossing over produces new grouping of genes and thus produces new phenotypic traits. Hence it is of great significance in breeding programme.
2. Crossing over provides physical proof for the linear arrangement of genes on the chromosomes.
3. Gene mapping on a chromosome is made possible by assessing the cross over value. The percentage of cross over is a function of distance between two gene loci.
4. Crossing over provides a new definition for the gene. According to this "Gene is the smallest section of the chromosome within which no cross over can take place.

CHROMOSOME MAPPING OR GENE MAPPING

One of the most useful aspects of the study of the phenomenon of linkage and crossing over is the help it provides in locating different genes on the chromosome. This is made possible by observing the products of a cross as to their phenotypic traits. The inheritance pattern of two gene loci and their corresponding phenotypic trait provides a clue as to the distance between them. A comparison between the parental combination and recombination among the progeny provides an understanding of the possible distance between the two genes in question. A higher parental combination with a relatively low recombination suggests that the two genes are very close, while an increased recombination value indicates that the genes are apart. The following are some of the salient features which are to be borne in mind as a preparation for the construction of linkage maps of chromosomes.

1. The number of genes far exceeds the number of chromosomes. Hence many genes are present on any single chromosome.
2. Genes are arranged in a linear fashion on the chromosome and they follow the chromosome during inheritance.
3. Generally speaking, all the genes on a chromosome behave as a single large linkage group and are inherited together. The number of linkage groups corresponds to the number of chromosomes.

4. Linked genes however may be assorted separately due to cross over taking place between their loci. The extent of separation is an index of the distance between two genes in question.
5. The strength of linkage is inversely proportional to the distance between two genes.
6. Each gene in a linkage group has a definite sequence or order as the cross over value between any two genes is always constant.

It was Sturtevant (1913), who thought that the constant value of cross over between any two genes can be exploited successfully to determine their location and relative distance on the chromosome. One can also determine the order of genes in a large linkage group. The first map of a chromosome was constructed for the X chromosome of *Drosophila* by Sturtevant and Morghan. They plotted the position of five genes.

Chromosome map may be defined as the graphic representation of genes in a linear fashion, with the distances between them (genes) being decided by cross over values.

Construction of linkage map The construction of linkage map involves several steps. At the first instance, several variants of the organism in question have to be studied and a pure breeding stock has to be obtained. Several breeding experiments are to be conducted in order to establish the linkage groups. The next step would be to identify the number of genes in the linkage group. The relative distance between these genes can be established by analysing the frequency of recombinations between these genes. The following is a detailed stepwise analysis of the construction of a linkage map.

1. **Determination of linkage groups :** Cross breeding should be conducted between two variants of a species involving at least three characters. The results of these crosses are to be analysed as to whether there is independent assortment or parental combinations far exceed the recombinations. When the parental combinations far out number recombinations, such traits or genes are said to form a linkage group.
2. **Determination of relative positions of genes :** Breeding experiments are to be conducted in such a way as to ascertain the value of recombination between one gene and the rest. For instance there are three genes A,B,and C. Cross over value for A and B and A and C has to be established first and then between B and C also has to be studied. Assuming the genes A,B,and C are linked in the same sequence, the cross over value between A and B will be less than A and C indicating, that the genes A andB are closer to each other than A and C. When the value of A and C is known, the value between B and C will help us to locate the gene in its exact position between A and C. If we assume the percentage cross over is 2% between A and B, then the distance between A and B is 2 units of map distance or 2 map unit morgan or 2 **centimorgan.**

If the F_1 hybrid (AB/ab) in a cross with the double recessive produces 16% of cross over individuals (Ab,aB) then the distance between A and B is 16 centimorgan. Each chiasma produces 50% of cross over products. 50% cross over is said to be equivalent to 50 centimorgans. If the mean number of chiasma are known for a pair of bivalents, the total length of the map may be predicted as follows.

Total length = mean number of chiasma x 50.

Two point test cross : In a two point test, cross over is known to occur at two points between two gene loci A and B. For example if an F_1 individual (AB/ab) is test crossed with the double recessive (ab/ab), among the progeny (F_2) we obtain 37% dominant genes (*AB*), 37% double recessive *ab*/*ab*, 13% first locus dominant second one recessive Ab/ab and 13% first gene recessive and second one dominant aB/ab. The last two Ab and aB are cross over gametes totalling to about 26%. Hence the distance between A and B is 26 centimorgan. Double cross over does not occur between genes less than 5 centimorgans. In order to asses the genes situated closer than this distance, a three point test cross is adopted.

Three point test cross : A three point test cross helps us not only to understand the relative distance between these genes, but also to analyse the linear order in which the genes are placed. Let us take an example in which three hypothetical genes are involved - A,B and C. These three are linked genes and they could be arranged in any one of the following sequences ABC, BAC or ACB. In the first sequence, B is in the middle, while in the second and third A and C are in the middle position.

Therefore in finding out the sequences the middle gene has to be found out first in order to locate the other two. Let us study the results of the following cross in order to understand the gene sequence from the cross over value. Pure line parents with the genotype AABBCC and aabbcc are crossed to produce F_1 hybrid with the genotype ABC/abc. When this F_1 individual is crossed with the double recessive parent abc/abc the following progenies are produced with the hypothetical frequencies given against them.

	State of cross over	Genotype	Phenotype	Supposed Frequency
1.	Non cross over	ABC/abc	ABC	36% - a
2.	Non cross over	abc/abc	abc	36% - b
3.	cross over (A - B)	Abc/abc	Abc	9% - c
4.	Cross over (A - B)	aBC/abc	aBC	9% - d
5.	Cross over (B - C)	ABc/abc	ABc	4% - e

6. Cross over (B - C)	ABc/abc	ABc	4% - f
7. Cross over A - B and B - C	abC/abc	abC	1% - g
8. Cross over A - B and B - C	aBc/abc	aBc	1% - h
			Total 100 - T

If we assume, that the cross over value between A and B = X and that of B and C = Y and that of A and C = Z, it can be calculated as follows.

1. Cross over between A - B (both double and single cross overs)

$$X = \frac{c+d+g+h}{T} \times 100$$

$$X = \frac{9+9+1+1}{T} \times 100$$

= 20 centimorgan (distance between A and B)

2. Cross over between B - C

$$Y = \frac{e+f+g+h}{T} \times 100$$

$$Y = \frac{4+4+1+1}{T} \times 100$$

= 10 centimorgan (distance between and B and C)

3. Cross over between A - C

$$Z = \frac{c+d+e+f}{T} \times 100$$

$$Z = \frac{9+9+4+4}{T} \times 100$$

= 30 centimorgan (distance between A and C)

To summarise the distance are as follows :

A - B = 20
B - C = 10
A - C = 30

Determining the gene sequence : To mark these genes on the chromosome in their proper order, let us suppose, gene A is in the middle -

I

B	20	A	30	C
B	10			C

(Actual distance)

In the above example if we assume A is in between B and C, since the distance between B and C is known to be only 10 and the distance now works to be 20 + 30 (B + A and A + C) = 50. This assumption is not correct. A cannot be in the middle.

II Let us assume that gene C is in the centre (ACB)

A	30	C	10	B
A	20			B

In this instance, the distance between A and B works out to 30 + 10 (A + C, + C + B), while the actual distance is 20. Hence C cannot be placed in the middle.

III Let assume that gene **B** is in the middle.

A	20	B	10	C
A	30			C

The above sequence of location of genes A-B-C is correct since the value of A + B and B + C (20 + 10) equals the actual value between A anc C (30). Hence the genes A, B, C are in the order in which they are noted below.

A B C

Linkage map in maize : We shall study one more example of linkage map with an illustration from maize, using a three point test cross. Three characters are chosen, all dealing with the endosperm.

1. Coloured aleurone (C) and colourless alleurone (c)
2. Full endosperm (Sh) and shrunken endosperm (sh)
3. Non waxy endosperm (Wx) and waxy endosperm (wx)

When a pure breeding coloured shrunken non waxy (C sh Wx / C sh Wx) is crossed with another pure breeding colourless full waxy individual (c Sh wx / c Sh wx) the F_1 will be coloured full and non waxy with the genotype C Sh Wx / c Sh wx. When this F_1 hybrid is crossed with the double recessive individual (c sh wx/c sh wx) the following kinds of progeny are obtained in F_2 generation. The data given below are those of Hutchinson (1922).

		Gametes C sh wx	Number of individuals in F_2
No cross over	C sh Wx	Csh Wx/c sh wx (coloured, shrunken non waxy)	5246

	c Sh wx	c Sh wx/c sh wx (colourless, full waxy)	
Single cross over	C Sh wx	C Sh Wx/c sh wx (coloured, full waxy)	229
	C sh Wx	c sh Wx/c sh Wx/c sh wx (colourless, shrunken non waxy)	
Single cross over sh - Wx	C sh wx	Csh wx/c sh wx (coloured, shrunken waxy)	1227
	c Sh Wx	C Sh wx/c sh wx (colourless, full waxy)	
Double cross over C - sh - and sh - Wx	C Sh Wx	C Sh Wx/c sh wx (coloured, full non waxy)	6
	c sh wx	c sh wx/c sh wx (colourless, shrunken waxy)	

As has been shown in the previous example involving hypothetical genes ABC, the cross over values for C - Sh, Sh - Wx and C - Wx can be worked out in order to find the linear distance between C, Sh and Wx.

1. Cross over value for C -sh $= \frac{229+6}{6708} \times 100 = 3.5\%$

2. Cross over value for Sh - Wx $= \frac{1227+6}{6708} \times 100 = 18..4\%$

3. Cross over value for C - Wx $= \frac{229+6}{6708} \times 100 = 21.7\%$

Here the cross over value C - Wx (21.7%) is approximately equal to the total value of C - Sh and Sh - Wx x (21.9%).

Hence Sh should be located between and C and Wx as follows.

C	3.5	Sh	18.4	Wx

The slight difference 0.2% (21.9 - 21.7) is due to the fact that double cross overs are not included in the C - wx cross over value.

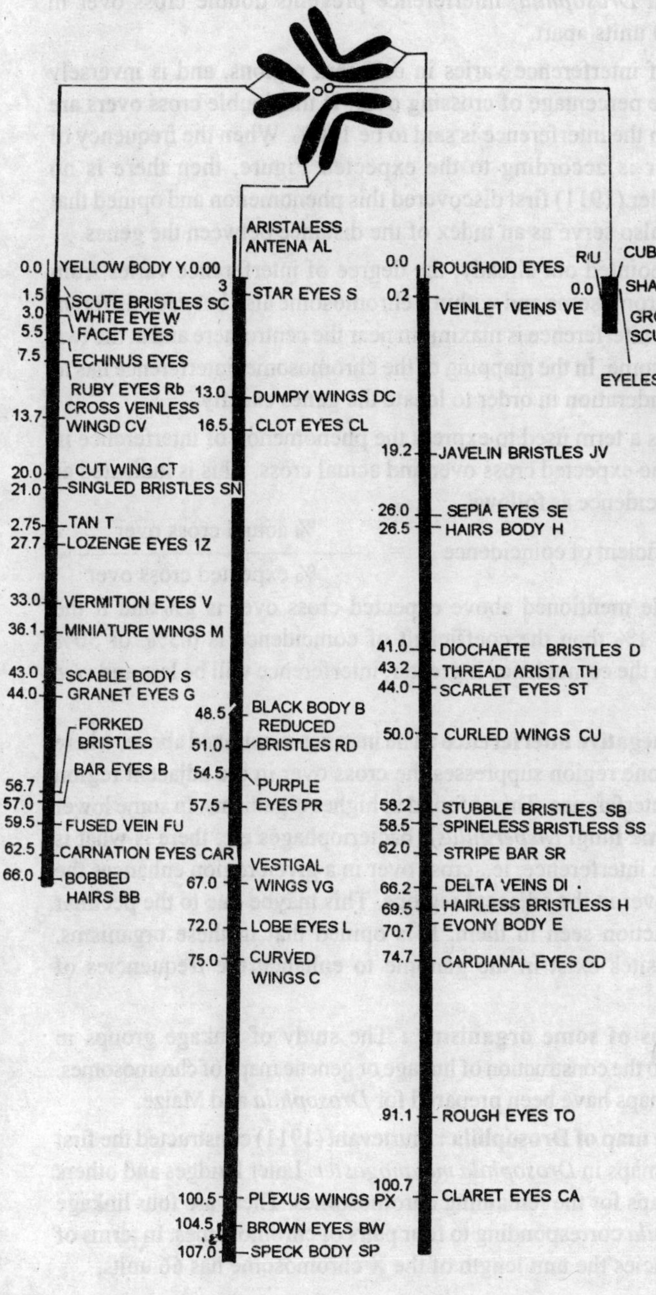

Fig. 12.12 Linkage and Crossing over

Interference. It is because of interference cross overs between close gene loci is not possible. In *Drosophila,* interference prevents double cross over in genes less than 10 units apart.

The degree of interference varies in different regions, and is inversely proportional to the percentage of crossing over. If the double cross overs are totally absent, then the interference is said to be 100%. When the frequency of double cross over is according to the expected Figure, then there is no interference - Muller (1911) first discovered this phenomenon and opined that interference may also serve as an index of the distance between the genes.

As has been pointed out already, the degree of interference varies from chromosome to chromosome and within a chromosome also between different regions. Generally interference is maximum near the centromere and at the two tips of the chromosome. In the mapping of the chromosome, interference has to be taken into consideration in order to locate the genes exactly.

Coincidence is a term used to express the phenomenon of interference in terms of ratio of the expected cross over and actual cross. This is indicated as coefficient of coincidence as follows.

$$\text{Coefficient of coincidence} = \frac{\text{\% actual cross over}}{\text{\% expected cross over}}$$

In the example mentioned above expected cross over is 2% and if the observed value is 1% then the coefficient of coincidence is 0.5%. or 50% coincidence, when the coincidence is greater, interference will be less and *vice versa.*

Positive and negative interference : The instance mentioned above, where cross over in any one region suppresses the cross over in the adjacent region is called Positive interference. This is found in higher organisms. In some lower organisms like some fungi *(Aspergillus),* bacteriophages etc, there is what is known as negative interference, ie., cross over in a given region enhances the chances of cross over in the adjacent regions. This maybe due to the peculiar modes of reproduction seen in them. It is opined that in these organisms, effective pairing sites exist in the genome to enhance the frequencies of recombination.

Linkage maps of some organisms : The study of linkage groups in organisms has led to the construction of linkage or genetic maps of chromosomes. Detailed linkage maps have been prepared for *Drosophila* and Maize.

Chromosome map of Drosopliila : Sturtevant (1911) constructed the first two chromosome maps in *Drosophila melanogaster.* Later Bridges and others constructed the maps for the remaining chromosomes. There are four linkage groups in *Drosophila* corresponding to four pairs of chromosomes. In terms of cross over frequencies the unit length of the X chromosome has 66 units, second chromosme 107.3 units, third chromosome has 106.2 umts and the fourth has only 0.2 units. These frequencies correspond fairly well with the observed length of the chromosomes.

Linkage maps in Maize : Among the plants,the chromosomes of Maize are quite extensively mapped. The pioneering works of Emerson, McClintock are worth mentioning in this regard. There are 10 linkage groups in maize corresponding to 10 chromosmes. Chromosome maps in man: A detailed map of all the chromosomes is highly desireable and perhaps will go a long way in combating genetic disorders and improving the health of man. But the construction of genetic maps of human beings is a very arduous task and is beset with many difficulties. This is not only because of a large number of genes (estimated to be about I lakh) involved but also due to many sociological constraints. It is not possible to choose mates for any desired experiment. Hence pedigree studies have to be made and results have to be computed. Sex chromosomes have been studied and tentative maps have been prepared. Recently attempts have been made by an international group of scientists to map the autosomes of human beings.

13

CHROMOSOMAL ABERRATIONS

The chromosomes - often called vehicles of heredity are sufficiently characteristic to provide the cytologist morphological criteria of an individual that can be used for identification much in the same way as a taxonomist uses phytographic (morphological characters of organs) characters to delimit the species. The constancy of karyotype is due to the fact-as Swanson (1957) observes "at a given stage of cell division and in a given tissue each cell of an organism has a constant number of chromosomes of reasonably definite volume, length and shape........" This constancy of chromosome number, volume etc., is determined in turn by the number arrangement, and sequence of genes in a definite order.

It becomes clear from the above discussion that constancy in chromosome number and structure are very essential for the individual not only to maintain its morphological characters, but many a time even for its own survival. Occasionally however, there occur variations in structure as well as number in the chromosomes producing alterations variations etc., in the individual. Depending on the magnitude of these changes, the individual undergoing these changes either will survive with little or no change or may barely survive with great changes. In some instances however, the changes are of such a magnitude that they may be lethal to the individual. These changes are called *chromosomal aberrations*. Very often these aberrations are also called *chromosoma mutations*, as they bring about change in genetic characters of the individual. Chromosomal mutations or aberrations should not be mistaken with gene mutations as the latter are involved with single genes, while the former affect either segments of chromosomes or sometimes entire chromosomes or even the entire set of genome (the total chromosomes present in an individual).

In natural populations, the chromosomal aberrations occur spontaneously but less frequently. In grass hoppers about one in a thousand individuals is known to carry some form of chromosome rearrangement. Naturally occuring chromosomal abberrations are found in human beings also. After the initial discovery of a naturally occurmg aberration, geneticists learnt that they can induce these aberrations artificially by using mustard gas, X ray, colchicine, etc.

Of what use are the aberrations to a geneticist? Chromosomal aberrations in an individual in comparison with the normal chromosomes provide a wealth of data to geneticists to understand the structure of a chromosome and also to compare any morphological expression with a particular segment of the chromosome. If the changes brought about are not good, the geneticist may try to find a way out to overcome these aberrations. On the other hand if these aberrations produce a new desirable trait, he may learn to produce those aberrations artificially.

For a cytologist, chromosomal aberrations, particularly structural are a gold mine as he can study by camparison (with the normal chromosome) the arrangement of genes, their order, frequency etc. It helps him to map the chromosome and locate a gene in a particular segment of the chromosome.

On a broad classification, chromosomal aberrations are of two kinds- *Structaral* and *Numerical.* Structural aberrations bring about a change in the chromosome structure in terms of length, shape or sometimes in the form of altered, or loss of genes and their arrangement. When the chromosomal changes affect a large segment, the changes may even be visible morphologically. Numerical changes on the other hand do not affect the structure of the chromosome, but are only involved in alteration in the number which may be either increase or decrease when compared with the normal number. This change also has many drastic genetic influences on the individual.

STRUCTURAL ABERRATIONS

As has been pointed out already, any change in the structure and architecture of the individual chromosome comes under the perview of structural changes.

The structural aberrations are of the two following categories -

(i) **Intra chromosomal aberrations**

(ii) **Inter chromosomal aberrations.**

The alterations of the genes or gene sequences influencing only one chromosome or taking place in a chromosome are called Intrachromosomal aberrations. This should not be mistaken to mean that more than one chromosome will not undergo alteration, but it means to say that all the events leading to a particular aberration take place within a chromosome.

Inter chromosomal structural aberrations on the other hand involve at least two chromosomes with reference to an aberration. Both intra and inter chromosomal aberrations are further classified as the chart below shows.

Chromosomal aberrations

- **Structural**
  - **Intra chromosomal**
    1. Deletions
    2. Duplications
    3. Inversions.
  - **Inter chromosomal**
    4. Translocations
- **Numerical** (Details given later in the chapter)

DELETIONS

Also known as deficiencies, these involve loss of a segment or loss of a portion of the chromosome, but the entire chromosome is never lost. The deleted portion which comes out of the remainder of the chromosome will not survive if it does not have a centromere. But in cases of chromosomes with a diffused centromere the fragment or fragments that come out of the chromosome do survive and this leads to increase in the chromosome number, but no loss of genes. The portion of the chromosome with the centromere (in the case of chromosomes with a single centromere) functions as a deficient chromosome.

How does a deletion occur? A break in the chromosome at any one point results in the tearing away of a part of the chromosome. This break may occur naturally or by the influence of external agencies like chemical gas, ionising radiation, drugs etc.

Types of Deletion

Two types of deletion have been noticed in organisms. These are *Terminal deletion* and *Interstitial deletion* (Intercalary deletion).

In a terminal deletion, a break occurs in a chromosome a little away from one end and a segment comes out. For instances if a chromosome has gene loci ABCDEFGH, the deleted portion will be GH and the chromosome will now have only ABCDEF. The terminal deletion produces two raw ends-in the main chromosome. The raw end heals up, while the deleted segment is lost in the next cell division as it lacks a centromere.

In an interstitial deletion, a segment comes out of the chromosome from the middle. This is brought about by two breaks at two points in the chromosome, removal of an intercalary segment and fusion of the two broken ends in the original chromosome. For instance, in a chromosome with gene loci ABCDEFGH, there will be two nicks-one between C and D and the other between E and F, with the result a segment of the chromosome carrying gene loci DE comes out, the two ends-at C and F unite to form the deleted chromosome ABCFGH.

Between the two types of deletions, terminal deletions are generally rare as they involve the loss of a telomere. The healing end of the chromosome must be such that, it should not fuse with any other chromosome.

Deletion Heterzygote

Deletion may affect one or both of the homologous chromosomes in a diploid organism. If both the homologues lose an identical segment, it will be known as deletion homozygote or as it happens quite frequently, one of the homologues may remain normal, while the other may undergo deletion, such an individual is said to be a deletion heterozygote. Generally individuals with homozygous deletion will not survive.

Occurrence

Deletions have been reported from a wide variety of organisms both in the plant and animal kindgom maize (McClintok 1941), *Drosophila,* Grasshopper, man etc, where several phenotypic traits have been associated with deletions.

Cytological identification

A homozygous deletion (if the individual survives) is difficult to identify morphologically at the chromosome level. However if the normal chromosome has been studied previously, the deleted chromosome can be pointed out because of its reduced length and also change in shape, since a terminal deletion in a metacentric chromosome would reduce the length of one arm and produces a sub metacentric chromosome.

Heterozygous deletion on the other hand is easy to identify cytologically during meiosis, when the normal and the deficient homologues undergo pairing. The pairing homologues produce specific configurations and can be easily located. In the case of a terminal deletion, the deleted chromosome will be shorter in length, hence there is no pairing in that portion and one can see a long and short chromosome undergoing linear pairing with the terminal part of the normal chromosome left out of pairing. In the case of interstitial deficiency, the normal chromosome forms a loop in that portion (of its chromosome) where the genes are lost in the other homologue.

Deletions can be easily identified by visual observation in instances like salivary gland chromosomes of *Drosophila,* where a deletion may lead to the loss of several bands. The same thing is possible in lampbrush chromosomes also where several specific loops will be lost.

Genetic consequences of Deletion

A deletion is involved in the loss of genetic material, hence it is imperative that there will be consequences which affect not only the individual but future generations also. The magnitude of genetic effect will however depend on *the size of the deletion and the quality of the genes so lost.* In many an instance the organism survives with little or no variation or may barely survive with drastic morphological and functional alteration or as in extreme cases, it may be lethal to the individual. In general however, homozygous deletions rarely survive as both the alleles are lost.

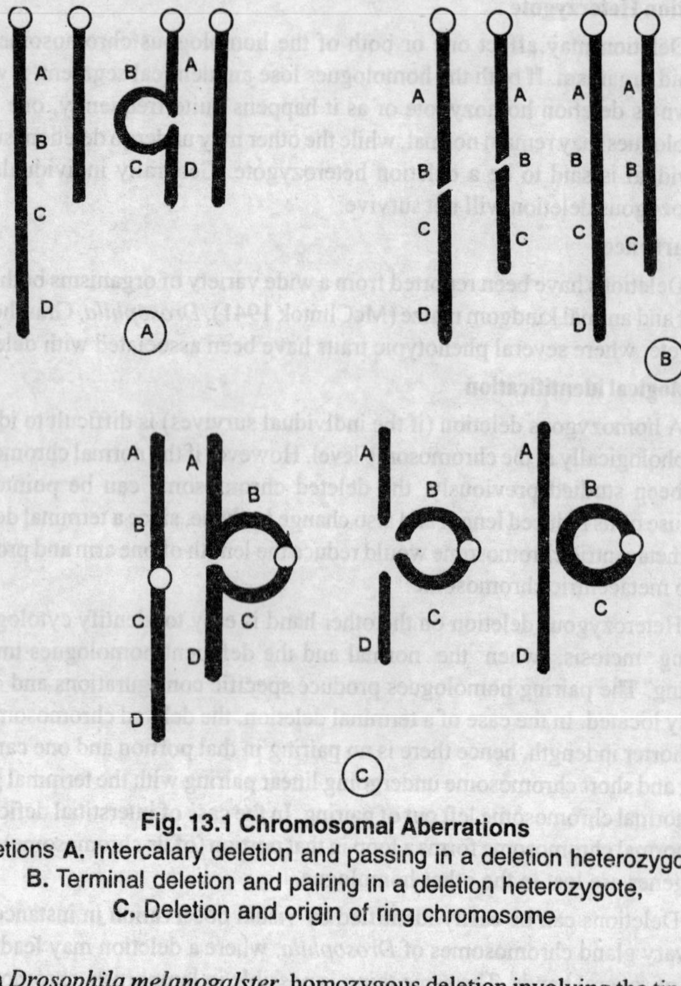

Fig. 13.1 Chromosomal Aberrations
Deletions A. Intercalary deletion and passing in a deletion heterozygote, B. Terminal deletion and pairing in a deletion heterozygote, C. Deletion and origin of ring chromosome

In *Drosophila melanogalster,* homozygous deletion involving the tip of X chromosome can survive provided they are very small characters such as yellow, scute etc, whose loss can be withstood without much problem. In *Drosophila,* several morphological traits are associated with deletions. These are blond, pale, beaded, carved, snipped and plexate. In addition to this, most of the *notch and.minutte* are also known to be due to deletions.

The phenomenon of *pseudodominance* seen in some plants by Stadler is also due to deletion. In a cross between a double dominant and double recessive, the F_1 should normally be heterozygous dominant. But if a deletion affects the gene D (dominant), then the recessive allele even in the single dose will express itself morphologically.

Pseudodominance has also been seen in the *waltzing mice.* The name is due to their (rat's) erratic movement.

In man, several chromosome deletions have been identified and associated with many congenital abnormalities in the heterozygous condition. Lejeunc and his coworkers discovered a chromosome deletion in man that has been associated with *cri-du-chat* syndrome. The babies affected with this disorder cry like cats. Other characters are microcephaly (small broad head), broad face, and saddle nose. There is general mental retardation. Generally such children die early. The symptoms are due to deletion in the short arm of chromosome no.5. Another abnormality in which chromosome no.22 has lost a fragment is called the *Philadelphia chromosome* named after the city where it was discovered by L.M. Tough and his collegues. This deletion brings-about chronic granulocytic (myeloid) leukemia and is identified with the loss of a part of long arm of chromosome no 22 as identified in bone marrow cells.

Deletion and chromosome mapping

In *Drosophila,* deletions have been employed to locate genes on the salivary gland chromosome. First of all a correlation is made between the loss of a band and the occurence of a particular morphological trait. For instance the red eye color is dominant over white eye. When the Band 302 is lost due to deletion, the eye color becomes white in the next progeny due to the loss of dominant allele, hence the white eye colour. It appears then that the loss of band 302 is near or on the 302 band. In the same way the *notch* wing appears, when ever the band 30, is lost due to deletion. In this way deletion can be employed as markers to locate the genes on the chromosome.

Evolutionary importance of Deletions

It is highly unlikely that deletions have a profound evolutionary impact, since the loss generally diminishes the potentiality of the organism. But theoritically speaking, if deletion removes some genes that suppress other genes, which can then express themselves, it may lead to new and desirable phenotypic variations being prefered by natural selection. In annual herbs, the loss of chromatin material is known to lead to specialization like adaptation to a habitat, to an insect pollinator etc. In *Crepis* (Comopositae) the loss seems to be restricted to only heterochromatin material.

DUPLICATIONS

This is another kind of structural chromosomal aberration affecting a single chromosome. In the affected chromosome, a segment of the chromosome containing several gene loci duplicates itself so that the chromosome has those genes in double dose. Duplication may occur at the tip of the chromosome or in the middle. In some instances the duplicated portion may attach to the chromosome *i.e* the same sequence or in a reverse sequence. Usually terminal duplications are rare due to the presence of the telomere. The following are the types of duplications in a chromosome with a normal gene complement ofABCDEFGHI. Let us assume that the segment DEF is duplicated.

Tandem duplication

In a tandem or sequential duplication the duplicated segment is attached to the chromosome in the same locus and in the same sequence. In the present example, the chromosome with tandem duplication will have the gene sequence *ABCDEFDEFGHI.*

Reverse tandem

Here the duplicated part will attach itself to the chromosome at the same locus but in a reverse sequence. For e.g. *ABCDEFFEDGHI*

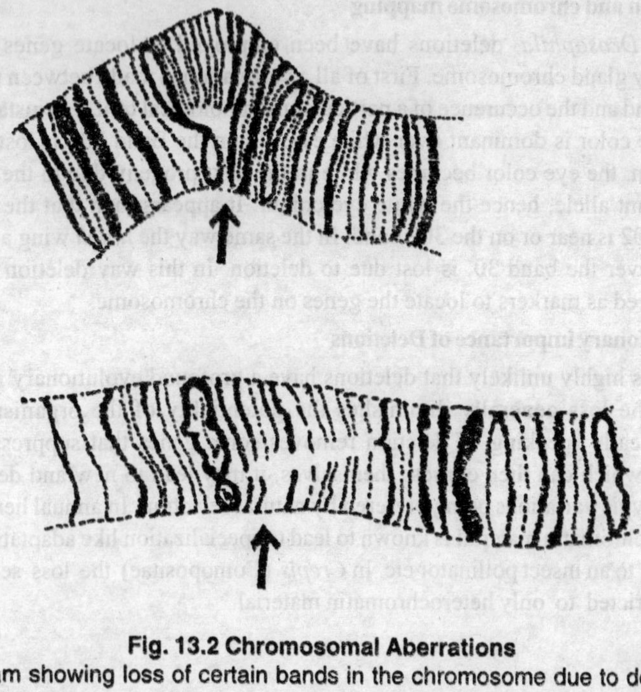

Fig. 13.2 Chromosomal Aberrations
Diagram showing loss of certain bands in the chromosome due to deletior

Displaced duplication

Here the segment DEF (duplicated portion) will attach to the same chromosome but in a different locus other than its original position. For e.g. RSTDEFUVW. Since the duplicated segment retains the sequence it is called tandem, and since it joins the chromosome in a different locus it is called *displaced tandem.* In a displaced reverse tandem duplication, the gene sequence would be RSTFEDUVW. Here the duplicated segment joins the chromosome in a different locus in the reverse order.

The displaced duplication (reverse or tandem) may take place on the same arm *(Homobrachial displacement)* or an a different arm *(Heterobrachial displacement).* Unusually, the duplicated segment may join another non homologous chromosome, in which case it is called *transposition.* (This is different from translocation where there is no duplication of any segment).

How duplication occurs

In a tandem duplication there is a break and the duplicated part joins the point of breakage. In a displaced duplication, a break occurs in a point releasing the duplicated part and then rejoins. In another part another break occurs and the duplicated segment is accommodated. Duplication may also be produced by unequal crossover between homologues. In the same way as in deletions, either one or both the homologues may undergo duplication. The former is called duplication heterozygote, while the latter is called duplication homozygote.

Duplications have been reported more frequently than deletions in a number of organisms. One reason for this is perhaps duplications are less deleterious than deletions.

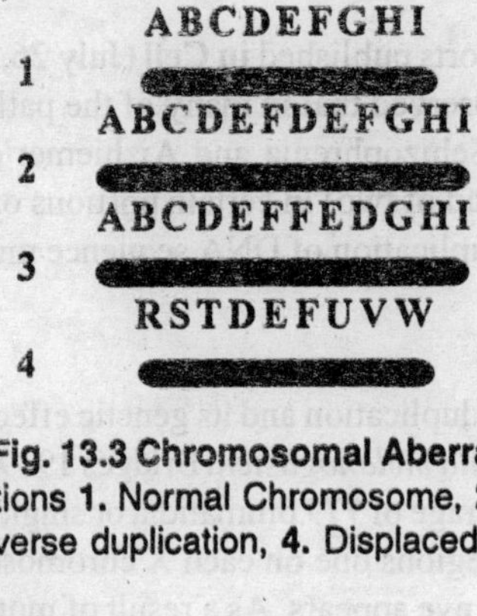

Fig. 13.3 Chromosomal Aberrations
Types of duplications **1.** Normal Chromosome, **2.** Tandem duplication, **3.** Reverse duplication, **4.** Displaced duplication

Cytological identification

The peculiar chromosome configuration observed during meiotic pairing in a duplication heterozygote provides a demonstrable visual evidence for duplication. When a normal homologous chromosome pairs with a duplicated chromosome, the duplicated fragment remains out of the synapsis in the form of loop. (In the case of deletion the normal chromosome forms a loop). Duplications are readily detected in *Drosophila* because of large size of salivary gland chromosomes, but in some plant species (haploid individuals) studies in meiosis have indicated pairing (in duplicated segments) where no pairing is expected.

Genetic effects of duplication

Generally speaking duplications are less harmful to the organism than deletions as additions are not expected to cause drastic changes. This is not true always. For instance an individual can survive a lethal gene in the heterozygous condition, but when the lethal gene is duplicated, it will be dangerous to the organism.

While deletions can be recognised genetically by the occurrence of recessive phenotypes, when they are not expected, similarly duplications can be inferred by the occurence of dominant character and total absence of recessive phenotype even in a cross between two heterozygous dominant individuals.

The phenotypic expression produced by duplication is illustrated by the attached X females in *Drosophila.* If we consider such flies which are homozygous for some reccesive, trait (located on both the X chromosomes), and these are fertilized by a male (having a *dominant gene),* the resultant individual will express the dominant character, eventhough two recessive genes are present. Evidently, the presence of dominant gene (in the same locus) in double dose on the same chromosome is sufficient to suppress the recessive allele.

In one of the recent reports published **in Cell** (July 26,1991) P.I. Patel and J.R. Lupski (1991) have observed that in many of the patients suffering from mental disorders such as Schizophrenia and Aizhiemer's disease, there are three copies of DNA (instead of two) in certain portions of Chromosomes no. 17. They believe that the duplication of DNA sequence might be the cause for mental disorders.

Position effect

A classical example of duplication and its genetic effect has been the study of *bar eye* locus in *Drosophila melanogaster* (Bridges 1919). The wild eye with the normal shape has an average of 779 ommatidia or single eyes. Genetically it is represented by two bar regions one on each X chromosome. However on a mutation in this locus a bar eye appears. As a result of mutation the bar region in one of the X chromosomes undergoes duplication, the size of the eye is reduced and it is called *bar eye* (BB,B). The individual heterozygous for bar eye has three bar regions and has 358 facets in the compound eye. In some bar individuals one more duplication may occur in the bar region (BBB, B) producing an *ultra bar* or *double bar* having totally four bar regions, each eye reduced to about 45 facets. Sturtevent and Bridges who studied these inheritance patterns of eye shape in *Drosophila* have concluded that during crossing over occuring at different levels in the chromosomes carrying bar regions, bar eyed individuals may give rise to normal individuals or even to *ultra bar.* Ultra bar is produced when crossing over joins together three bar regions. The fly having four bar regions and is homozygous bar has both the X chromosomes with two bars. Even though there are four bars, the eye will not be *ultrabar* (it has 68 facets in comparison to 45 for ultra bar), where as in heterozygous individuals with four bars, (BBB, B) there are three bar regions produced by double duplication on one X chromosome and the other normal X chromosone has a single bar locus. This individual will be *ultra bar,* because the three bar regions are located together on a single chromosome (BBB, B) while the single bar region is on another X chromosome. In the previous instance even though there are four bar regions (BB, BB) they did not produce an ultra bar, because on any one chromosome there are only two bar regions. This is due to a phenomenon called *position effect,* which is brought about by juxtaposition of bar sections (3,1) in the same chromosome having a stronger effect than the same quantity of bar regions (2,2) distrbuted on two chromosomes.

Catcheside, has found similar position effect due to duplications in *Oenothera.*

An instance of duplication and position effect in human beings is represented by the variants in haemoglobin. Haemoglobin A having 2 alpha and 2 beta chains is the most commonly found one. Haemoglobin B differs in the beta chain being replaced by delta chains. An abnormal haemoglobin (Hb-Lepore) has 2 alpha chains and one delta and one beta chain. Balgioni who studied this, reports that unequal crossing over taking place in beta and delta loci have deprived one chromosome of both the loci, while the other got both, resulting in the production of beta and delta chains.

Evolutionary significance of duplication

In contrast to deletions, duplications are certainly of some evolutionary consequence as they can produce viable and heritable morphological traits. Duplication seems to provide a feasible method for the acquisition of new genes and hence new functions. Duplications also provide the organism the survival of the original gene (in case it has any selective value) in case mutation occurs changing the original gene.

INVERSIONS

This is a kind of intrachromosomal aberration where there will be no change in quality or quantity of genes, but there will be alteration in their sequence. In order to bring about an inversion, there will be two breaks in the chromosome, with a segment getting separated from the chromosome. The segment joins back the chromosome but in a reverse sequence. For eg., in a chromosome with a gene sequence ABCDEFG, the segment BCDE is inverted, after the inverted portion rejoins the chromosome the altered chromosome would have the gene sequence AEDCBFG

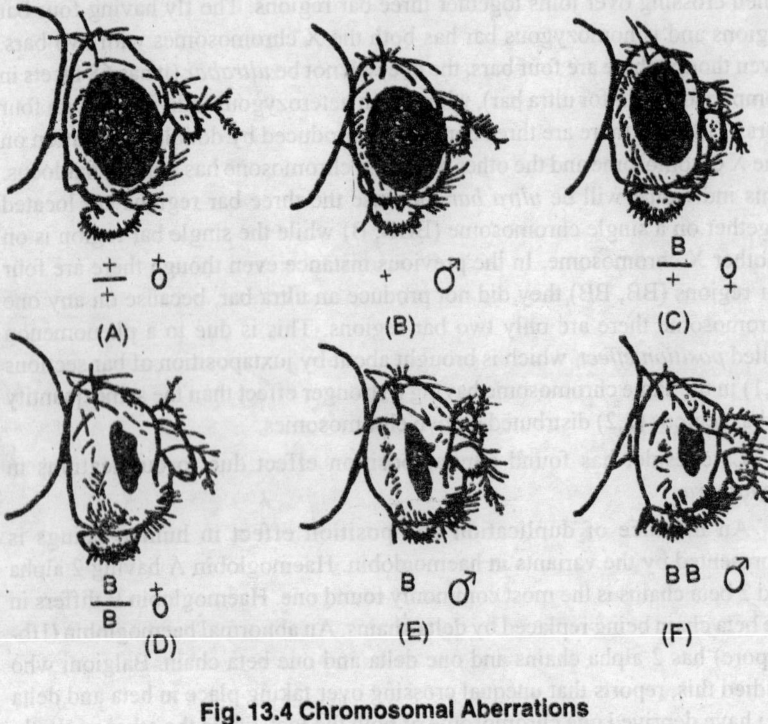

Fig. 13.4 Chromosomal Aberrations
Effect of duplication on the size in Drosophila
A. Normal female, **B.** Normal male, **C.** Normal (Heterozygous) female, **D.** Bar eyed female, **E.** Bar eyed male, **F.** Double bar male.
(Note : In D even though two bar loci are present, it is only bar eyed as the bar loci are present on different chromosome. In F, two bar loci are present on the same chromosome and the eye becomes double bar due to position effeect).

Inversions have also been reported in a wide variety of plants and animals like deletions and duplications. These perhaps have a wider distribution than the other two aberrations in natural populations. Among plants, inversions have been reported in *Paris, Tradescantia* etc. Among animal species, the favourite fruit fly, *Drosophila* exhibits quite a number of inversion types. *D.willstoni* and *D.psuedo obscura* show a variety of inversions that can be studied in great detail in salivary gland chromosomes. Several types of inversions have also been reported in maize (McClintock 1938).

Anopheles mosquito, urodeles Amphibia and many species of grasshoppers are other organisms where inversions have been reported.

Types of inversion

Inversions are classified into two basic categories viz, *paracentric inversion* and *pericentric inversion.* Paracentric inversion does not involve the centromere. In other words, paracentric inversion does not alter the morphological shape of the chromosome. In pericentric inversion however, the centromere is included and as a result quite frequently leads to changes in the appearance of chromosome in somatic cells. For e.g. as it happens *Drosophila melanogasiner,* a metacentric chromosome (2nd and 3rd) undergoing pericentric inversion, will have two breaks at different distances from the centromere. When the segment rejoins the chromosome in the reverse order, the metacentric chromosome becomes submetacentric because the centromere is shifted a little away from the centre (see diagram for details). Inversions can be cytologically detected; inversion homozygotes with some difficulty, and inversion heterozygotes easily. An inversion homozygote is one in which both the homologues have undergone inversion, while in inversion heterozygote one of the homologue would have undergone inversion while the other is normal.

Identification of inversion is possible in inversion homozygotes, only by means of genetic tests as meiosis would show normal pairing between homologous chromsomes. If the inverted portion contains known genes and is linked to genes outside inversion, the linkage will break and thus results in changed phenotype. In inversion heterozygotes on the other hand, cytological demonstration is possible visually as the normal and inverted homologues pair in a peculiar loop type configuration. While the inverted chromosome forms an overlapping loop, the normal chromosome forms a loop surrounding the overlapping loop.

Paracentric inversions : As has already been pointed out, here the centromere does not figure in the inversion. Let us assume a chromosome with gene loci *abcdefg* with the centromere between *a* and *b.* After inversion in the segment *bcde* the inverted chromosome will be *aedcbfg,* the centromere however will not be altered and it is located even now between *a* and *b.* How does pairing occur between a normal (ABCDEFG) and an inverted chromosome of this type? While the normal chromosome forms .a loop outside, the inverted chromosome will form an overlapping loop (ie one thread overlaps the other) so that all gene loci are matched. In such a configuration a cross over might take place wherein the two non sister chromatids exchange segments. In the given example let us assume a crossover occuring between C and D. What will be the type of separation and the fate of chromatids? Will all of them have same loci or will they be different? When the anaphasic movement begins there will be an unusual sight at the equatorial plate. Out of the four chromatids -two will move to the opposite poles, one will remain in the centre without moving to any of the poles while another will strecth between the two poles forming what is classically known as the *anaphasic bridge.* The gene loci of the above chromatids will be as follows:

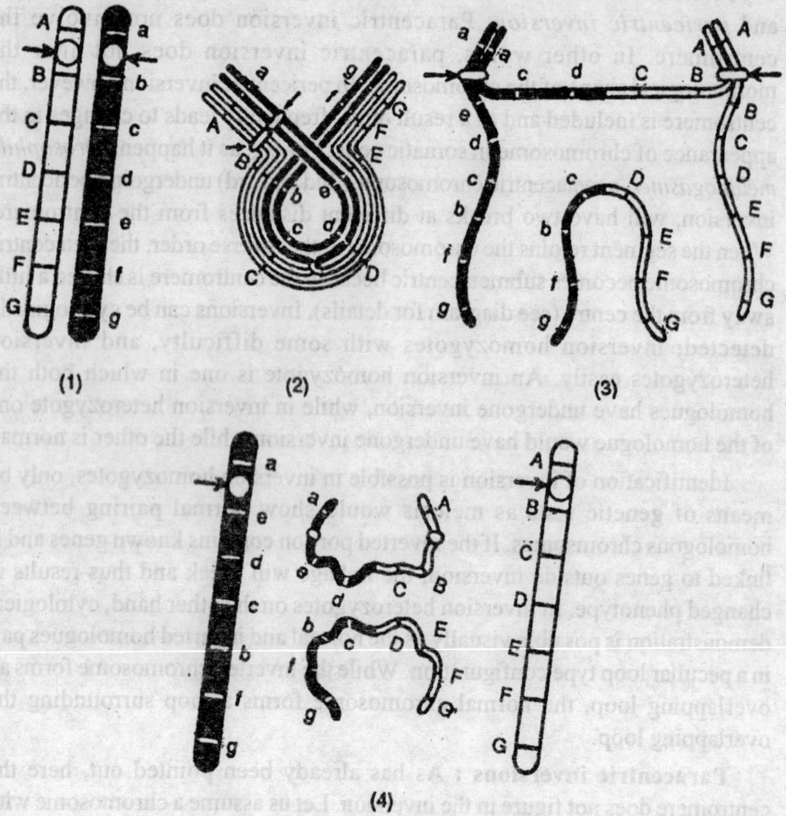

Fig. 13.5 Chromosomal Aberrations

Consequence of crossing over in a heterozygote with a paracenteric inversion

1. Homologus chromosome (Dark chromosome) shows inversion
2. Pairing and overlapping loop forming in an inversion heterozygote
3. Movement of chrosomes during anaphase (Note in the chrosome bridge formed by a dicentric chrosome and an acentric chromosome lying in the centric
4. Meiotic products showing duplications and deficiencies

Chromatids at	1 aedcbfg
Either poles	2 ABCDEFG
Chromatid at the centre	3 gfbcDEFG
Chromatid streching between poles	4 aedCBA

(If one carefully traces the diagram (at Diplotene) following the chromatid with different colours, it will be easy to find out the gene loci that the four different chromatids will get.)

In the chromatids mentioned above, No. 1 will be an inverted chromosome; it will have a centromere and the gamete that encloses this chromatid will be viable as it has all the gene loci. Chromatid two is a perfectly normal one. The gamete that gets this will also survive. Chromatid 3 will not survive because not only it has duplications and defficiencies for genes (deficiency for A and duplication for G and F) but it also lacks a centromere. Hence the gamete which will include this chromatid will not be viable. In fact the chromatid will normally be not included in the gamete as it fails to move during division. The chromnatid No. 4 will have two centromeres and is streched between the two poles. It has duplications and deficiencies (deficiency for A and duplication for A) This chromatid also is not viable. Normally however the oocyte where there is inversion and a cross over occurs within the inversion, it (egg) will not survive due to meiotic irregularities mentioned above; because after meiosis I the products may not participate in the second division. The chances of survival of gametes however is only 50% in the case of pollen mother cells.[In the female side after meiosis, only one cell *ie* egg will be produced where as in the male four cells- sperms-{ pollen in plants} are produced.

McClintock, Darlington and Stebbins have observed the acentric fragments and chromosome bridges in a variety of plants and these serve as good cytological indications of inversion when the chromosomes are not as large as salivary gland chromosomes. Sturtevant and Beadle (1936) opine that only the non cross over chromatids will survive provided the cross over is suppressed. *In Drosophila* males inversion does not pose a problem as cross over is generally absent. In females with inversion, whose eggs survive it is either due to lack of cross over or due to non inclusion of anaphase bridge in the egg nucleus.

When there are two cross overs within an inverted segment the survival will depend upon the number of chromatids involved. Double and multiple cross overs are however rare.

Pericentric inversions

The inversions that include the centromere are called pericentic inversions. Very often pericentic inversions change the morphology of the chromosome. However if the two breaks in the chromosome occur at equal distance from the centromere, there will not be any morphological change in the chromosome. While there is no estimate of natural populations where pericentric inversions have drastically altered chromosome morphology, in certain Orthoptera and species of *Drosophila,* centromeric shifts brought about by inversion have been responsible in producing new karyotypes.

Pairing in an inversion heterozygote produces loop formation in pericentric inversion also. Let us study the example of a pericentric inversion in a chromosome with the gene loci *abcdefg* with the centromere being situated between *d* and *e* . The chromosome obviously is a metacentric one with inversion taking place in the segment *bcde*, the resultant submetacentric chromosome will have gene loci *aedcbfg*. During meiosis pairing occurs in a heterozygous individual between a normal {ABCDEFG} and an inverted *{aedcbfg}* chromosome. Similar type of loop formation seen in paracentric inversion is seen here also. If a cross over takes place {see Fig. 1.90 in the inversion region between gene loci Cand D and the cross over products separate during anphase what will be the chromosome compliment of each one of the chromatids? Unlike in the case of paracentric inversion no anphase bridge or acentric fragment is produced; all the chromatids will have centromere but they {50%} suffer from duplications and deficienicies. The chromosome compliment of each of the chromatid is as follows.

	↓	
Chromatid 1 (non cross over)	ABCDEFG	(Metacentric)
	↓	
Chromatid 2 (cross over)	ABCdea	(Submetacentric)
	↓	
Chromatid 3 (cross over)	gfbcDEFG	(Submetacentric)
	↓	
Chromatid 4 (non cross over)	aedcbfg	(Submetacentric)

(Arrows among gene sequence indicate the position of the centromere)

In the above instance, if the inversion has occured on the male side 50% of the pollen (in plants) or 50% of the sperms (in animals) would survive. Balance would abort due to duplications and deficiencies. In the case of female however, it depends as to which of the chromatid would be included in the egg. The egg would survive if it gets chromatid 1 or 4, if it gets 2 or 3 it would not survive. In the case of embryo sacs of higher plants generally the intact chromatids (1 and 4) are so oriented that they will be included in the outer megaspores and the irregular chromatids (2 and 3) will be in the central megaspores. Consequently, the basal megaspore receives either a normal or an inverted chromosome and the gametophyte survives.

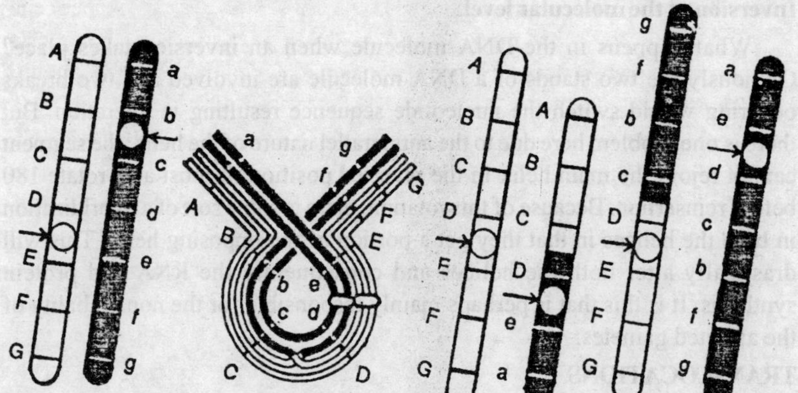

Fig. 13.6 Chromosomal Aberrations

Consequence of crossing over in a heterozygote, with pericentric inversion

1. Homologous chromosomes
2. Crossing over with loop formation
3. Meiotic products

Overlapping inversions

In some instances a second inversion may take place within the first inversion; such a type of inversion is said to be overlapping inversion. For example the normal chromosome is ABCDEFG and the inverted chromosome is AEDCBFG; a second inversion taking place within the inverted segment EDCB between B and E will result in a chromosome of the gene loci - AECDBFG.

Role of inversions in evolution : Interest in inversion having a bearing on evolution centres round paracentric inversions. Overlapping inversions studied in *Drosophila pseudoobscura* and *D.persimilis* have shown that the inversions are interrelated. An analysis of the natural populations at mountain ranges and elsewhere has shown that, populations in cool weather have a high frequency of inversions than those found elsewhere. This indicates that inversions are not only tolerated but may actually have selective value in natural populations. For the inversion to have any selective value, it must not be too long to have cross over effecting the allelic sequence but must be long enough to accumulate a genetic difference. It is possible according to Swanson (1967) that inversions can serve as foci for species divergence given sufficient time to accumulate genetic differences and the erection of barriers to prevent breeding within the population.

Inversion at the molecular level

What happens in the DNA molecule when an inversion takes place? Obviously the two stands of a DNA molecule are involved and two breaks occuring would switch the nucleotide sequence resulting in alteration. But there is one problem here due to the antiparallel nature of the helix; the segment cannot rejoin the main helix in the reverted position; it must also rotate 180 before reinsertion. Because of this rotation, there is some sort of a hybridization in both the helices in that they get a portion of the opposing helix. This will drastically alter both the helices and consequently the RNA and protein synthesis. It is this that is perhaps mainly responsible for the non viabality of the affected gametes.

TRANSLOCATIONS

This is a type of structural aberration affecting at least two chromosomes. Hence it is also known as inter chromosomal structural aberration. As the very name indicates, in *translocation* a segment of the chromosome is removed and it attaches itself to another non homologous chromosome. Let us assume that there are two non homologous chromosomes 1) ABCDEF and 2) GHIJKL; the DEF portion from the first chromsome will attach itself to the second chromosome. After traslocation, the two chromosomes would be 1) ABC and 2) GHIJKL DEF. There would be deficiency in the first chromosome, but there are no duplications, in the second chromosome. This type of translocation is theoretically possible but it will not normally take place because of the presence of telomere in the other chromosome (GHIJKL) which will not allow any attachment at the tip. Normally however two non homologuous chromosomes exchange their segments. For e.g. ABCDEF and GHIJKL exchange segments DEF and JKL. After translocation the two chromosomes would be ABCJKL and GHIDEF. In order to have this type of exchange, there will be breaks in both the chromosomes. Such a type of translocation in which two non homologous chromosomes mutually exchange segments is known as **reciprocal translocation**. The two new chromosomes thus produced will survive provided they have only one centromere. If the reunion is such that as to produce dicentric chromosomes or acentric fragments, they will generally not survive due to failure of proper seggregation during anaphase.

Like inversions, translocations may exist either in a homozygous state or in a heterozygous state. Homozygous translocations behave perfectly normal during meiosis, exhibit linear pairing except for the fact of new linkage group of genes. It they survive, they give rise to new chromosome race. These are found sporadically in many organisms, but more commonly in *Oenothera, Paeonia, Datura* etc., in plants and scorpions and roaches among animals.

Translocation heterozygotes have only one of the homologues (of 2 chromosomes) affected while the other is normal. For e.g. considering the above example the pollen mother cell or MMC (or oocyte or spermatocyte) will have the following chromosomes.

Homologue I	ABCDEF and ABCJKL
Homologue II	GHIJKL and GHIDEF.

Homologue I **ABCDEF and ABCJKL**
Homologue II **GHIJKL and GHIDEF.**

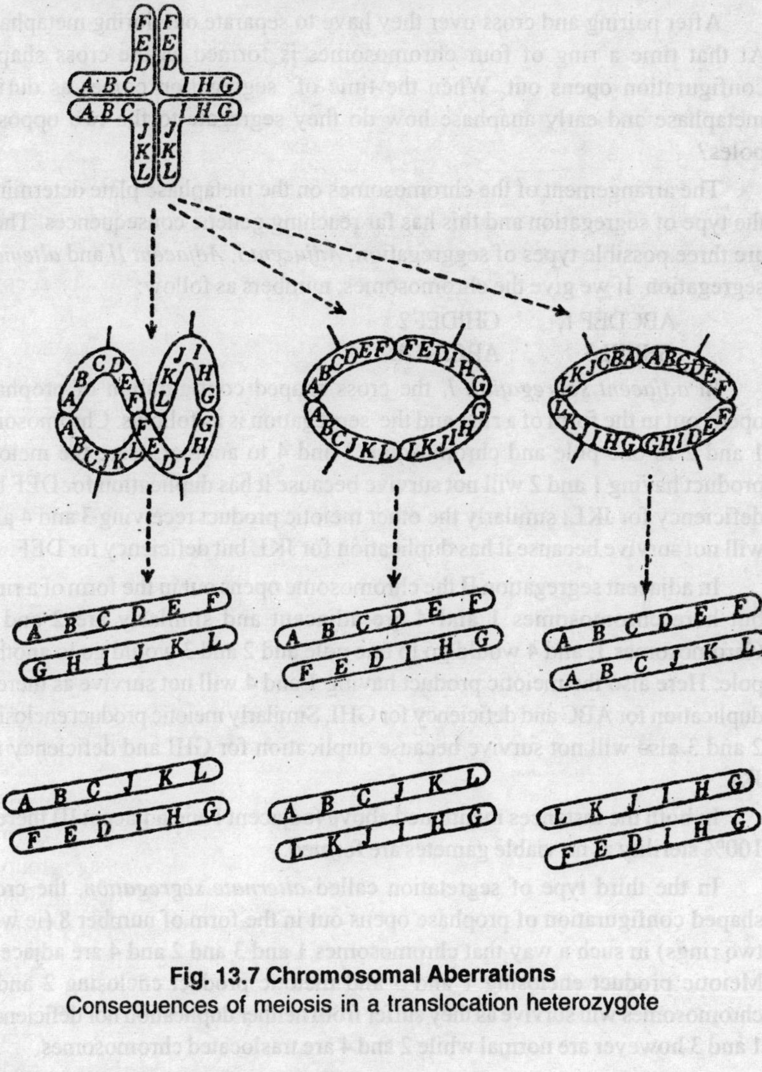

Fig. 13.7 Chromosomal Aberrations
Consequences of meiosis in a translocation heterozygote

In such a situation how can pairing take place between ABCDEF and ABCJKL; GHIJKL and GHIDEF. If there is a linear pairing roughly 50% of the length of the chromosome in both cases will remain out of synopsis. This is not conducive to the individual. Hence in order to have pairing, at all gene loci, all the four chromosomes combine and form a peculiar cross shaped or + shaped configuration at pachytene with the pairing partners being changed at the point of translocation break. In other words one half of the chromosome pairs with its homologue and the other half pairs with the allelic loci attached to a non homologous chromosome.

After pairing and cross over they have to separate out during metaphase. At that time a ring of four chromosomes is formed as the cross shaped configuration opens out. When the time of segregation comes as during metaphase and early anaphase how do they segregate to the two opposite poles?

The arrangement of the chromosomes on the metaphase plate determines the type of segregation and this has far reaching genetic consequences. There are three possible types of seggregation. *Adjacent I, Adjacent II* and *altemale* segregation. If we give the chromosomes, numbers as follow:

ABCDEF 1, GHIDEF 2
GHIJKL3 ABCJKL4.

In *adjacent segregation I,* the cross shaped configuration of prophase opens out in the form of a ring and the segregation is as follows. Chromosome 1 and 2 to one pole and chromosome 3 and 4 to another. Here the meiotic product having 1 and 2 will not survive because it has duplication for DEF but deficiency for JKL; similarly the other meiotic product receiving 3 and 4 also will not survive because it has duplication for JKL but deficiency for DEF.

In adjacent segregation II the chromosome opens out in the form of a ring; but here chromosomes 1 and 4 are adjacent and similarly are 2 and 3. Chromosomes 1, and 4 would go to one pole and 2 and 3 would go to another pole. Here also the meiotic product having 1 and 4 will not survive as there is duplication for ABC and deficiency for GHI. Similarly meiotic product enclosing 2 and 3 also will not survive because duplication for GHI and deficiency for JKL.

In both the instances mentioned above (adjacent I and adjacent II) there is 100% sterility as no viable gametes are formed.

In the third type of segretation called *alternate segregation,* the cross shaped configuration of prophase opens out in the form of number 8 (ie with two rings) in such a way that chromosomes 1 and 3 and 2 and 4 are adjacent. Meiotic product enclosing 1 and 3 and meiotic product enclosing 2 and 4 chromosomes will survive as they suffer from neither duplication nor deficiency. 1 and 3 however are normal while 2 and 4 are traslocated chromosomes.

The translocations discussed above deal with the involvement of only two non homologous chromosomes and in the ring at metaphase only four chromosomes are found. However if the arm of one of the translocated chromosomes is involved in exchange with a third non homologous chromosome, a ring of 6 chromosomes is found at the metaphase. If there is a third exchange a ring of eight chromosomes would be formed. This process can go on ultimately to involve all the chromosomes in an individual to form a *translocation complex.* This phenomenon is found in *Rheo discolor* a member of the family Commelinaceae of monocots, where a ring of 12 chromosomes are found at metaphase. This is called *chromosome ring.* (different from ring chromosome seen in some prokaryotes) and is due to the whole genome being involved in translocation. The plants *Oenothera* and *Paeonia* represent different stages (grades) of translocation complex. *Paeonia californica* with 5 pairs of chromosomes has the following combinations

(a) Ring of 4 Chromosomes and 3 bivalents
(b) Ring of 6 Chromosomes and 2 bivalents
(c) Ring of 8 Chromosomes and 1 bivalent
(d) Ring of 10 Chromosomes and no bivalent.

In *Oenolhera* also, different grades of complexes are found; while *O.hookeri* has all 7 normal bivalents, *O.lamarckiana* has all the 14 chromosomes forming a ring. (more details of *Oenothera* are discussed later)

Genetic detection of translocation

Heterozygous translocation may be detected by crossing two such individuals. The resultant progeny is of three types *viz.* Normal homozygotes, Interchange heterozygotes and Interchange homozygotes.

Translocation both in the homozygous and in the heterozygous condition affects the gene linkage. In homozygotes, the linkage is disrupted, if the translocated segment is not linked to its original group. Study of this linkage can be used to detect translocation and identify the chromosome involved. In the translocation heterozygote all the genes in the involved chromosomes are linked. This is because only the perfectly balanced gametes survive and take part in fertilization. Heterozygotes can also be detected by their semisterility as in the male side only 50% would survive. In the female side however there can be partial or complete sterility.

Evolutionary implication of translocation

The evolutionary implications of translocation are quite perplexing in that they produce new chromosome linkages. Among plants *Datura* and *Oenothera* have provided a wealth of information. In a translocation heterozygote involving two chromosomes, a ring of four is formed at the metaphase plate. The four chromosomes do not have an independent assortment. Sheer pressure of survival, forces on them alternate seggregation, resulting in two normal chromosomes forming a group and so are the two translocated chromosomes. In other words the two chromosomes which were initially independent, now are linked in spite of the fact that chromosomes exist as independent entities. As the translocation complex increases in number, the independent behaviour of the chromosomes decrease and as it happens in *Oenothera lamarckiana,* all the 14 chromosomes form one large translocation complex (a large chromosome ring is formed at metaphase involving all the 14 chromosomes). In this plant, the translocation complexes are called the **Renner complexes** named after the discoverer. Of the 14 chromosomes found at the metaphase plate, once again the chromosomes can not afford the luxury of free or independent seggregation. Survival instinct again forces them into two groups (or complexes) of 7 each. In effect there are two linkage groups with 7 chromosomes each. In a seggregation of this type there is no lethality. These two chromosome groups are called *Renner complexes.* Each ring having 14 chromosomes is actually a dual entity, because each has two complexes which may differ in genetic content. In *O.lamarckiana,* the two Renner complexes called **Gaudens** and **Velans** yield entirely different projenies when crossed with other forms. The individuals of *O.lamarckiana* are generally self pollinated and breed true and retain the structural complexity. A further refinement of the Renner complex is the *Renner effect,* when the two complexes are trasmitted in different ways ie. one *via.* egg and the other *via* sperm.

The evolutionary path in *Oenothera* is via the translocation complexes. The translocation having once arisen in the course of evolution has been stabilized in the gene pool by making it workable in the genetic machinery. The *Oenothera* is a clear example as to how a genetic irregularity deleterious to the organism to a certain degree has been incorporated into the system and made to function well. Similarly, the absence of cross pollination and retention of only self pollination, which to a certain extent is disadvantageous to the organism, is actually beneficial here as it helps in maintaining the complex, while cross pollination can break the complex by bringing in new and unknown genes into picture.

NUMERICAL ABERRATIONS

In the biological heirarchy, species form the basic genetic unit capable of surviving and maintaining its structure. This constancy (subject of course to mutational influences) of character in a species has as its basis, the chromosomal background. The number and kinds of chromosomes in a species are sufficiently constant to delimit it from others. But as the genes mutate or change, so also chromosomes are subject to variations. The process of alteration in chromosome numbers seen very often in plants and to a less extent in animals is called numerical aberrations.

Numerical aberrations in chromosomes produce two types of cells or individuals. These are 1) those whose somatic genome (chromosome complement) is the exact multiple of the basic number characteristic of the species and 2) those in which the somatic number is an irregular multiplication of the basic number. Occasionally there could be reduction in the basic number also. The first category is said to be *Euploid* and the second one is called *Aneuploid.* Euploids may be Haploid (monoploid), diploid, triploid, tetraploid and polyploid. Haploid is the basic number of the species with one set of chromosomes in the body cells, diploids have 2 sets, triploids have 3 sets, tetraploids have 4 sets and further multiplication series are refered to as polyploid.

Individuals whose genome complement is aneuploid have one or two extra chromosomes added or one or two chromosomes deleted from the normal number of diploid in the body cells. For instance a single chromosome may be missing from the diploid complement when it is called 2n -1: The following is the classification of numerical aberrations.

Numerical aberrations

- Euploidy
  - Haploidy
  - Diploidy
  - Polyploidy
- Aneuploidy
  - Hyperploidy
    - (a) Trisomy
    - (b) Tetrasomy
  - Hypoploidy
    - (a) Monosomy
    - (b) Nullisomy

ANEUPLOIDY

In aneuploidy, there is either reduction or partial increase of the chromosome number. Reduction is said to be hypoploidy, while increase is said to be hyperploidy. Hypoploidy is due to loss of one or two chromosomes. If a single chromosome is lost from a diploid complement it is said to be Monosomy (individuals are called *monosomics)* and is symbolically designated as 2n-l. If two chromosomes are lost (ie. both the homologues in a pair) it is said to be

Nullisomy (2n-2). Similarly the addition of a third chromosome to one of the bivalents is called *Trisomy* (2n + 1) and addition of 2 chromosomes to a bivalent is said to be *tetrasomy* (2n + 2).

Monosomy : In monosomic diploid individuals, (in the genome) one of the pairs of homologues would undergo a loss of one chromosome and it becomes a univalent while all others are bivalents. The diploids cannot generally withstand such a loss, while polyploids can withstand it. This is due to the fact that polyploids (see later in the same chapter) have more than two homologues for every pair. Theoretically the number of monosomics possible in an individual equals its haploid number. In common wheat where there are 21 pairs of chromosomes; there are 21 possibilities for monosomics. E.R. Sears has artificially induced these 21 monosomics in a wheat variety called *Chinese spring*. Monosomics have been induced in cotton (Endrizzi *et al)* and tobacco (Clausen and Cameron).

While diploids with monosomics do not survive, there are instances as in tomato, where monosomics could be produced. Here the chromosome number is 2n = 24. Double and triple monosomics (2n - 1 - 1) have two homologues lost from two different bivalents. During meiotic prophase, together with the normal bivalents there will be two separate univalents indicating that both these have lost their partners. Similarly in triple monosomy one homologue each is lost from three different bivalents and three univalents appear at the meiotic prophase.

Monosomy can be detected by comparing it with the normal individual and associated traits. In a cross between two monosomics, we will get normal, monosomics and nullisomics in a mixture. Nullisomics (if they survive) will not exhibit any of the morphological traits seen on the missing chromosome; there by specific traits and genes can be located on this chromosome.

Aneuploidy has been noticed in human beings also. A classical instance is the *Turner's syndrome* named after the discoverer H.H. Turner (1938). It occurs in 0.23 infants for every 1000 births. Here there is a loss of one of the X chromosomes and the females would be XO instead of XX. Adult females with Turner's syndrome have no ovaries and no secondary sexual characters and are sterile. They have abnormal jaw, hypoplasia and webbed neck. Turner's syndrome arises by the fertilization of an egg with a defective sperm (having no X chromosome)

Nullisomy : In this, an entire pair of homologous chromosomes will be missing (2n - 2). This should not be mistaken with double monosomy where 2 chromosomes are lost from two different pairs. In nullisomics, the metaphase plate at meiosis I will apparently show no abnormality and only a comparison with the normal can detect the loss of one whole pair of chromosomes. Nullisomics have been identified in wheat and other plants.

Trisomy : In trisomic individuals one of the pairs of homologues will have an additional member. Trisomic individuals exhibit a trivalent together with the normal bivalents during meiotic metaphase. These are designated as 2n + 1. There could be a double trisomy in which case two of the bivalents will have one additional member. At the metaphase of meiosis I, there will be two trivalents. The number of possible trisomy in individuals equals the number of haploid set. For eg. in barley where the haploid number is 7, seven trisomics are possible.

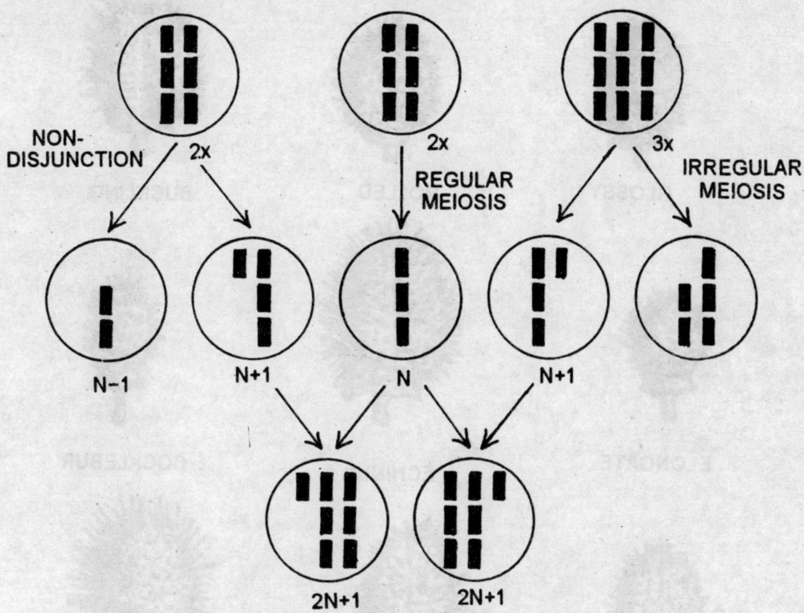

Fig. 13.8 Chromosomal Aberrations
Origin of different types of aneuploidy

Trisomics are classified into primary, secondary and tertiary based on the type of the additional chromosome. If the extra member is identical to the other two homologues, the trisomy is said to *be primary trisoinm* (there is a triplicate of a chromosome). In secondary trisomy, while two are normal chromosomes, the third member is an isochromosome (with two arms genetically identical). In tertiary trisomy, the extra chromosome is the product of translocation.

In plants, trisomy has been studied extensively *in Datura,* maize, tomato and *Nicotiana.* In animals, trisomy has been reported in *Drosophila* and others including human beings.

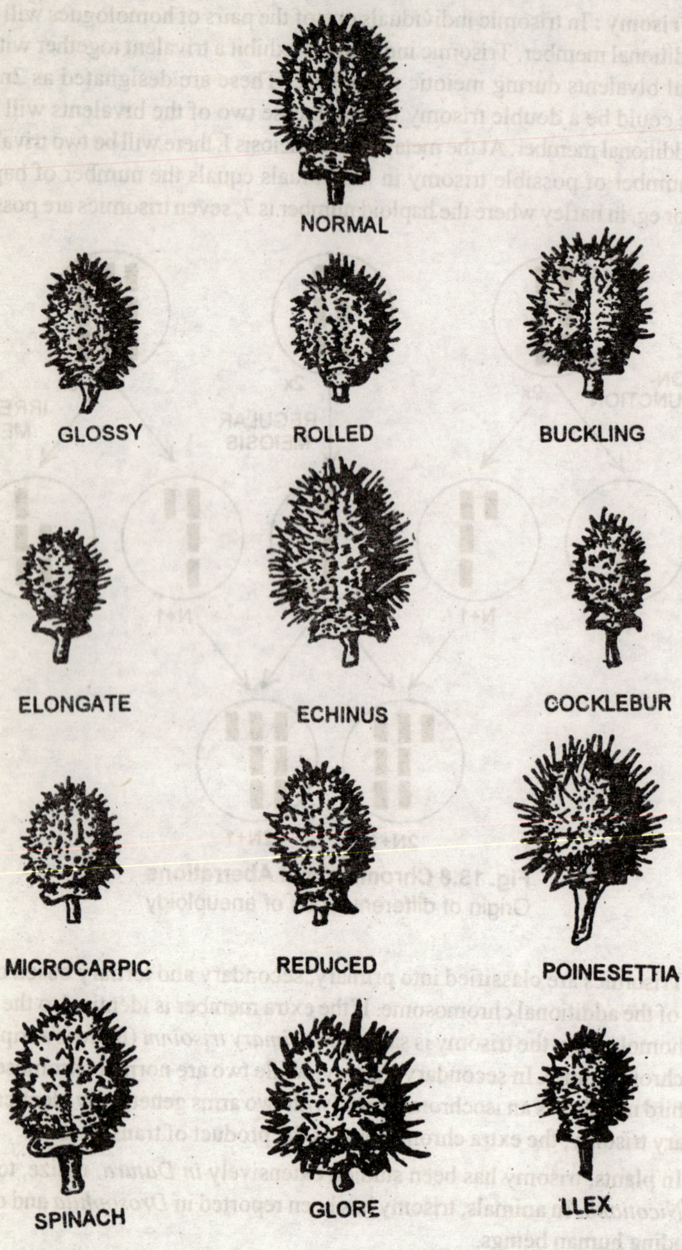

Fig. 13.9 Chromosomal Aberrations
Trisomy in Datura, resulting in different types of fruit shapes

Blackslee and his associates have described a series of trisomy in the jimson weed *(Datura stramonium)* belonging to the family Solanaceae. The base chromosome number of this species is 12 (n = 12) and 12 primary trisomics are possible. In this case instead of 12 bivalents at the metaphase of meiosis I, there will be 12 trivalents. Each one of this is associated with a distinct fruit character and can be separated from the others. Blackslee has named these trisomics (as represented by the fruit character) as follows - *Rolled, Glossy, Buckling, Elongate, Echinus Cocklebur, Microcarpic, Reduced, Poinsettia, Spinach, Globe* and *Ilex*. Secondary and tertiary trisomics have also been reported from *Datura.*

In *Drosophila,* trisomy has been reported for the 4th chromosome. The fly can survive this and breed also.

Trisomy has also been reported in human beings leading to many abnormalities, both in the physical features as well as in mental faculties. *Down's syndrome* (mongoloid idiocy) was first described by Langdon Down in 1866. Affected individuals are short in stature, have slanting eyes, broad skulls, and stubby hands. They have low mental faculties. It was Lejuene (1959) who first associated this disorder with a chromosomal abnormality. Cytological investigations have revealed that in Down's syndrome, chromosome 21 has an additional member. Trisomy in chromosome 21 is produced due to primary non disjunction occuring at meiosis in the mother. The two chromatids do not separate and as a result the egg has both the chromatids of 21, while the first polar body receives none. The incidence of Down's syndrome occurs quite frequently (1:700) among European people.

Other trisomic abnormalities in human beings include the *Patau syndrome* (Patau 1960), *Edward's syndrome* (Edward J.H. 1960) and *Kleinfeter syndrome* (H.F. Kleinfelter 1942). The first two are autosomal trisomics while the third one is a trisomy for the sex chromosomes.

Kleinfelter syndrome occuring one in every 500 births, results in sexual malfunctioning. Individuals are morphologically male, but have small testes and are effiminate in their behaviour. Enlarged breasts, lack of body hair are other characters. Cytologically the cells show a Y chromosome in addition to 2X (XXY). This also indicates that Y chromosome determines the maleness in human beings.

Tetrasomy : In tetrasomy, a particular bivalent has two extra members; in other words there will be four homologous chromosomes in one pair, Symbolically tetrasomy is represented as 2n+2. Tetrasomics have been identified in wheat (E.R. Sears).

EUPLOIDY

Euploid individuals could be haploids, diploids or polyploids and represent the multiplication of the entire genome.

Haploids : Haploid or *n* number of chromosomes in the body cells may be normal or abnormal. In lower plants and animals, haploid condition is normal in the sense, here the individuals have only one set of chromosomes in the body cells, while diploid condition is represented only in the zygote. This is of normal occurrence. Haploidy is of interest when it occurs in diploid individuals. Here normally the body cells have 2 sets of chromosomes. Reduction division takes place and haploid gametes are formed. These gametes however have no morphological expression ie. they cannot directly develop into new individuals. They fuse with another gamete resulting in the formation of a diploid individual. Rarely however haploid individuals are produced as during parthenogenesis when the egg directly develops into a new individual without fertilization. Haploids are of two kinds *euhaploids* and *aneuhaploids.* While the former are derived from euploids the latter are derived from aneuploids.

Origin and production of Haploids : In some insects haploid individuals are produced naturally due to parthenogenesis. In plants also parthenogenetic development of egg results in haploid individuals. Such haploids have been seen in cotton, tomato etc.

In rare instances, pollen tube, or microspores may develop into new individuals. Such individuals are called *androgenic haploids* as they are produced from the male tissue. Occasionally synergids, antipodals etc may also develop into haploid embryos.

Haploids can artificially be obtained by tissue culture of the anther, pollen, X ray treatment, colchicine treatment, distance hybridization etc.

Anther culture has yielded excellent haploids in *Nicotiana* and Potato. Similarly Inter specific crosses may also result in haploid individuals. According to Kasha (1970), in a cross between *Hordeum vulgare* X *H. bulbosum'*. Chromosomes *of H. bulbosum* are eliminated in early zygotic division. Embryosacs of these can be cultured to obtain haploid individuals.

Morphology of haploids : There is general reduction in size and stature of the individual. This is also reflected in their physiology and generally they are not vigorous.

Cytology of haploids : In a haploid set all the chromosomes are non homologous and hence at metaphase I, they retain as univalents. Consequently these are distributed randomly at anaphase I for e.g. in maize a haploid ($n = 10$) individual produces gametes whose chromosome number ranges from 0-10. As a result there will be partial or complete sterility. Restitution nucleus may be formed.

Use of Haploidy : In normal diploids, haploid individuals help us to understand the functioning of recessive genes and also the expression of other genes in the haploid state. Doubling of haploid chromosomes will also bring about a completely homozygous diploid. This is of particular interest in plant breeding as this would bring about the effect of several generations of inbreeding.

POLYPLOIDS

In polyploid individuals, the body cells have more than two sets of chromosomes and consequently multivalents are formed at the metaphase I in meiosis.

There are mainly three categoreis of polyploids - *Autopolyploids, Allopolyploids* and *Autoallopolyploids*.

The following chart gives the derivation of these from three diploids with genomes AA and $B_1 B_1$ and $B_2 B_2$.

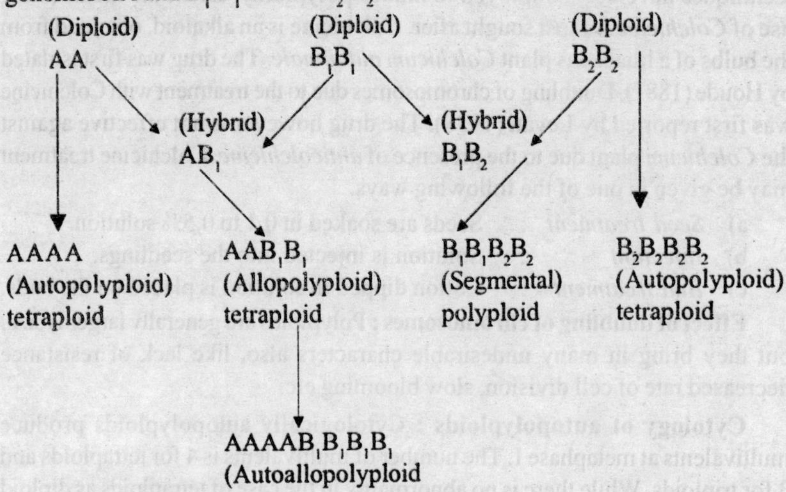

Autopolyploids : Here the same basic genome set is multiplied. For instance if the diploid genome is AA, autotriploid would be AAA, autotetraploid would be AAAA, autooctoploid - AAAAAAAA etc. An auto tetraploid would result either by doubling of somatic chromosomes or by the union of two unreduced gametes. Autotriploid can arise by a cross between a tetraploid and a diploid as shown below.

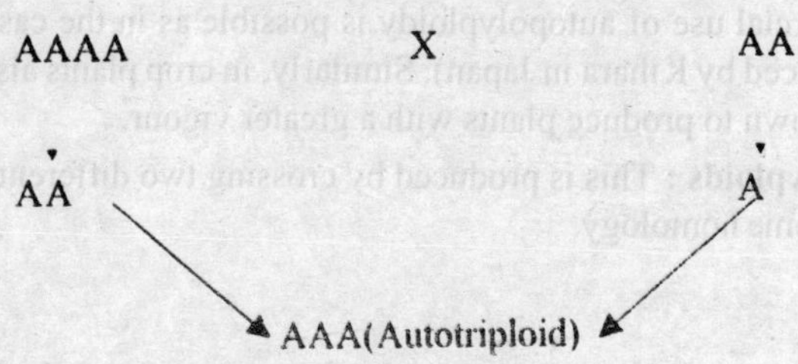

Distribution ofAutopolyploids : According to Muntzing (1936), these are widely distributed among the plants. But recent studies have shown (Clausen, Stebbins etc) that they are not as frequent as they were thought to be. Autopolyploidy has also been reported in the common Indian grass *(Cynodon dactylon)*. This is known to be an autotriploid and survives mainly due to vegetative propagation. Autotriploids are known in water melons, sugar beet, tomato etc. Similarly autotetraploids are known in rye, corn, red clover, marigolds, snapdragons, flox, grapes etc.

Autopolyploidy can be artificially induced by various treatments. These include, ionizing radiation, chemicals, mustard gas etc. Since polyploid individuals have large stature, bigger fruits, increased yield etc, various techniques have been employed to induce polyploidy. Of all the methods the use of *Colchicine* is most sought after. Colchicine is an alkaloid, obtained from the bulbs of a liliaceous plant *Colchicum autuinnale*. The drug was first isolated by Houde (1887). Doubling of chromosomes due to the treatment with Colchicine was first reported by Levan (1938). The drug however is not effective against the *Colchicum* plant due to the presence of *anticolchicine*. Colchicine treatment may be given in one of the following ways.

a) *Seed treatment* : Seeds are soaked in 0.1 to 0.5% solution.
b) *Injection* : Solution is injected into the seedlings.
c) *Bud treatment* : Cotton dipped in solution is placed on the buds.

Effect of doubling of chromosomes : Polyploids are generally larger is size, but they bring in many undesirable characters also, like lack of resistance decreased rate of cell division, slow blooming etc.

Cytology of autopolyploids : Cytologically autopolyploids produce multivalents at metaphase I. The number of multivalents is 4 for tetraploids and 3 for triploids. While there is no abnormality in the case of tetraploids as diploid gametes can be formed, the meiosis is irregular in triploids. During disjunction, the trivalents separate as 2:1, or 1:1 with one chromosome left at the centre. This depends on the type of cross over. Since there are three homologues one chromosome may have cross over with two others at different levels. This leads to irregularity in disjucntion and deficiencies in the gamete. In tetraploids, the disjunction is usually 2:1 or rarely 3:1. Hence autotetraploids may show partial sterility, while triploids are generally sterile.

Commercial use of autopolyploidy is possible as in the case of seedless fruits (Produced by Kihara in Japan). Similarly, in crop plants also polyploidy has been known to produce plants with a greater vigour.

Allopolyploids : This is produced by crossing two different species with no chromosome homology.

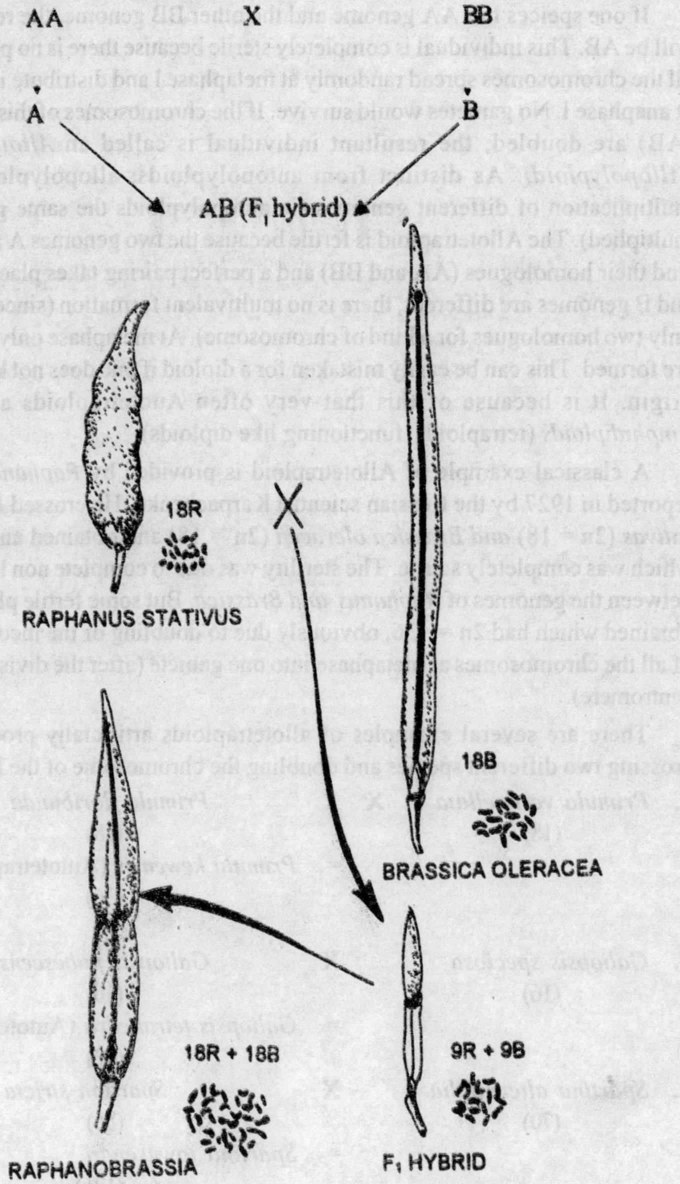

Fig. 13.10 Chromosomal Aberrations
Origin of Raphanobrassica

If one speices has AA genome and the other BB genome, the resultant F_1 will be AB. This individual is completely sterile because there is no pairing and all the chromosomes spread randomly at metaphase I and distribute irregularly at anaphase I. No gametes would survive. If the chromosomes of this F_1 hybrid (AB) are doubled, the resultant individual is called an *Allotetraploid (Allopolyploid).* As distinct from autopolyploids, allopolyploids have multiplication of different genomes (in autopolyploids the same genome is multiplied). The Allotetraploid is fertile because the two genomes A and B will find their homologues (AA and BB) and a perfect pairing takes place Since A and B genomes are different, there is no multivalent formation (since there are only two homologues for a kind of chromosome). At metaphase only bivalents are formed. This can be easily mistaken for a diploid if one does not known the origin. It is because of this that very often Auotetraploids are called *Ainphidiploids* (tetraploids functioning like diploids).

A classical example of Allotetraploid is provided by *Raphanobrassica* reported in 1927 by the Russian scientist Karpachenko. He crossed *Raphanus sativas* (2n = 18) *and Brassica oleracea* (2n = 18) and obtained an F_1 hybrii which was completely sterile. The sterility was due to complete non homology between the genomes of *Raphanus and Brassica.* But some fertile plants were obtained which had 2n == 36, obviously due to doubling or the incorporation of all the chromosomes at metaphase into one gamete (after the division of the centromere).

There are several examples of allotetraploids artificially produced by crossing two different species and doubling the chromosome of the F_1 hybrid.

1. *Primula verticellata* (18) X *Primula floribunda* (18)
   = *Primula kewensis* (Autotetraploid) (36)

2. *Galiopsis speciosa* (16) X *Galiopsis pubescens* (16)
   = *Galiopsis tetrahedra* (Autotetraploid) (32)

3. *Spartina alternifolia* (70) X *Spartina stricta* (56)
   = *Spartina townsendii* (126)

(Numbers in parenthesis indicate the 2n chromo number)

Examples of Amphidiploidy (Allotetraploid) among cultivated plants are seen in cotton, wheat, barely etc.

Segmental allopolyploids : In certain allopolyploids, the genomes coming from different parents are not altogether different. As a result there is some degree of pairing and multivalents are formed. This means that segments of chromosomes and not entire chromosomes are homologous. Such allopolyploids are called segmental allopoyploids and they are intermediate between auto and allopolyploids. The common bread wheat is regarded as a segmental allopolyploid.

Auto allopolyploid : This is a higher series of polyploid (octoploid). It can be produced as follows. A cross between two species (AA, BB) would yield a F_1 sterile hybrid (AB); on doubling, this becomes an amphidiploid (AA BB). If there is doubling of chromosomes in an amphidiploid it results in an autoallopolyploid (AA AA BB BB). The name autoallopolyploid is due to the fact that allopolyploid genomes are doubled to give rise to a polyploid having characters of both auto and allopolyploids.

ROLE OF POLYPLOIDY IN EVOLUTION

Polyploid individuals are different from their diploid relatives in many characters; yet they retain some degree of genetic semblance (allopolyploids) with the diploids. Polyploidy influences not only the morphological attributes of the individual but also has a profound effect on the reproductive potentiality and genetic variability of the progeny. All these obviously play an important role in evolution.

Comparison of natural polyploids with their diploid relatives and artificially produced polyploids provides us an indepth knowledge into the genetic machinery of polyploids. The resynthesis of polyploids from their diploid ancestors has provided additional proof about the origin of polyploids. Natural hybridization or introgression between species and doubling of the chromosomes has contributed a great lot in the evolution of species.

Polyploidy is widely distributed among the various groups of plant kingdom, but they are relatively scarce among animals. Even though this seems to be a mystery a careful analysis of the karyotype of animals and plants will readily provide an answer for the paucity of polyploidy among animals. In majority of the animals, there is a sharp contrast of two sexes resting upon a delicate balance between autosomes and sex chromosomes. This does not permit the addition or deletion of any member of the genome as is easily done in the case of hermaphroditic plants. In animals, wherever polyploidy has been reported it is associated with parthenogenetic development. There is one exception however as in the case of the saw fly, where an authentic case of polyploidy has been reported without the accompanying parthenogenesis.

Effect of polyploidy on structure and function of the organisms

A polyploid individual arising from a diploid population can survive and stabilize itself in the population provided the genetic alterations are such that it can fit into the existing genetic pool, while still showing variations. Between allopolyploids and autopolyploids, the former have a greater chance of survival. The effect of polyploidy on the morphology and physiology of the individual to a great extent will determine, whether the individual in question will survive in the environment. One of the favourite effects of polyploidy is gigantism, but it may not be always so. The other morphological and physiological effects of polyploidy have already been discussed.

Distrubiton ofpolyploids in the plant kingdom : Polyploidy has a wide distribution in the plant kingdom except perhaps for lower plants. Among the lower plants, tetraploidy perhaps exists in baker's yeast *(Saccharomyces cereviseae)*. Tischler (1950) records a number of Algae as being polyploid. According to Manton (1952) the Horsetails, Clubmosses and the Psilotales represent the remanants of an ancient polyploid system. The Filicales also record a very high chromosome number *(Ophioglossum valgatum* has 500 chromosomes). Among the Gymnosperms polyploids are encountered in Gnetales.

Among the flowering plants however, polyploids are quite extensively distributed. According to Stebbins (1950), Tischler (1950), Darlington andJanakiammal (1945), about 30-35% of flowering plants are polyploids (allopolyploids). In grasses about 75% are polyploids. Rosaceae, Polygonaceae, Malvaceae, Crassulaceae among Dicots and Cyperaceae, juncaceae,Iridaceae and Orchidaceae among Moncots exhibit polyploid individuals.

The propensity of allopolyploids and the apparent paucity of autopolyploids in flowering plants is a clear indication that natural hybridization or Introgression and the subsequent doubling of the chromosomes have been going on in plant populations since a long time and have greatly contributed to evolution.

Polyploidy and speciation : In the evolutionary heirarchy, the lowest stable morphological unit is species. Two species are delimited not only on the basis of morphological traits, but underlying this is the chromosomal divergence. When two individuals have sufficient non homology among their chromosomes as to make a cross impossible, they are then said to have evolved into new species. If this statement is true completely; after the establishment of a species, the only factor for variation would be mutation. But evolution cannot depend on only one mechanism for its functioning. Even though species cannot be crossed generally, it does happen in nature and nature itself (or the selective forces) has found an answer to the resultant sterility in the F_1 hybrid. The chromosomes are doubled and the individual becomes fertile; has basic morphological and genetic characters sufficient enough to raise it to the status of a new species. It has happened in the case of *Primula kewensis*. Ever since then several new polyploid species have been created such as *Galeopsis tetrahedra*, *Spartina townsendii* etc.

Several of our cultivated crops represent a polyploid series indicating their origin from diploid ancestors. The two classical examples are those of wheat and cotton.

The wheat represents an allopolyploid series of diploid, tetraploid and hexaploid species. These three groups are called *Einkorn*, *Emmer* and *Vulgare* groups. The einkorn group has two diploid (2n = 14) species viz. *Triticum monococcum* and the wild *I. boeticum*. The einkorn speices are single seeded and not of much use for human consumption, because the grains are tightly held in glumes. In some parts of middle east and Europe, they are used for making dark bread.

The emmer group consists of seven species of tetraploid wheats of which the important ones are *Triticum dicoccum* and *T.durun*. The emmer wheat has arisen as a cross between an einkorn wheat and a wild grass *Aegilops* as shown below.

Triticum boeticum (2n = 14) AA X *Aegilops speltoides* (2n = 14) BB

↓

AB (Diploid sterile hybrid)

↓ Doubling

AABB

Triticum dicoccooides (2n = 28) Fertile alloplyploid

↓ evolved

Triticum dicoccum (2n = 28) A'A'B'B'

The vulgare group consists of five species of hexaploid wheats (2n = 42) including the familiar bread wheat *Triticum aestivum*. This is said to have been originated as a result of a cross between *Triticum dicoccum* (A'A'B'B') and another species of goat grass *viz. Aegilops squamosa* (diploid DD). The bread wheat therefore has genomes from 3 different sources. All the three wheat groups are interrelated and it is very obvious as to how polyploidy and associated chromosome doubling would have resulted in different species of wheat. But for polyploidy, the other evolutionary force (mutation) could not have produced such variety among wheat.

In cotton, the upland cotton *Gossypicum hirsutum* is a tetraploid and is a result of crossing (and subsequent chromosome doubling) of *G.herbaceum* and *G.raimondii*.

Several such polyploid series are found in *Bromus, Sanicula* etc. Among the Rosaceae the entire tribe pomoideae seems to have been arisen through poplyploidy. According to Sax (1931), the group Pomoideae with n = 17 seems to have been the result of an ancient cross between Spiraeoideae (n = 9) and Prunoideae (n = 8).

Among the polyploid series, the allotetraploid perhaps represents a stage very advantageous to the individual. Allotetraploidy combines the best of both worlds; it has the advantage of higher chromosome numbers, but the genetic and pairing behaviour of a diploid. No wonder then among the polyploids which survive, allotetraploids rank first. The higher series of polyploids however have gone too far in polyploidy and have committed themselves to a polyploid existence with its attendant advantages and disadvantages. Thus they lack the plasticity that is noticed in allotetraploids.

According to Swanson (1975), the sequence of evolution where polyploidy has been an important consideration is as follows. Gene mutation, aberrations and recombination will provide the principal raw material at the diploid level for natural selection. This results in a number of ecotypes from which arise divergent lines leading to higher taxonomic categories. In this scheme of things, autopolyploids have little role to play for they are no different genetically from their diploid parents.

Introgression and doubling of chromosomes however will lead to allopolyploidy and at once creates (evolves) a new species.

Triticale - a man made cereal : This is produced by a cross between wheat and rye. *Triticum durum* (2n = 28) *andSecalecereale* (2n = 14) are crossed producing a F_1 sterile (2n = 21). On doubling, this becomes a hexploid triticale (2n = 42). An octoploid triticale can also be produced by a cross between *Triticum aestivum* (2n = 42) and *Secale cereale* (2n = 14). The resultant F_1 hybrid (2n = 28) is sterile; on chromosome doubling this will lead to an octoploid triticale (2n =-56).

There is some acceptance for the hexaploid triticale from the point of view of grain quality and taste. These hexaploid Triticales are crossed again with wheat to bring about improvement. The products of these crosses are called *secondary Triticales* and are known to be better than *primary Triticales* (produced from wheat and rye). In producing a new species Triticale, man has just imitated the forces of nature which have brought in,an extensive array of biological diversity on earth.

14

CYTOPLASMIC INHERITANCE

Inheritance of characters from parents to offspring is primarily dependant on the nucleus. Oscar, Hertwig etc., had established in 1870 itself the importance of nucleus in the heriditary transmission of characters. The discoveries of Mendel further provided the genetic evidence for nucleus being the seat of all heriditary characters. The nucleus forms specific number of chromosomes during cell division and these duplicate, maintaining their number in future cell generations. Chromosomal aberrations, both structural and numerical leading to phenotypic abnormalities provided further proof for chromosomes albeit nucleus to be the only organelle concerned with the transmission of characters. The discovery of sex linked characters gave further impetus for the chromosome theory of inheritance as it was firmly established that a particular trait followed a particular chromosome. An additional proof for chromosomes being the only link between one generation and the other is the fact the gametes, atleast the male gamete consists of mainly the nucleus and very little amount of cytoplasm. With all these overwhelming evidences in favour of nuclear inheritance it was but natural for geneticists to beleive that nucleus is the only seat where all the genetic potentialities are located. But there are several instances of a peculiar type of transmission of characters both in the animal and plant kingdom which seemed to defy every known genetic law assuming the whole of inheritance to be through the nucleus. The inheritance of flower colour in 4° Clock plants, Kappa particles in *Paramecium* etc, can not be explained on the basis of nuclear inheritance. What then is the secret of their .inheritance pattern? Scientists started looking for other cell organelles which could perhaps play a role in inheritance. The gametes have some amount of cytoplasm surrounding the nucleus. The amount of cytoplasm is more in the case of female gametes, than in the male gametes. Geneticists started investigating, whether the cytoplasm also could help in the transmission of hereditary characters. If inheritance could be brought about by cytoplasm, there should be more chances of female or maternal inheritance as the egg contributes more amount of cytoplasm than male gamete to the offspring. Further, one of the important requirements of a transmission system (if it exists in cytoplasm) is the capacity of self reproduction and it should have the basic molecules of inheritance viz. nucleic acids - particularly DNA. Further investigations did provide enough evidence of the existence of transmission systems in the cytoplasm (other than the nucleus). Plastids, mitochondria, ribosomes etc., have DNA of their own and many evidences have indicated that they replicate on their own, independent of the replication of nuclear DNA. These evidences clearly indicated a nuclear independent transmission machinery in cytoplasm and geneticists were able to explain the mechanism of inheritance in some apparently paradoxical instances in both plant and animal kingdom. Some of these are inheritance of flower colour in 4° clock plants, inheritance of killer particles (Kappa) in *Paramecium* etc. These were explained on the basis of cytoplasmic inheritance. Even though there are many organelles in cytoplasm possessing DNA, only the role of plastids in inheritance has been proved in plants. In animals, kappa particles and other traits seem to rest in cytoplasm but not in any definite organelle. The salient features of cytoplasmic inheritance are.

1. Certain somatic characters do not follow the Mendelian pattern, nor are they following the chromosomes in inheritance.
2. The genes or heriditary particles often called plasmagenes are located in cytoplasm.
3. In some instances, plasmagenes may be traced to definite organelles like plastids.
4. The most salient feature of cytoplasmic inheritance is female dominance. Since the egg contributes more amount of cytoplasm, offspring are more likely to get the maternal characters than the paternal. This has been established by reciprocal crosses.
5. Characters dependent on cytoplasm for transmission are never located on the chromosome. Hence the inheritance is also known as extra chromosomal inheritance.

In this chapter we shall study a few examples of cytoplasmic inheritance from both plant and animal kingdom.

Plastid inheritance in Mirabilis

Mirabilis jalapa, of the family Nyctaginaceae is popularly called the 4° clock plant, as the flowers bloom in the evening. The plant is unique in several respects. 1. The leaves are variegated or some times completely green. 2. The branches are of three types deep green, pale green and variegated. 3. The flowers are of various colours on the same plant in different branches they are either pink, yellow, or white with pink dots. The mechanism of inheritance in this plant was worked out by Correns, who opined that the flower and branch colour inheritance is not through the nucleus, but through the plastids. The phenotype of the progeny depends from which part of the plant the seed is obtained. If a seed is obtained from a flower on a green branch, the progeny would alway produce a green branch. Again a seed obtained from the same plant, but from a variegated branch would produce offspring with a variegatci green and pale branches. The same is true for flowers also. In thes inheritances, the source of the pollen does not seem to have any effect on the phenotype of the progeny.

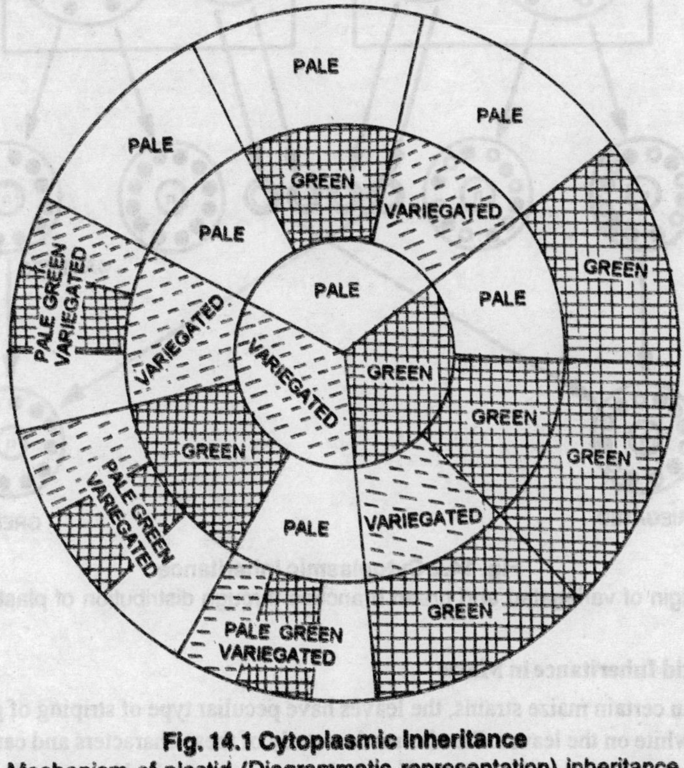

Fig. 14.1 Cytoplasmic Inheritance
Mechanism of plastid (Diagrammatic representation) inheritance

The mechanism of inheritance is as follows- The colour of the branch and flowers depend on specific plastids. The female gamete from a particular source has one or more types of plastids. The pollen carries no plastids at all as it has a limited cytoplasmic boundary. When the zygote divides, the plastids are distributed irregularly. Some cells may get plastids for green, some may get plastids for pink etc. These cells develop into branches, leaves flowers etc.

Hence the colour of the branch and flower depends on what types of plastids they have received during development,

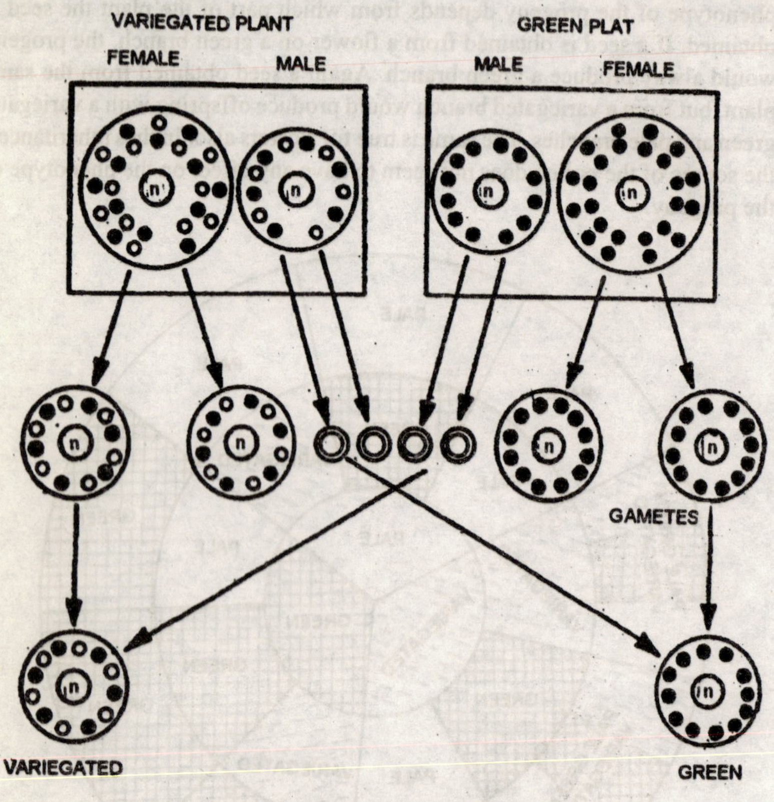

Fig. 14.2 Cytoplasmic Inheritance
Origin of variegated and green branches through distribution of plastids

Plastid Inheritance in Maize

In certain maize strains, the leaves have peculiar type of striping of green and white on the leaves. Rhodes made a study of these characters and came to the conclusion, that the **iojap** (striped) character is controlled by a gene located on chromosome number 7. The dominant gene *Ij* would produce green plastids, and its recessive allele *ij* would produce the iojap striping, *IjIj* is green and *ijij* is striped. The *ijij* plants would carry both green as well as abnormal white plastids in their cytoplasm, while *IjIj* plants would carry only the green plastids. In a cross between iojap and normal plant, if iojap parent is used as female, the female gametes carry either only green plastids, only white plastids or both green and white plastids. Consequently when iojap is the female parent and *ljlj* is the male parent, the offspring could be either green, variegated or white. The white seedlings die. If a striped offspring from the F_1 *(ljij)* is used as the female parent and the *ljlj* as the male parent and a cross carried out, two genotypes are obtained *ljlj - ijij*. But phenotypically they are of three types - either green, white or striped, ie. each one has all the three .phenotypic expressions.

Fig. 14.3 Cytoplasmic Inheritance
Variegated leaf of mirablis jalapa

The above experiments clearly show, that the iojap character, once inherited through the female parent maintains itself in the cytoplasm, irrespective of the genotype of the plant. Hence the striped phenotype (iojap) may have any one of the genotypes - *ljlj, ljij, ijij*.

Male sterility in plants : Sterility in male may be influenced by nuclear genes, cytoplasmic genes or by both cytoplasmic and nuclear genes acting in co-ordination. We shall briefly discuss all these three.

Male sterility by nuclear genes : A single dominant gene on the chromosome controls the fertility, while its recessive allele brings about sterility. In a cross between the two, the F_1 individuals are heterozygous fertile. Selfing of the F_1 individuals to obtain F_2 shows a typical 3:1 seggregation of fertile and sterile plants.

Cytoplasmic male sterility : In many o. the crop plants notably maize, pollen grains degenerate even though the ovules are normal and fertile. Marcus Rhodes made a detailed study of the male sterility in maize. He conducted breeding experiments extensively and was successful in replacing all the ten chromosomes of a male sterile strain with those of a male fertile strain. Inspite of this replacement of all the chromsomes, the progeny of this plant continued to be male sterile. This indicated that the nuclear genes did not play any role in determining the male sterility. Since male sterility was more when male sterile plant was used as the female parent in any cross, it is obvious, that the male sterility factor must be present in cytoplasm. A reciprocal cross in which the male sterile plant was used as the male parent and the male fertile pant as the female parent proved this tact (Although it may seem a paradox to use a male sterile plant as the male parnet, some pollen grains are fertile even in a male sterile plant).

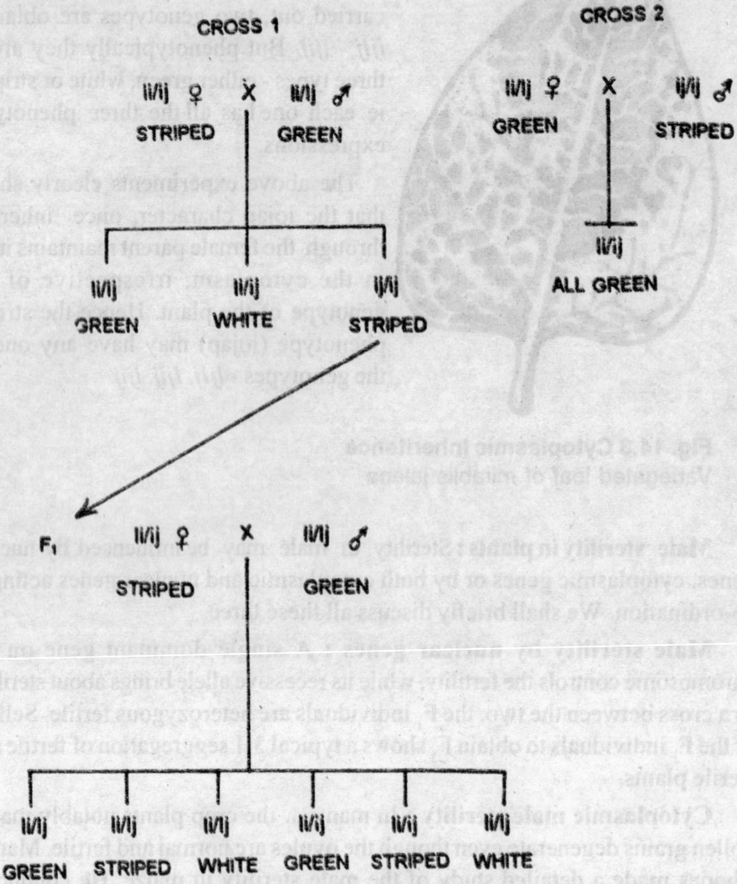

Fig. 14.4 Cytoplasmic Inheritance
Results of reciprocal cross between iojap and normal in maize

Cytoplasmic and nuclear male sterility : In certain instances cytoplasmic induced male sterility may be reversed by a fertility restorer gene in the nucleus. For example, if the female parent in a cross is male sterile, then the genotype (nuclear) of the male parent will decide the phenotype of the F_1 progeny. The male sterile female parent will have the recessive genes (rr) while the restorer

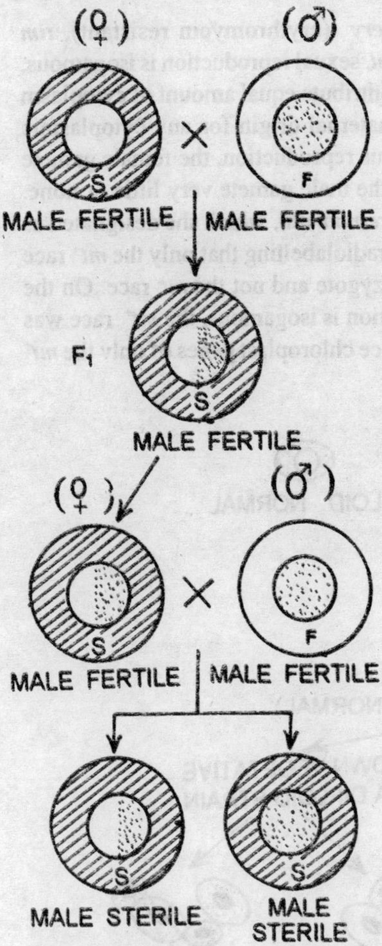

Fig. 14.5 Cytoplasmic Inheritance
Maternal inheritance of male sterility in maize

gene is dominant (R). If the male parent is homozygous dominant, the F_1 progeny will be heterozygous (Rr) and fertile. On the other hand if the male parent is rr all the F_1 progeny will be also male sterile. A test cross between F_1 hybrid (Rr) and double recessive male parent will produce 50% male sterile and 50% fertile plants.

Cytoplasinic inheritance in Chlamydomonas

Chlamydomonas is a fresh water green alga belonging to the order Volvocales. Genetic investigations in *Chylamydomonas* have shown the existence of three independent and operationally distinct genetic systems (Hudock and Rosen, 1976). These are

1. A Mendelian system comprising 16 linkage groups and residing in the nucleus.
2. A non Mendelian uniparental transmission of genes through the mating type plus *(mt⁺)* parent which is traccable to DNA in the chloroplast.
3. A non Mendelian biparental trasmission, the genes for which are thought to reside in the mitochondrial DNA (Alexander *et al*, 1974).

We will study the second category of inheritance which is cytoplasmic in nature. The pioneer in the field of plasma gene studies in *Chylamydomonas* is Ruth Sager (1954, 1964,1970) who has studied the inheritance of streptomycin resistance in *C. reinhardtii.* Seven gene pairs have been identified on the chloroplast genome. Ruth Sager (1972) has proposed a circular model for the chloroplast. DNA. Cytoplasmic inheritance in *Chlamyomonas reinhardtii* is studied using mutations for various chemicals; such as *ac* (acetate requiring mutant), *Sniz* (streptomycin resistant), *ery* (erythromycin resistant), *tim* (temperature sensitive) etc. In *C. rcinhardtii,* sexual reproduction is isogamous, hence both the gametes are supposed to contribute equal amount of cytoplasm and as such it is difficult to assess the maternal origin for any cytoplasmic factor. In higher organisms, with oogamous reproduction, the female gamete always contributes more cytoplasm, and the male gamete very little or none. Ruth Sager has identified two races of C. *reinhardtii,* which she designated as mt^+ and mt^-. Of these, it was found out by radiolabelling that only the mt^+ race can donate its chloroplast genome to the zygote and not the mt^- race. On the basis of this even though sexual reproduction is isogamous, the mt^+ race was called maternal and mt^- race paternal. Hence chloroplast genes of onlv the mt^+ race are inherited.

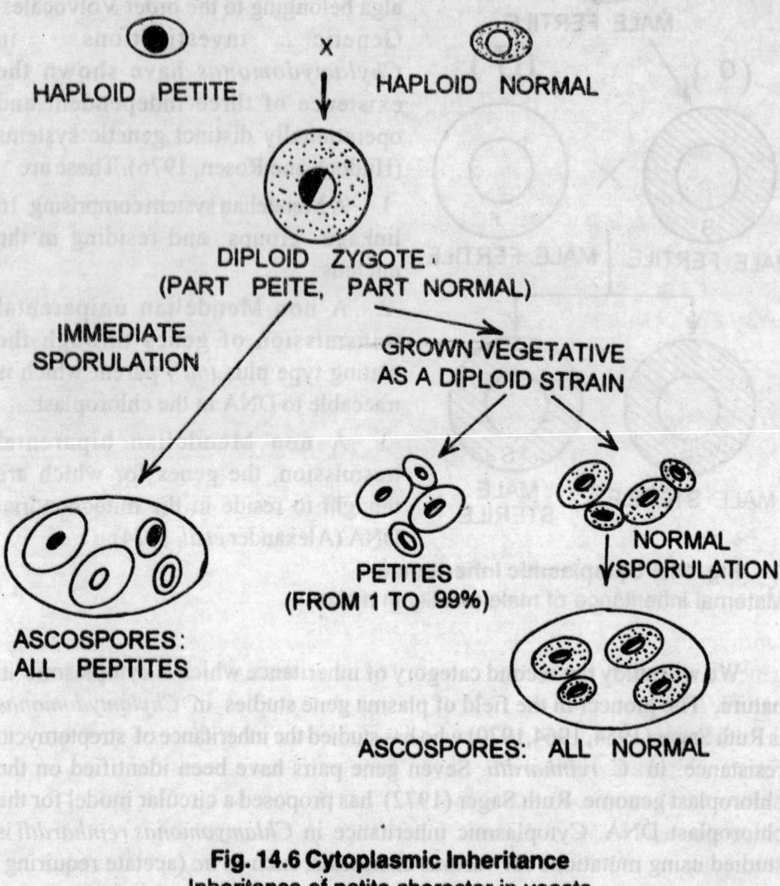

Fig. 14.6 Cytoplasmic Inheritance
Inheritance of petite character in yeasts

Petite character inheritance in Yeasts

Petite character in Yeasts refers to slow growth of mutants in. aerobic medium. The petite yeasts develop into small colonies when compared with the normal one, if the medium is oxygenated. The petite yeasts have a defective aerobic respiratory mechanism. Some of the characters of petite yeasts are as follows a) they are insensitive to cyanide poisining (affecting the aerobic path way) b) Absence of cytochrome a and cytochrome a-?) Incompletely formed mitochondria d) non stainability of petite mitochondria with Janus green etc. The genetic basis of petite character is a cytoplasmic factor , which is either defective or absent in petite yeasts. Thus in petite yeasts p is p or p' (defective or suppressive). The p' cytoplasmic factor is known to be transmitted through the mitochondrial DNA. Evidence for petite factor to be linked with mitochondrial DNA comes from the fact that ethidium bromide, inducing petite mutations disturbs the *mit* - DNA. *Mit* DNA with suppressive petite factor have alteration in their base sequence when compared with non petite *mit* DNA (N.W. Gilham, 1974).

Kappa particles in Paramecium

There are two strains in *Paramecium,* the *killer strain* and the *sensitive strain.* T.M. Sonneborn, USA has conducted extensive studies on these strains in *P. aurelia.* The killer strains have certain bodies in cytoplasm called kappa particles. These kappa particles are composed of DNA. The killer strains produce a substance called *paramecin,* which is toxic to the sensitive strains. When this substance is released into the medium, the sensitive strains die. But the paramecin has no effect on the killer strain even though they are produced by it. The killing capacity of the killer strain is influenced by temperature, food and repeated cell division. Low temperature, inadequate nutrition and frequent cell divisions would reduce the killing ability. This is due to the fact, that with rapid cell division, the kappa particles do not multiply at the same rate, and the daughter cells in subsequent generations receive less and less kappa particles, hence reduction of killing ability. It has been estimated that atleast 400 kappa particles are necessary to maintain the killing ability. Kappa particles are self reproducing, but their multiplication depends on the presence of the dominant gene *K.* But there is a peculiar relationship between the gene and the kappa particle. *K* gene is necessary for the multiplication of the kappa particles, but if the cytoplasm lacks the kappa particles, *K* gene cannot produce them *de now.* Hence even though the genotype may be KK, the strain will not be a killer. Similarly, in the absence of *K,* kappa particles even if they are present in cytoplasm will not reproduce and gradually their number diminishes with every cell division. The genotypes of killer and sensitive strains are as follows.

Killer strain *KK* or *Kk*
Sensitive strain *kk*

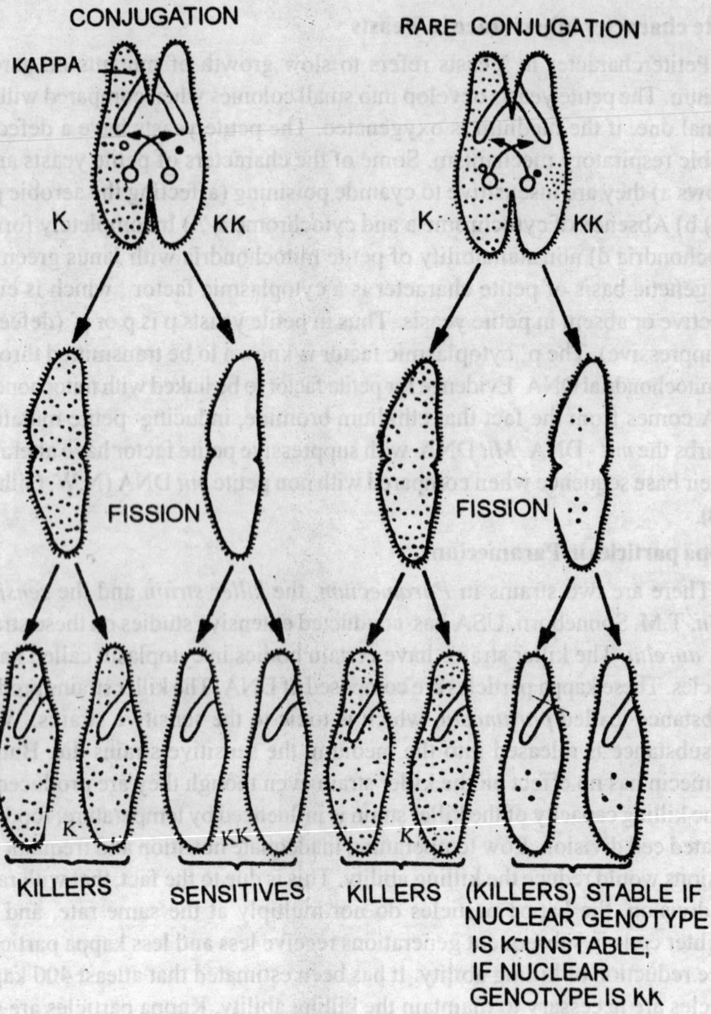

Fig. 14.7 Cytoplasmic Inheritance
Inheritance of kappa particles in Paramecium

By repeated culturing, it is possible to induce a rapid cell division in the killer strains with either *KK* or *Kk* genotypes. The kappa particles can not keep pace with this division and consequently after several generations, the progeny would be without kappa particles, even though they are genotypically KK or Kk and their parents were killers. These progenies (*KK* or *Kk*) would now be sensitive strains. This is not possible if the killer character were to be located on the nuclear gene. For, with every cell division, nucleus also divides and even after repeated divisions, the killer strain should have retained its killer gene if it is in the nucleus.

When the killer *(KK)* strain with kappa particles is placed along with the sensitive strain *(kk)* on a fresh medium without paramecin, conjugation occurs and the genotype of the exconjugants (products of conjugation) will be *Kk*. The phenotype (ie. killer or sensitive) however will depend on the duration for which conjugation takes place. If the conjugation is over in a short period of time the *exconjugants* will be killers and sensitive strains like their parents. Killer strain will remain killer and sensitive strain will remain sensitive even after conjugation. This is not possible if the killer factor were to be located on the nucleus. All the heterozygotes (*Kk*) of a cross (*KK* and *kk*) should have been killers. But if the conjugation persists for a long time, the sensitive exconjugant will now turn a killer for, with longer conjugation it would have received cytoplasm and together with it kappa particles also from the killer strain.

Recent discoveries have indicated that kappa particles could be parasites permanently housed in the cell of *Paramecium.*

Coiling of shells in Snails : The coiling pattern in snails *(Limnaea peregra)* has been studied in detail with reference to cytoplasmic inheritance (Boycott, Diver and Garstang). *Two* types of coiling have been observed in this species during the development. In some strains the coiling is dextral (coiling is to the right) and in some it is sinistral (coiling is to the left). The direction of the coiling is supposed to be controlled by the gene *D* for right hand coiling and its recessive allele *d* controls left hand coiling. But in crosses involving the dextral and sinistral snails, the inheritance pattern does not always follow the genotype. When dextral females are mated to sinistral males all the F_1 offspring are dextral. In the F_2 generation however the expected ratio of 3:1 for dextral and sinistral is not obtained. All the individuals -are dextral and there is no recessive trait. The phenotypes of all the progeny resemble that of the mother. When the double recessive *(dd)* individuals are inbred they are all sinistral. On the other hand when *DD* (dextral) was crossed with Dd (dextral), there appears to be no seggregation and all the offspring are dextral. When a sinistral (recessive) female is crossed to a dextral (dominant) male, the offspring are all like the mother - sinistral.

Investigations have shown that, the coiling is influenced by the orientation of the spindles formed during the cleavage of first metaphase. The spindle is shifted to the right in dextral and to the left in sinistral. This is controlled by the maternal genes (not nuclear) and the paternal genes do not seem to have any influence.

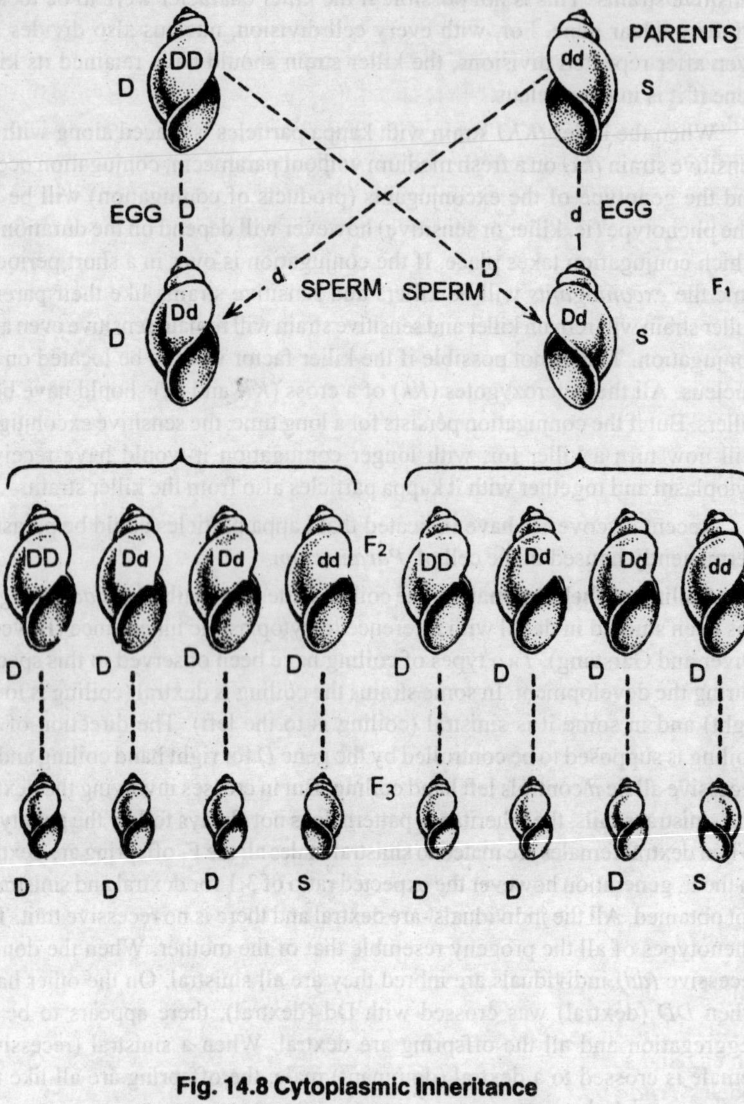

Fig. 14.8 Cytoplasmic Inheritance
Coiling of shells in snails (Limnea peregra)

Carbon dioxide sensitivity in Drosophila : Heritier and Teissier have described a strain of *Drosophila,* which is very sensitive to carbon dioxide than others. Such strains get immobilized on exposure to a high concentration of carbon dioxide. By breeding experiments, Heriteir and Teissier obtained a true breeding strain for carbon sensitivity and established the trait to be heritable. They further proved that this heritable trait does not follow the chromosomal inheritance. When sensitive female strains were crossed to normal male strains, all the progeny were sensitive. However one can assume this to be the dominance of sensitivity over normal. But a reciprocal cross soon disproved this. When a sensitive male was crossed with a normal female, all the progeny were normal. This indicated that, sensitivity trait is maternal in transmission. The progeny could get it only through the mother.

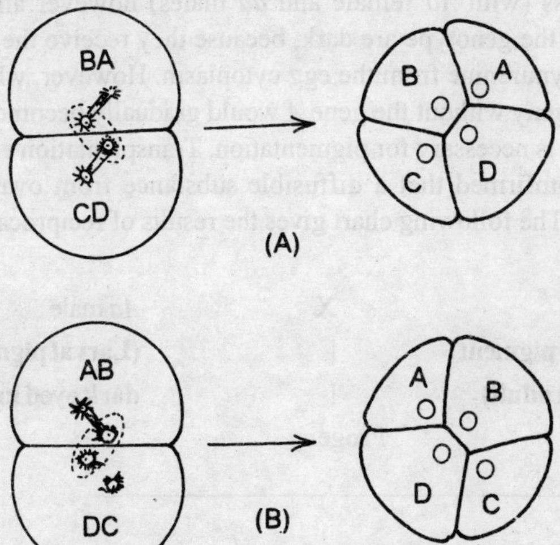

Fig. 14.9 Cytoplasmic Inheritance
Oreintationo of spindle determining the direction of coiling,
A. Cleavage leading to sinistral, **B.** Cleavage leading to dextral

Merogenic hybrids in Urodeles : Merogenic hybrids are obtained from a cross between two different species in which, the individuals have the cytoplasm of the egg, but the nucleus of the sperm. Obviously before fusion, the egg nucleus is removed, hence the zygote will be haploid, but has the cytoplasm of both egg and sperm. Hadom, conducted his experiments with two species of urodeles *Triton palmatus* and *T.cristatus* which differed in the nature of the epidermal cells. The egg of *T.palmatus* was fertilised by the sperm of *T.cristatus.* The zygote was haploid as it had only the sperm nucleus and developed up to the blastula stage. When a portion of the tissue (which would develop into epidermis) from this embryo was grafted on to the larva of another species, the tissue developed into epidermis characteristic of *T.palmatus,* which contributed cytoplasm, through the egg. There was no resemblance to the epidermis of *T.cristatus,* which contributed the nucleus.

Cytoplasmic inheritance in Ephestia : Ephestia is the common flour moth, that occurs in old and stale flour. In one of the species, *E.kuhniella* a clear instance of maternal inheritance has been observed. The dark and the light skin colour is known to be transmitted through the cytoplasm of the mother. The dark skin pigmentation is known to be governed by the gene *A,* while its recessive trait *a* brings about light skin character. Homozygous *AA* individuals are dark, while homozygous *aa* individuals are light. In a cross between, *aa* females with males, half of the progeny are dark and the other half are light skined. In a reciprocal cross (with *Aa* female and *aa* males) however all the progeny irrespective of the genotype are dark, because they receive the precursors of the pigment Kynurenine from the egg cytoplasm. However, when the larvae moult, the progeny without the gene *A* would gradually become pigmentless, as the gene *(A)* is necessary for pigmentation. Transplantation experiments of ovary have comfirmed that a diffusible substance from ovary influences pigmentation. The following chart gives the results of reciprocal crosses.

1 *aa* female **(No larval pigment red eyed in adult)** X *Aa* male **(Larval pigment present dark eyed in adult)**

Progeny

aa (50%) **Unpigmented** — *Aa* (50%) **Pigmented**

2 *Aa* female X *aa* male

Progeny (All with larval pigment)

aa (50%) **Red eyes in adult)** — *Aa* (50%) **(Dark eyes in adult)**

Other instances of cytoplasmic transmission : Besides the above mentioned examples, which are studied in detail there are also a number of less known instances of cytoplasmic transmission of characters.

Epilobium is a member of the family Onagraceae and has been studied in some detail by Lehmann, Renner, Michelis and others. There are two species of

Epilobium, E.hirsutum and *E.roseum,* which differ in some morphological characters. Reciprocal crosses between these two species produces some interesting results. In a cross where *E.hirsutum* is the female parent and *E.roseum,* the male parent, the progeny are nearly sterile with reduced anther size. In a reciprocal cross, where the *E.hirsutum* is the male parent and *E.roseum,* the female parent, the offspring are normal with little or no sterility. Michelis produced a progeny [by repated back crossing of the F_1 female plant *E.roseum* (female) x *E.hirsutum* (male)] with *E.hirsutum* males, where the cytoplasm was derived from *E. roseum,* while all the chromosomes belonged to *E. hirsutum.*

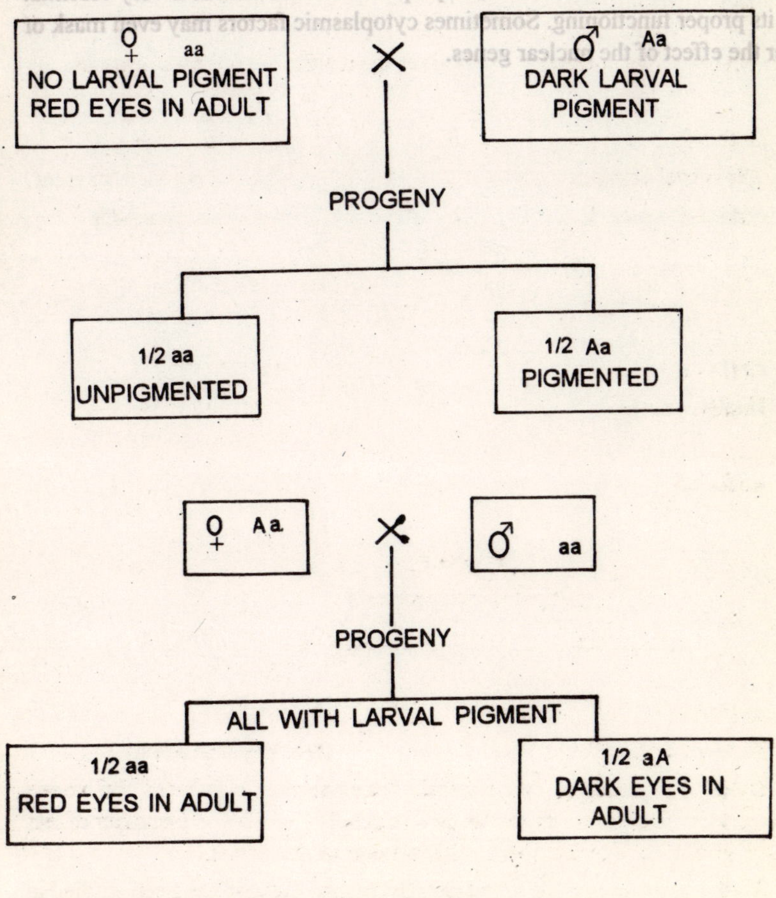

Fig. 14.10 Cytoplasmic Inheritance
Maternal inheritance in flour moth (Ephistia Kuhniella)